新工科建设·电子信息类系列教材

通信与电子信息专业英语

English for Communication, Electronics and Information Technology

古海云　任　蕾　主编

電子工業出版社
Publishing House of Electronics Industry
北京·BEIJING

内 容 简 介

本书内容涵盖电子、通信、电信及计算机等相关专业方向，具有较强的先进性和实用性。本书内容分为两部分，共 18 个单元。专业阅读部分包括 14 个单元：电路原理、嵌入式系统、集成电路、信号与系统、数字信号处理、数字图像处理、通信原理、移动通信、光通信、电信网络、电磁学、微波通信、计算科学基础、新技术趋势。实践指导部分包括 4 个单元：专业学习方法、研究方法、求职指导及沟通技巧。本书提供配套的电子课件 PPT、词汇表、习题参考答案、教学大纲、教学指南等。

本书可作为电子工程、通信工程、电信工程等专业大学本（专）科生的科技英语和专业英语课程教学用书，也可供相关领域的工程技术人员学习参考。

图书在版编目（CIP）数据

通信与电子信息专业英语 / 古海云，任蕾主编. —北京：电子工业出版社，2024.2

ISBN 978-7-121-47165-0

Ⅰ. ①通…　Ⅱ. ①古…　②任…　Ⅲ. ①通信工程－英语－高等学校－教材②电子信息－英语－高等学校－教材　Ⅳ. ①TN91②G203

中国国家版本馆 CIP 数据核字（2024）第 022204 号

责任编辑：王晓庆　　文字编辑：韩玉宏
印　　刷：保定市中画美凯印刷有限公司
装　　订：保定市中画美凯印刷有限公司
出版发行：电子工业出版社
　　　　　北京市海淀区万寿路 173 信箱　　　邮编：100036
开　　本：787×1 092　1/16　印张：18　　　字数：599 千字
版　　次：2024 年 2 月第 1 版
印　　次：2024 年12月第 2 次印刷
定　　价：59.00 元

凡所购买电子工业出版社图书有缺损问题，请向购买书店调换。若书店售缺，请与本社发行部联系，联系及邮购电话：（010）88254888，88258888。

质量投诉请发邮件至 zlts@phei.com.cn，盗版侵权举报请发邮件至 dbqq@phei.com.cn。

本书咨询联系方式：（010）88254113，wangxq@phei.com.cn。

前　　言

在信息科技日新月异的时代，工程师已经成为一个全球性的职业，需要与国内外同行进行英语交流。能够用专业英语清晰地读写、解释技术和商业文件，并能够将其理解的工程概念有效地传达给同行、主管和公众，是工程师和专业技术人员的必备技能。专业英语课程的学习和训练将培养工科大学生阅读和翻译专业英语文献的能力、专业英语写作能力、专业英语交流能力和基本的专业科研能力，以提高学生在全球领域的求职竞争力。

本书内容分为两部分。第一部分是 Academic Reading，共包括 14 个单元，分别是 Electric Circuits、Embedded Systems、Integrated Circuits、Signals and Systems、Digital Signal Processing、Digital Image Processing、Principles of Communications、Mobile Communication、Optical Communication、Telecommunication Networks、Electromagnetics、Microwave Communication、Basics of Computer Science、New Technology Trends，涵盖电子、通信、电信及计算机等相关专业方向。每个单元由课文、词汇、注释、习题 4 个部分组成。课文选材广泛，既包括专业基础内容，也包括最新的研究热点；词汇部分按文中出现的顺序列出相关专业术语，并尽量保持术语在文中的固定搭配形式，便于学生更好地掌握术语的含义和用法；注释部分包括课文中的特色句型，并配有翻译；习题部分提供多种形式的练习，包括专业词汇释义、英汉互译、专业材料阅读总结和实用语法，力求学以致用。

第二部分是 Practical Guidance，共包括 4 个单元，分别是 Learning Skills、Research Skills、Careers in ECE、Communication Skills，旨在培养学生的学习能力、科研能力、求职能力和沟通能力。每个单元包括相关主题的实用指导与建议，以及针对性训练。

本书可作为电子工程、通信工程、电信工程等专业大学本（专）科生的科技英语和专业英语课程教学用书。本书能够满足大多数高校 32～48 学时的教学计划安排，教师可根据学生的专业方向和课程的学时适当选取教学内容。**本书提供配套的电子课件 PPT、词汇表、习题参考答案等，请登录华信教育资源网（www.hxedu.com.cn）注册后免费下载，也可联系本书编辑（wangxq@ phei.com.cn，010-88254113）索取。**教学大纲、教学指南，请发送电子邮件至作者（271435904@qq.com）索取。

本书由多年担任专业英语课和专业双语课、全英语课教学的教师编写。由古海云主持制定编写大纲，并对全书进行统稿，古海云和任蕾担任主编。其中，Unit 1～3、Unit 7～13、Unit 15～18 由古海云编写，Unit 4～6 和 Unit 14 由任蕾编写。

工程师是一个需要终身学习的职业，随着新技术的不断涌现，我们的学习、工作和生活方式也都在与时俱进。“授之以鱼，不如授之以渔”，希望本书在能力培养方面做出的努力对读者有所助益。

编　者

目　　录

Part Ⅰ　Academic Reading

Part Ⅱ Practical Guidance

Part I Academic Reading

Unit 1 Electric Circuits

1.1 Circuit Elements

An element is the basic building block of a circuit. An electric circuit is simply an interconnection of the elements. Circuit analysis is the process of determining voltages across (or the currents through) the elements of the circuit.

There are two types of elements found in electric circuits: passive elements and active elements. An active element is capable of generating energy while a passive element is not. Examples of passive elements are resistors, capacitors, and inductors. Typical active elements include generators, batteries, and operational amplifiers. Our aim in this section is to gain familiarity with some important active elements.

The most important active elements are voltage or current sources that generally deliver power to the circuit connected to them. There are two kinds of sources: independent and dependent sources.

An ideal independent source is an active element that provides a specified voltage or current that is completely independent of other circuit elements. **In other words, an ideal independent voltage source delivers to the circuit whatever current is necessary to maintain its terminal voltage.** Physical sources such as batteries and generators may be regarded as approximations to ideal voltage sources. Figure 1.1 shows the symbols for independent voltage sources. Notice that both symbols in Figure 1.1(a) and (b) can be used to represent a DC voltage source, but only the symbol in Figure 1.1(a) can be used for a time-varying voltage source. Similarly, an ideal independent current source is an active element that provides a specified current completely independent of the voltage across the source. That is, the current source delivers to the circuit whatever voltage is necessary to maintain the designated current. The symbol for an independent current source is displayed in Figure 1.2, where the arrow indicates the direction of current i.

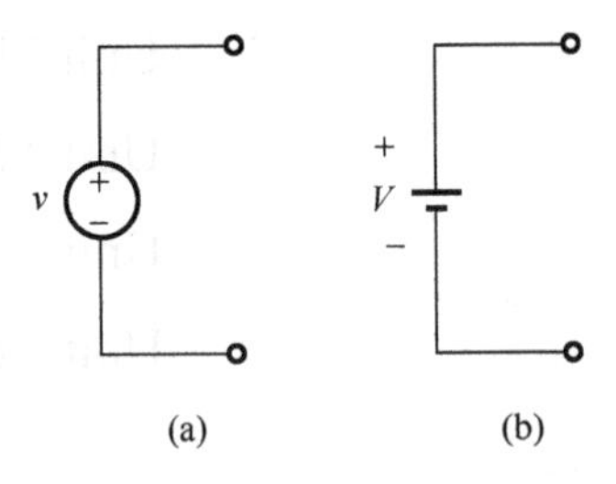

Figure 1.1 Symbols for independent voltage sources: (a) used for constant or time-varying voltage; (b) used for constant voltage (DC)

An ideal dependent (or controlled) source is an active element in which the source quantity is controlled by another voltage or current. Dependent sources are usually designated by diamond-shaped symbols, as shown in Figure 1.3. Since the control of the dependent source is achieved by a voltage or current of some other element in the circuit, and the source can be voltage or current, it follows that there are four possible types of dependent sources, namely:

1. A voltage-controlled voltage source (VCVS).
2. A current-controlled voltage source (CCVS).
3. A voltage-controlled current source (VCCS).
4. A current-controlled current source (CCCS).

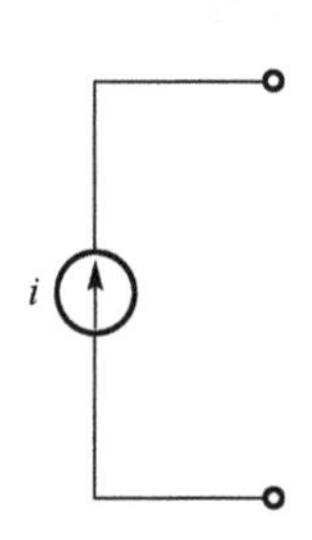

Figure 1.2 Symbol for an independent current source

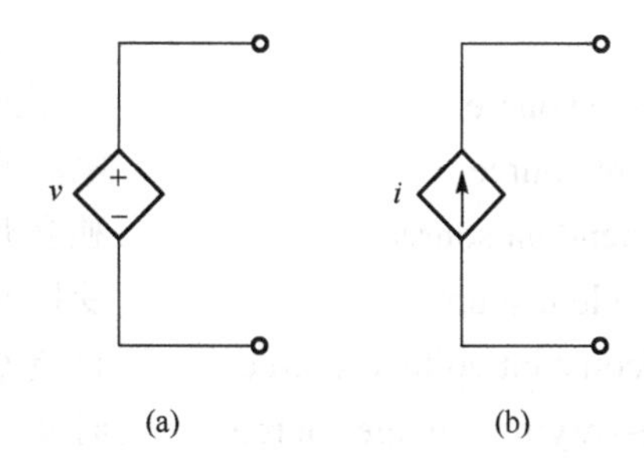

Figure 1.3 Symbols for dependent sources: (a) dependent voltage source; (b) dependent current source

Dependent sources are useful in modeling elements such as transistors, operational amplifiers, and integrated circuits. An example of a current-controlled voltage source is shown on the right-hand side of Figure 1.4, where the voltage $10i$ of the voltage source depends on the current i through element C. Students might be surprised that the value of the dependent voltage source is $10i$ V (and not $10i$ A) because it is a voltage source. The key idea to keep in mind is that a voltage source comes with polarities (+ −) in its symbol, while a current source comes with an arrow, irrespective of what it depends on.

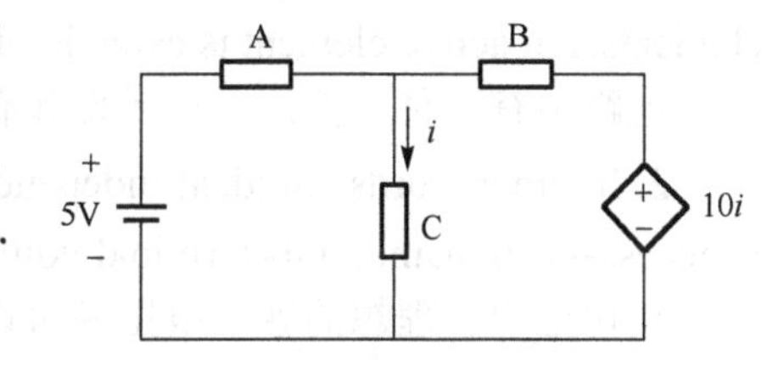

Figure 1.4 An example of a current-controlled voltage source

It should be noted that an ideal voltage source (dependent or independent) will produce any current required to ensure that the terminal voltage is as stated, whereas an ideal current source will produce the necessary voltage to ensure the stated current flow. Thus, an ideal source could in theory supply an infinite amount of energy. **It should also be noted that not only do sources supply power to a circuit, they can absorb power from a circuit too.** For a voltage source, we know the voltage but not the current supplied or drawn by it. By the same token, we know the current supplied by a current source but not the voltage across it.

Words and Expressions

circuit element	电路元件
passive element	无源元件
active element	有源元件
resistor [rɪ'zɪstər]	*n.* 电阻器
capacitor [kə'pæsɪtər]	*n.* 电容器

inductor [ɪn'dʌktər] n. 电感器；电感线圈；感应器（体，物，元件，线圈）
generator ['dʒenəreɪtər] n. 生成器；发电机
battery ['bætəri] n. 电池
operaional amplifier 运算放大器
gain [geɪn] v. 获得
n. 增加；好处；收益；增益
voltage source 电压源
current source 电流源
independent source 独立源
dependent source 受控源，非独立源
DC/constant voltage source 直流/恒定电压源
time-varying voltage source 时变电压源
transistor [træn'zɪstər] n. 晶体管
integrated circuit IC，集成电路

Notes

1. There are two types of elements found in electric circuits: passive elements and active elements. An active element is capable of generating energy while a passive element is not.

电路中有两种元件：无源元件和有源元件。有源元件能产生能量，而无源元件不能。

2. In other words, an ideal independent voltage source delivers to the circuit whatever current is necessary to maintain its terminal voltage.

换句话说，理想的独立电压源可以为电路提供维持其终端电压所需的任何电流。

3. An ideal dependent (or controlled) source is an active element in which the source quantity is controlled by another voltage or current.

理想非独立（或受控）电源是有源元件，其电压或电流的大小由另一个电压或电流控制。

4. Dependent sources are useful in modeling elements such as transistors, operational amplifiers, and integrated circuits.

非独立源可用于晶体管、运算放大器和集成电路等元件的建模。

5. It should also be noted that not only do sources supply power to a circuit, they can absorb power from a circuit too.

还应注意的是，电源不仅向电路供电，也可以从电路中吸收功率。

1.2 Basic Laws

1.2.1 Ohm's Law

Materials in general have a characteristic behavior of resisting the flow of electric charge. **This physical property, or ability to resist current, is known as *resistance* and is represented by the symbol *R*.** The resistance of any material with a uniform cross-sectional area A depends on A and its length l, as shown in Figure 1.5(a). We can represent resistance (as measured in the laboratory), in mathematical form:

$$R = \rho \frac{l}{A} \tag{1.1}$$

where ρ is known as the *resistivity* of the material in ohm-meters. **Good conductors, such as copper and aluminum, have low resistivities, while insulators, such as mica and paper, have high resistivities.** Table 1.1 presents the values of ρ for some common materials and shows which materials are used for conductors, insulators, and semiconductors.

Table 1.1 Resistivities of common materials

Material	Resistivity(Ω·m)	Usage
Silver	1.64×10^{-8}	Conductor
Copper	1.72×10^{-8}	Conductor
Aluminum	2.8×10^{-8}	Conductor
Gold	2.45×10^{-8}	Conductor
Carbon	4×10^{-5}	Semiconductor
Germanium	0.47	Semiconductor
Silicon	6.4×10^{2}	Semiconductor
Paper	10^{10}	Insulator
Mica	5×10^{11}	Insulator
Glass	10^{12}	Insulator
Teflon	3×10^{12}	Insulator

The circuit element used to model the current-resisting behavior of a material is the *resistor*. For the purpose of constructing circuits, resistors are usually made from metallic alloys and carbon compounds. The circuit symbol for the resistor is shown in Figure 1.5(b), where R stands for the resistance of the resistor. The resistor is the simplest passive element.

Georg Simon Ohm (1787–1854), a German physicist, is credited with finding the relationship between current and voltage for a resistor. This relationship is known as Ohm's law. That is,

$$v \propto i \tag{1.2}$$

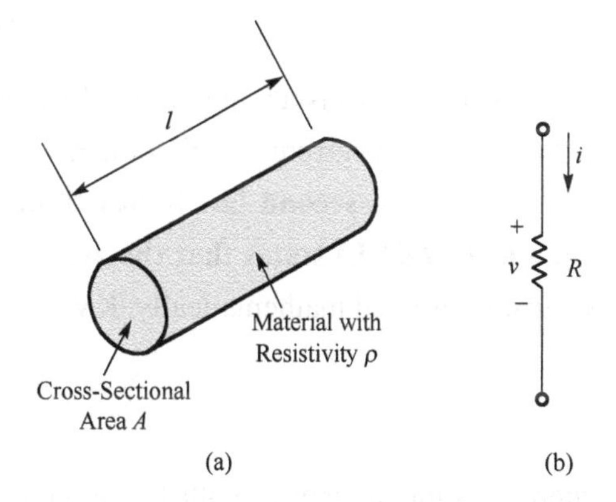

Figure 1.5 (a) Resistance; (b) circuit symbol for the resistor

Ohm defined the constant of proportionality for a resistor to be the resistance, R. (The resistance is a material property which can change if the internal or external conditions of the element are altered, e.g., if there are changes in the temperature.) Thus, Equation (1.2) becomes

$$v = iR \tag{1.3}$$

which is the mathematical form of Ohm's law. R in Equation (1.3) is measured in the unit of ohms, designated Ω.

1.2.2 Kirchhoff's Laws

Ohm's law by itself is not sufficient to analyze circuits. **However, when it is coupled with Kirchhoff's two laws, we have a sufficient, powerful set of tools for analyzing a large variety of electric circuits.** Kirchhoff's laws were first introduced in 1847 by the German physicist Gustav Robert Kirchhoff (1824–1887). These laws are formally known as Kirchhoff's current law (KCL) and Kirchhoff's voltage law (KVL).

Kirchhoff's first law is based on the law of conservation of charge, which requires that the algebraic sum of charges within a system cannot change.

Kirchhoff's current law (KCL) states that the algebraic sum of currents entering a node (or a closed boundary) is zero. Mathematically, KCL implies that

$$\sum_{n=1}^{N} i_n = 0 \tag{1.4}$$

where N is the number of branches connected to the node and i_n is the nth current entering (or leaving) the node. By this law, currents entering a node may be regarded as positive, while currents leaving the node may be taken as negative or vice versa.

Consider the node in Figure 1.6. Applying KCL gives

$$i_1 + (-i_2) + i_3 + i_4 + (-i_5) = 0 \tag{1.5}$$

since currents i_1, i_3, and i_4 are entering the node, while currents i_2 and i_5 are leaving it. By rearranging the terms, we get

$$i_1 + i_3 + i_4 = i_2 + i_5 \tag{1.6}$$

Equation (1.6) is an alternative form of KCL: the sum of the currents entering a node is equal to the sum of the currents leaving the node.

Kirchhoff's second law is based on the principle of conservation of energy: Kirchhoff's voltage law (KVL) states that the algebraic sum of all voltages around a closed path (or loop) is zero. Expressed mathematically, KVL states that

$$\sum_{m=1}^{M} v_m = 0 \tag{1.7}$$

where M is the number of voltages in the loop (or the number of branches in the loop) and v_m is the mth voltage.

To illustrate KVL, consider the circuit in Figure 1.7. The sign on each voltage is the polarity of the terminal encountered first as we travel around the loop. We can start with any branch and go around the loop either clockwise or counterclockwise. Suppose we start with the voltage source and go clockwise around the loop as shown; then voltages would be $-v_1$, $+v_2$, $+v_3$, $-v_4$, and $+v_5$, in that order. For example, as we reach branch 3, the positive terminal is met first; hence, we have $+v_3$. For branch 4, we reach the negative terminal first; hence, we have $-v_4$. Thus, KVL yields

$$-v_1 + v_2 + v_3 - v_4 + v_5 = 0 \tag{1.8}$$

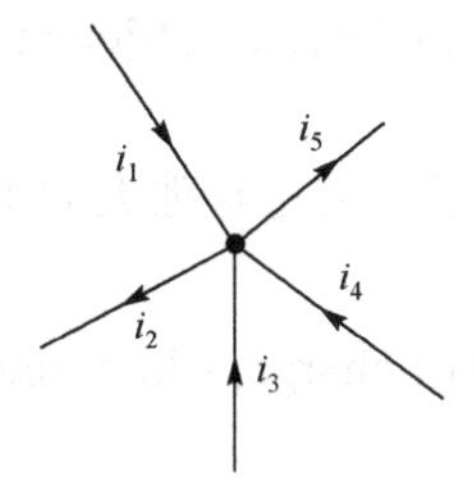

Figure 1.6 Currents at a node illustrating KCL

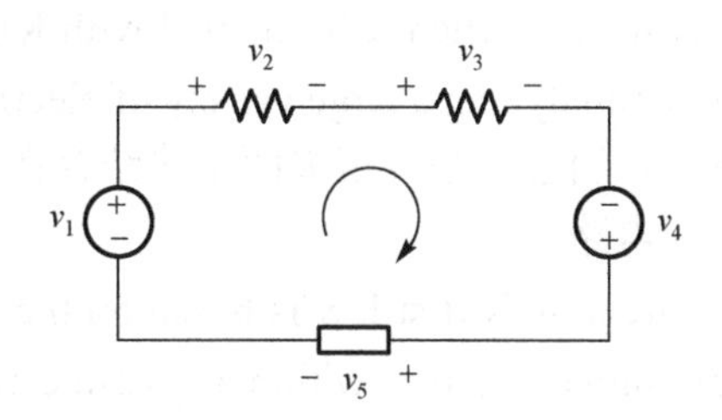

Figure 1.7 A single-loop circuit illustrating KVL

Rearranging terms gives

$$v_2 + v_3 + v_5 = v_1 + v_4 \tag{1.9}$$

which may be interpreted as: sum of voltage drops = sum of voltage rises. This is an alternative form of KVL. Notice that if we had traveled counterclockwise, the result would have been $+v_1$, $-v_5$, $+v_4$, $-v_3$, and $-v_2$, which is the same as before except that the signs are reversed. Hence, Equations (1.8) and (1.9) remain the same.

Words and Expressions

Ohm's law	欧姆定律
resistance [rɪ'zɪstəns]	*n.* 电阻，电阻值；抵抗力；阻力
conductor [kən'dʌktər]	*n.* 导体
insulator ['ɪnsəleɪtər]	*n.* 隔热（或绝缘、隔音等的）材料（或装置）
semiconductor ['semikəndʌktər]	*n.* 半导体；半导体装置
metallic alloy	金属合金
carbon compound	碳化合物
aluminum [ə'lumənəm]	*n.* 铝
germanium [dʒɜːr'meɪniəm]	*n.* 锗
silicon ['sɪlɪkən]	*n.* 硅
mica ['maɪkə]	*n.* 云母
teflon ['teflɑːn]	*n.* 特氟隆，聚四氟乙烯（不粘锅涂层材料）
Kirchhoff's current law (KCL)	基尔霍夫电流定律
Kirchhoff's voltage law (KVL)	基尔霍夫电压定律

Notes

1. This physical property, or ability to resist current, is known as *resistance* and is represented by the symbol *R*.

这种物理特性或抵抗电流的能力称为电阻，用符号 *R* 表示。

2. Good conductors, such as copper and aluminum, have low resistivities, while insulators, such as mica and paper, have high resistivities.

良导体（如铜和铝）的电阻率较低，而绝缘体（如云母和纸）的电阻率较高。

3. However, when it is coupled with Kirchhoff's two laws, we have a sufficient, powerful set of tools for analyzing a large variety of electric circuits.

然而，当它与基尔霍夫两定律结合在一起时，我们就有了一套足够强大的工具来分析各种各样的电路。

4. Kirchhoff's first law is based on the law of conservation of charge, which requires that the algebraic sum of charges within a system cannot change.

基尔霍夫第一定律基于电荷守恒定律，该定律要求系统内电荷的代数和不能改变。

5. Kirchhoff's second law is based on the principle of conservation of energy: Kirchhoff's voltage law (KVL) states that the algebraic sum of all voltages around a closed path (or loop) is zero.

基尔霍夫第二定律基于能量守恒原理：基尔霍夫电压定律（KVL）规定闭合路径（或回路）上电压的代数和为零。

1.3 Circuit Analysis

Having understood the fundamental laws of circuit theory (Ohm's law and Kirchhoff's laws), we are now prepared to apply these laws to develop two powerful techniques for circuit analysis: nodal analysis, which is based on a systematic application of Kirchhoff's current law (KCL), and mesh analysis, which is based on a systematic application of Kirchhoff's voltage law (KVL).

1.3.1 Nodal Analysis

Nodal analysis provides a general procedure for analyzing circuits using node voltages as the circuit variables. **Choosing node voltages instead of element voltages as circuit variables is convenient and reduces the number of equations one must solve simultaneously.** Given a circuit with n nodes without voltage sources, the nodal analysis of the circuit involves taking the following three steps:

1. Select a node as the reference node. Assign voltages v_1, v_2,…,v_{n-1} to the remaining $n-1$ nodes. The voltages are referenced with respect to the reference node.

2. Apply KCL to each of the $n-1$ nonreference nodes. Use Ohm's law to express the branch currents in terms of node voltages.

3. Solve the resulting simultaneous equations to obtain the unknown node voltages.

Consider, for example, the circuit in Figure 1.8(a). Node 0 is the reference node which is commonly called the ground since it is assumed to have zero potential ($v = 0$), while nodes 1 and 2 are assigned voltages v_1 and v_2, respectively. Keep in mind that the node voltages are defined with respect to the reference node. As illustrated in Figure 1.8(a), each node voltage is the voltage rise from the reference node to the corresponding nonreference node or simply the voltage of that node with respect to the reference node.

As the second step, we apply KCL to each nonreference node in the circuit. To avoid putting too much information on the same circuit, the circuit in Figure 1.8(a) is redrawn in Figure 1.8(b), where we now add i_1, i_2, and i_3 as the currents through resistors. At node 1, applying KCL gives

$$I_1 = I_2 + i_1 + i_2 \tag{1.10}$$

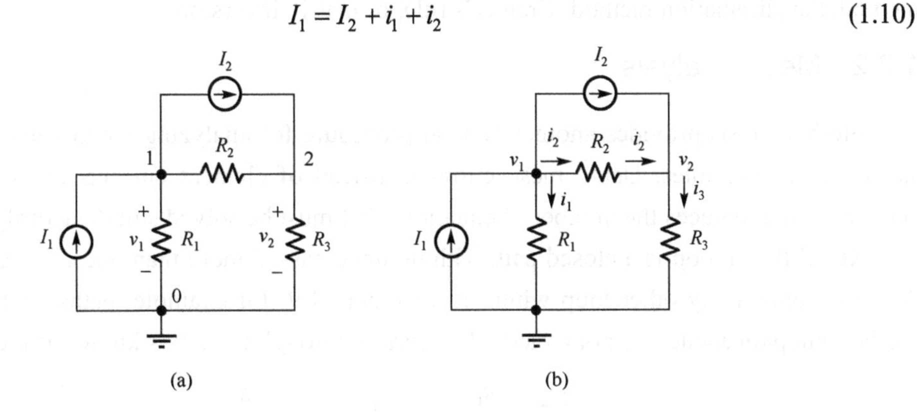

Figure 1.8 Typical circuit for nodal analysis

At node 2,

$$I_2 + i_2 = i_3 \tag{1.11}$$

We now apply Ohm's law to express the unknown currents i_1, i_2, and i_3 in terms of node voltages. **The key idea to bear in mind is that, since resistor is a passive element, by the passive sign convention, current must always flow from a higher potential to a lower potential.** With this in mind, we obtain from Figure 1.8(b)

$$i_1 = \frac{v_1 - 0}{R_1} \quad \text{or} \quad i_1 = G_1 v_1$$

$$i_2 = \frac{v_1 - v_2}{R_2} \quad \text{or} \quad i_2 = G_2(v_1 - v_2) \tag{1.12}$$

$$i_3 = \frac{v_2 - 0}{R_3} \quad \text{or} \quad i_3 = G_3 v_2$$

Substituting Equation (1.12) in Equations (1.10) and (1.11) results, respectively, in

$$I_1 = I_2 + \frac{v_1}{R_1} + \frac{v_1 - v_2}{R_2} \tag{1.13}$$

$$I_2 + \frac{v_1 - v_2}{R_2} = \frac{v_2}{R_3} \tag{1.14}$$

In terms of the conductances, Equations (1.13) and (1.14) become

$$I_1 = I_2 + G_1 v_1 + G_2(v_1 - v_2) \tag{1.15}$$

$$I_2 + G_2(v_1 - v_2) = G_3 v_2 \tag{1.16}$$

The third step in nodal analysis is to solve for the node voltages. If we apply KCL to $n - 1$ nonreference nodes, we obtain $n - 1$ simultaneous equations such as Equations (1.13) and (1.14) or (1.15) and (1.16). For the circuit in Figure 1.8, we solve Equations (1.13) and (1.14) or (1.15) and

(1.16) to obtain the node voltages v_1 and v_2 using any standard method, such as the substitution method, the elimination method, Cramer's rule, or matrix inversion.

1.3.2 Mesh Analysis

Mesh analysis provides another general procedure for analyzing circuits using mesh currents as the circuit variables. Using mesh currents instead of element currents as circuit variables is convenient and reduces the number of equations that must be solved simultaneously.

Recall that a loop is a closed path with no node passed more than once. A mesh is a loop that does not contain any other loop within it. In Figure 1.9, for example, paths abefa and bcdeb are meshes, but path abcdefa is not a mesh. The current through a mesh is known as mesh current.

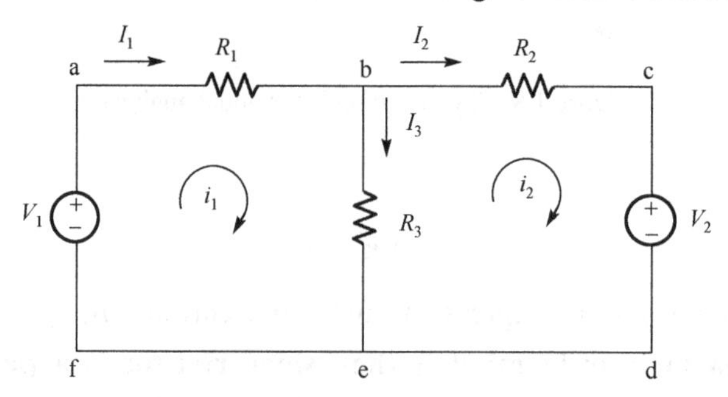

Figure 1.9 A circuit with two meshes

Nodal analysis applies KCL to find unknown voltages in a given circuit, while mesh analysis applies KVL to find unknown currents. In the mesh analysis of a circuit with n meshes, we take the following three steps:

1. Assign mesh currents $i_1, i_2, \ldots, i_n$ to the n meshes.

2. Apply KVL to each of the n meshes. Use Ohm's law to express the voltages in terms of the mesh currents.

3. Solve the resulting n simultaneous equations to get the mesh currents.

To illustrate the steps, consider the circuit in Figure 1.9. The first step requires that mesh currents i_1 and i_2 are assigned to meshes 1 and 2. **Although a mesh current may be assigned to each mesh in an arbitrary direction, it is conventional to assume that each mesh current flows clockwise.**

As the second step, we apply KVL to each mesh. Applying KVL to mesh 1, we obtain

$$-V_1 + R_1 i_1 + R_3(i_1 - i_2) = 0 \quad \text{or} \quad (R_1 + R_3)i_1 - R_3 i_2 = V_1 \tag{1.17}$$

For mesh 2, applying KVL gives

$$R_2 i_2 + V_2 + R_3(i_2 - i_1) = 0 \quad \text{or} \quad -R_3 i_1 + (R_2 + R_3)i_2 = -V_2 \tag{1.18}$$

Note in Equation (1.17) that the coefficient of i_1 is the sum of the resistances in the first mesh, while the coefficient of i_2 is the negative of the resistance common to meshes 1 and 2. Now observe that the same is true in Equation (1.18). This can serve as a shortcut way of writing the mesh equations.

The third step is to solve for the mesh currents. Putting Equations (1.17) and (1.18) in matrix form yields

$$\begin{bmatrix} R_1+R_3 & -R_3 \\ -R_3 & R_2+R_3 \end{bmatrix}\begin{bmatrix} i_1 \\ i_2 \end{bmatrix}=\begin{bmatrix} V_1 \\ -V_2 \end{bmatrix} \tag{1.19}$$

which can be solved to obtain the mesh currents i_1 and i_2. We are at liberty to use any technique for solving the simultaneous equations. If a circuit has n nodes, b branches, and l independent loops or meshes, then $l = b - n + 1$. Hence, l independent simultaneous equations are required to solve the circuit using mesh analysis.

Notice that the branch currents are different from the mesh currents unless the mesh is isolated. To distinguish between the two types of currents, we use i for a mesh current and I for a branch current. The current elements I_1, I_2, and I_3 are algebraic sums of the mesh currents.

It is evident from Figure 1.9 that:

$$I_1 = i_1, I_2 = i_2, I_3 = i_1 - i_2 \tag{1.20}$$

Words and Expressions

nodal analysis 节点分析法
mesh analysis 网孔分析法；网格分析法
reference node 参考节点；参考点
node voltage 节点电压
branch current 支路电流
ground[graʊnd] *n.* 地；地电位
potential [pə'tenʃl] *n.* 电位；电势；电压
simultaneous equations 联立方程；联立方程组；联立方程式
substitution method 替换法；代入法；代换法；替代法；置换法
elimination method 消除法；消元法
Cramer's rule 克拉默法则
matrix inversion 矩阵求逆
mesh current 网孔电流；网格电流
loop [lu:p] *n.* 回路；环路
coefficient [ˌkoʊɪ'fɪʃnt] *n.* 系数

Notes

1. Having understood the fundamental laws of circuit theory (Ohm's law and Kirchhoff's laws), we are now prepared to apply these laws to develop two powerful techniques for circuit analysis: nodal analysis, which is based on a systematic application of Kirchhoff's current law (KCL), and mesh analysis, which is based on a systematic application of Kirchhoff's voltage law (KVL).

理解电路理论的基本定律（欧姆定律和基尔霍夫定律）后，我们现在准备应用这些定律来开发两种强大的电路分析技术：基于基尔霍夫电流定律（KCL）系统应用的节点分析法和基于基尔霍夫电压定律（KVL）系统应用的网孔分析法。

2. Choosing node voltages instead of element voltages as circuit variables is convenient and reduces the number of equations one must solve simultaneously.

选择节点电压而不是元件电压作为电路变量很方便，并且减少了联立方程的数量。

3. The key idea to bear in mind is that, since resistor is a passive element, by the passive sign convention, current must always flow from a higher potential to a lower potential.

要记住的关键思想是，由于电阻器是无源元件，根据无源符号惯例，电流必须始终从较高的电势流向较低的电势。

4. Although a mesh current may be assigned to each mesh in an arbitrary direction, it is conventional to assume that each mesh current flows clockwise.

尽管每个网孔的电流可以指定为任意方向，但通常假定网孔电流顺时针方向流动。

5. Note in Equation (1.17) that the coefficient of i_1 is the sum of the resistances in the first mesh, while the coefficient of i_2 is the negative of the resistance common to meshes 1 and 2.

注意在式（1.17）中，i_1 的系数是第一个网孔中电阻的总和，而 i_2 的系数是网孔 1 和 2 共有电阻的负值。

Exercises

1. Match the terms (1)–(6) with the definitions A–F.

(1) node	A. a chain of components with a single current path
(2) branch	B. an arbitrary reference for a given circuit that cannot necessarily be equated with earth ground
(3) ground	C. a circuit through which no current flows
(4) source	D. a point in a circuit where multiple branches in the circuit join
(5) open circuit	E. a circuit across which no voltage can be developed
(6) short circuit	F. a portion of a circuit capable of generating power

2. Translate into Chinese.

(1) There are two quantities that we like to keep track of in electronic circuits: voltage and current. These are usually changing with time; otherwise nothing interesting is happening.

(2) Currents flow through circuit elements, and voltages are applied across circuit elements. So you've got to say it right: always refer to the voltage between two nodes or across two nodes in a circuit; always refer to the current through a device or connection in a circuit.

(3) Power goes into heat (usually), or sometimes mechanical work (motors), radiated energy (lamps, transmitters), or stored energy (batteries, capacitors, inductors).

3. Translate into English.

（1）一个完美的电流源是一个双端的黑匣子，无论负载电阻或外加电压为何，都能向外部电路提供一个恒定电流。

（2）铁芯电感器通常被称为扼流圈，因为在交流电路中，它具有扼流圈效应，限制流过它的电流。

（3）使用基尔霍夫定律分析电路的主要优点之一是，我们可以在不改变电路原始结构的情况下分析电路；这种方法的一个主要缺点是，对于大型复杂电路，计算很烦琐。

4. Read the following article and write a summary.

Using electronics today is so much a part of our daily lives we hardly think of the way the world would be without electronics. Everything from cooking to music uses electronics or electronic components in some way. Our family car has many electronic components, as does our cooking stove, laptop and cell phone. Children and teenagers carry mobile phones with them everywhere and use them to take and send pictures, videos, and to play music. They send text messages on the cell phone to other phones and to their home computers.

Wireless internet is becoming more common all the time, with laptops set up in cyber cafes where people can drink coffee and check their e-mails all at the same time. The computer user can do all the web searching in relative privacy thanks to the electronic accessories which can be added to the computer. Conversely, more and more transactions are being sent electronically across the airwaves so security is becoming a larger issue than ever before. Merchants who sell products online must be able to assure their customers that information submitted at a website is not being accessed by unauthorized personnel.

Music is a prime user of electronics, both in recording mode and in playback mode. Stereos, record players, tape decks, cassette players, CD drives and DVD players were all the result of advances in electronics technology in the last few decades. Today people can carry a playlist of hundreds of songs around with them easily in a very small device—easily portable. When you add Bluetooth or headphones the music can be heard by the user, but does not disturb those nearby.

Electronics technology in cameras has increased dramatically. A digital camera is available to most Americans at a price they can afford and a cell phone often includes a fairly sophisticated digital camera that can capture still pictures or even video pictures and store them or transfer them to a computer where they can be saved, shared digitally with family or friends or printed out in hard form with a photo printer device. Pictures obtained through a camera or by means of a scanner can be edited, cropped, enhanced or enlarged easily through the marvel of electronics.

Electronics devices are being used in the health field, not only to assist in diagnosis and determination of medical problems, but to assist in the research that is providing treatments and cures for illnesses and even genetic anomalies. Equipment such as MRI, CAT and the older X-rays, tests for diabetes, cholesterol and other blood component tests all rely on electronics in order to do their work quickly and accurately. Pacemakers and similar equipment implanted in the body are now almost routine.

Literally thousands of everyday devices that we use constantly make use of electronics technology in order to operate. These are products ranging from automotive engines to automated equipment in production settings. Even artistic efforts benefit from computer modeling prior to the committing of valuable artistic media to create the finished product.

5. Language study: Describing block diagrams and circuits

We can describe a block diagram or a circuit like this: "It **consists of/is composed of** two resistors, one capacitor, and a voltage source. "

Also, we can describe the links between each building block like this: "The resistor **is connected to/ is linked** to a voltage source."

Finally, we can describe the values of the components like these: "R_1 a two-hundred-and-twenty-kilohm resistor" and "C_2 a hundred-picofarad capacitor".

Describe the circuits in Figure 1.8 and Figure 1.9 to your partner.

*This table provides the terms you might need.

Prefix	Symbol	Multiple	Example
giga	G	10^9	GHz gigahertz
mega	M	10^6	MΩ megohm
kilo	k	10^3	kV kilovolt
deci	d	10^{-1}	dB decibel
milli	m	10^{-3}	mW milliwatt
micro	μ	10^{-6}	μH microhenry
nano	n	10^{-9}	nF nanofarad
pico	p	10^{-12}	pF picofarad

Unit 2 Embedded Systems

2.1 Introduction to Embedded Systems

An embedded system can be broadly defined as a device that contains tightly coupled hardware and software components to perform a single function, forms part of a larger system, is not intended to be independently programmable by the user, and is expected to work with minimal or no human interaction. Two additional characteristics are very common in embedded systems: reactive operation and heavily constrained.

Most embedded systems interact directly with processes or the environment, making decisions on the fly, based on their inputs. This makes necessary that the system must be reactive, responding in real time to process inputs to ensure proper operation. **Besides, these systems operate in constrained environments where memory, computing power, and power supply are limited.** Moreover, production requirements, in most cases due to volume, place high cost constraints on designs.

2.1.1 Early Forms of Embedded Systems

The concept of an embedded system is as old as the concept of an electronic computer, and in a certain way, it can be said to precede the concept of a general purpose computer. If we look at the earliest forms of computing devices, they adhere better to the definition of an embedded system than to that of a general purpose computer. **One of the earliest electronic computing devices credited with the term "embedded system" and closer to our present conception of such was the Apollo Guidance Computer (AGC).** Developed at the MIT Instrumentation Laboratory by a group of designers led by Charles Stark Draper in the early 1960s, the AGC was part of the guidance and navigation system used by NASA in the Apollo program for various spaceships. In its early days it was considered one of the riskiest items in the Apollo program due to the usage of the then newly developed monolithic integrated circuits.

The AGC incorporated a user interface module based on keys, lamps, and seven-segment numeric displays (see Figure 2.1); a hardwired control unit based on 4,100 single three-input RTL NOR gates, 4 KB of magnetic core RAM, and 32 KB of core rope ROM. The unit CPU was run by a 2.048-MHz primary clock, had four 16-bit central registers and executed eleven instructions. It supported five vectored interrupt sources, including a 20-register timer-counter, a real-time clock module, and even allowed for a low-power standby mode that reduced in over 85% the module's power consumption, while keeping alive all critical components.

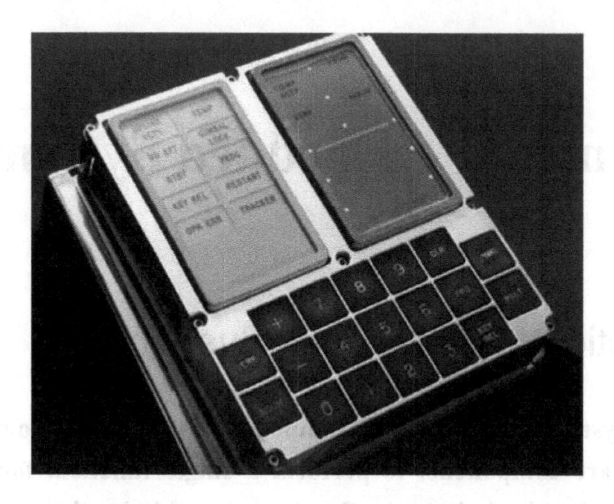

Figure 2.1 AGC user interface module (public photo EC96-43408-1 by NASA)

The system software of the Apollo Guidance Computer was written in AGC assembly language and supported a non-preemptive real-time operating system that could simultaneously run up to eight prioritized jobs. The AGC was indeed an advanced system for its time. As we enter into the study of contemporary applications, we will find that most of these features are found in many of today's embedded systems. But the flourishing of embedded systems in commercial applications had to wait until another remarkable event in electronics: the advent of the microprocessor.

2.1.2 Birth and Evolution of Modern Embedded Systems

The beginning of the decade of the 1970s witnessed the development of the first microprocessor designs. By the end of 1971, almost simultaneously and independently, design teams working for Texas Instruments, Intel, and the US Navy had developed implementations of the first microprocessors.

Gary Boone from Texas Instruments was awarded in 1973 the patent of the first single-chip microprocessor architecture for its 1971 design of the TMS1000 (see Figure 2.2). **This chip was a 4-bit CPU that incorporated in the same die 1 KB of ROM and 256 bits of RAM to offer a complete computer functionality in a single-chip, making it the first microcomputer-on-a-chip (a.k.a. microcontroller).** The TMS1000 was launched in September 1971 as a calculator chip with part number TMS1802NC.

The 4004 (see Figure 2.3) recognized as the first commercial, stand-alone single chip microprocessor, was launched by Intel in November 1971. The chip was developed by a design team led by Federico Faggin at Intel. This design was also a 4-bit CPU intended for use in electronic calculators. The 4004 was able to address 4 KB of memory, operating at a maximum clock frequency of 740 kHz. Integrating a minimum system around the 4004 required at least three additional chips: a 4001 ROM, a 4002 RAM, and a 4003 I/O interface.

The third pioneering microprocessor design of that age was a less known project for the US Navy named the Central Air Data Computer (CADC). This system implemented a chipset CPU for the F-14 Tomcat fighter named the MP944. The system supported 20-bit operands in a pipelined, parallel multiprocessor architecture designed around 28 chips. Due to the classified nature of this design, public disclosure of its existence was delayed until 1998, although the disclosed documentation indicates it was completed by 1970.

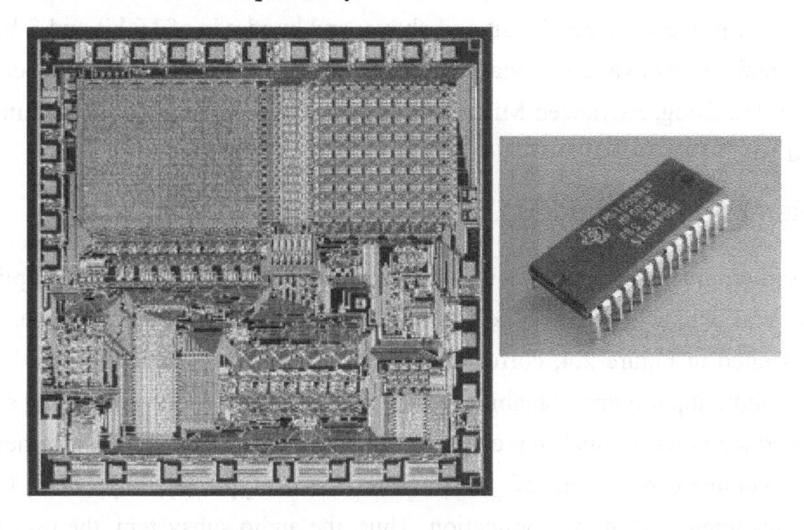

Figure 2.2 Die microphotograph (left) and packaged part for the TMS1000 (TI)

Figure 2.3 Die microphotograph (left) and packaged part for the 4004 (Intel)

After these developments, it did not take long for designers to realize the potential of microprocessors and their advantages for implementing embedded applications. Microprocessor designs soon evolved from 4-bit to 8-bit CPUs. By the end of the 1970s, the design arena was dominated by 8-bit CPUs and the market for microprocessor-based embedded applications had grown to hundreds of millions of dollars. The list of initial players grew to more than a dozen of chip manufacturers that, besides Texas Instruments and Intel, included Motorola, Zilog, Intersil, National Instruments, MOS Technology, and Signetics, to mention just a few of the most renowned. Remarkable parts include the Intel 8080 that eventually evolved into the famous 80x86/Pentium series, the Zilog Z-80, the Motorola 6800 and the MOS 6502. The evolution in CPU sizes continued through the 1980s and 1990s to 16-bit, 32-bit, and 64-bit designs, and now-a-days even some specialized CPUs crunching data at 128-bit widths. In terms of manufacturers and availability of processors, the list has grown to the point that it is possible to find over several dozens of different choices for processor sizes 32-bit and above, and hundreds of 16-bit and 8-bit processors. Examples of manufacturers available today include Texas Instruments, Intel, Microchip, Freescale (formerly Motorola), Zilog, Advanced Micro Devices, MIPS Technologies, ARM Limited, and the list goes on and on.

2.1.3 Contemporary Embedded Systems

Nowadays microprocessor applications have grown in complexity, requiring applications to be broken into several interacting embedded systems. To better illustrate the case, consider the application illustrated in Figure 2.4, corresponding to a generic multi-function media player. The system provides audio input/output capabilities, a digital camera, a video processing system, a hard drive, a user interface (keys, a touch screen, and a graphic display), power management and digital communication components. Each of these features are typically supported by individual embedded systems integrated in the application. Thus, the audio subsystem, the user interface, the storage system, the digital camera front-end, and the media processor and its peripherals are among the systems embedded in this application. Although each of these subsystems may have their own processors, programs, and peripherals, each one has a specific, unique function. None of them is user programmable, all of them are embedded within the application, and their operation require minimal or no human interaction.

The above illustrated the concept of an embedded system with a very specific application. Yet, such type of systems can be found in virtually every aspect of our daily lives: electronic toys; cellular phones; MP3 players, PDAs; digital cameras; household devices such as microwaves, dishwasher machines, TVs, and toasters; transportation vehicles such as cars, boats, trains, and airplanes; life support and medical systems such as pace makers, ventilators, and X-ray machines; safety-critical systems such as anti-lock brakes, airbag deployment systems, and electronic surveillance; and defense systems such as missile guidance computers, radars, and global positioning systems. These are only a few examples of the long list of applications that depend on embedded systems. **Despite being omnipresent in virtually every aspect of our modern lives, embedded systems are ubiquitous devices almost invisible to the user, working in a pervasive way to make possible the "intelligent" operation of machines and appliances around us.**

Words and Expressions

embedded system	嵌入式系统
coupled ['kʌpld]	*adj*. 耦合的；结合的
function ['fʌŋkʃn]	*n*. 功能；函数
programmable ['proʊgræməbl]	*adj*. 可编程的；程控的
interaction [ˌɪntər'ækʃn]	*n*. 交互作用（影响）；交相感应；干扰（涉）

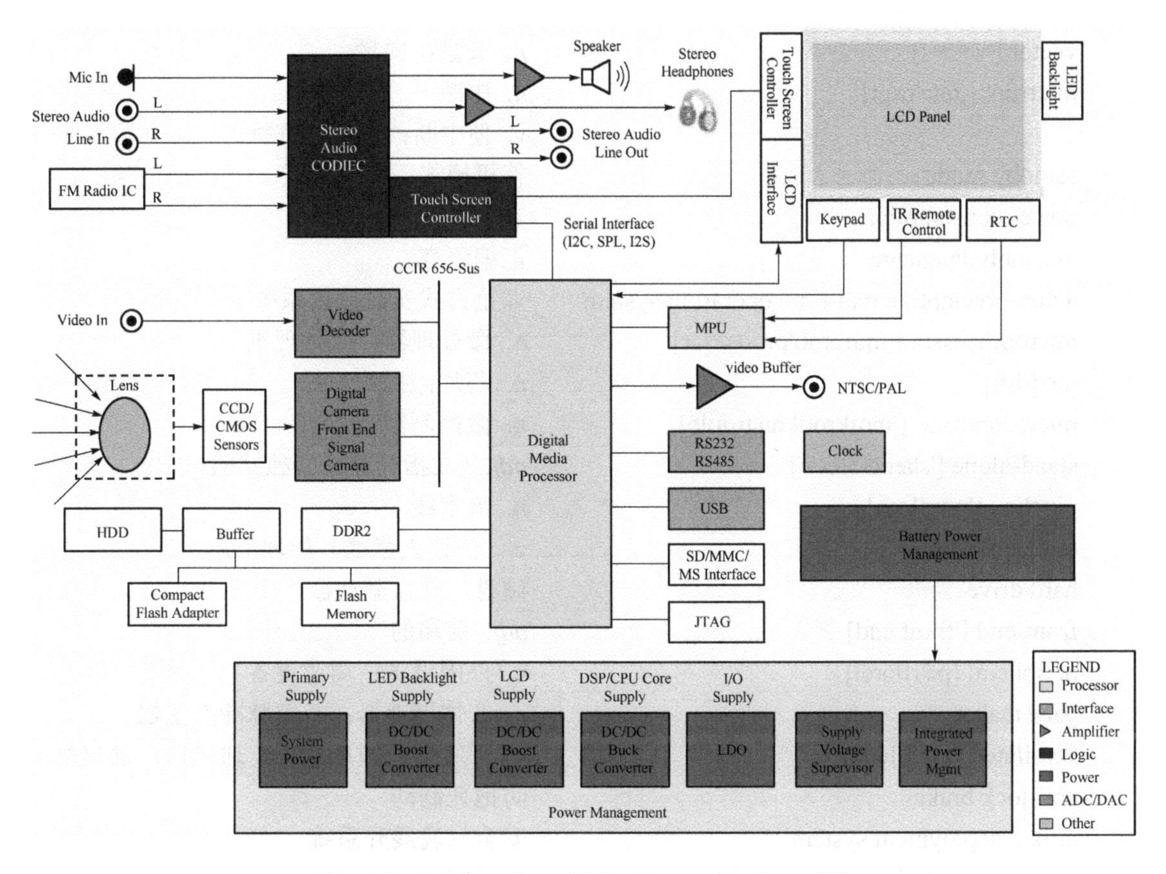

Figure 2.4 A generic multi-function media player (TI)

characteristic [ˌkærəktə'rɪstɪk] *n.* 特征；特点；品质

reactive [ri'æktɪv] *adj.* 反应的；回应的；反应式的

process [prə'ses] *n.* 进程；流程；工艺流程；过程；步骤
v. 处理；加工；审阅；审核；数据处理

on the fly 匆忙地；赶紧地

memory ['meməri] *n.* 存储器；记忆力；回忆

general purpose 通用的；普遍的；万能的；一般用途

navigation system 导航系统

monolithic [ˌmɑːnə'lɪθɪk] *adj.* 庞大而单一的；巨大而单调的

user interface 人机交互界面；操作界面；用户界面

module ['mɑːdʒuːl] *n.* 模块；功能块；程序块；组件；配件

RTL: resistor-transistor logic 电阻-晶体管逻辑

NOR gate 或非门

magnetic core 磁铁芯；磁芯

RAM：random-access memory 随机存取存储器，随机存储器

ROM：read-only memory 只读存储器

register ['redʒɪstər] *n.* 寄存器

instruction [ɪn'strʌkʃn] *n.*（计算机的）指令；用法说明；操作指南

vector ['vektər] *n.* 矢量；向量

interrupt [ˌɪntə'rʌpt] *n.* 中断
v. 使中断；打断；打扰

standby mode 待机模式；备用模式

power consumption 功耗

assembly language 汇编语言

a non-preemptive real-time operating system 非抢占式实时操作系统

microprocessor [ˌmaɪkroʊ'proʊsesər] *n.* 微处理器；微处理机

die [daɪ] *n.* 芯片；裸片

microcontroller[ˌmaɪkroʊkən'troʊlə] *n.* 微控制器

stand-alone ['stænd əloʊn] *adj.*（尤指计算机）独立的

pipeline ['paɪplaɪn] *n.* 流水线

crunch [krʌntʃ] *v.*（快速大量地）处理信息

hard drive 硬盘；硬盘驱动器

front-end ['frʌnt end] *adj.* 前端的

peripheral [pə'rɪfərəl] *n.* 外围设备；周边设备

pace maker *n.* 心律调整器；心房脉冲产生器

ventilator ['ventɪleɪtər] *n.* 呼吸器；通风设备；通风口；通气机

anti-lock brake 防抱死制动

airbag deployment system 安全气囊展开系统

electronic surveillance 电子监视（侦察）

radar ['reɪdɑ:r] *n.* 雷达

global positioning system GPS，全球定位系统

omnipresent [ˌɑ:mnɪ'preznt] *adj.* 无所不在的；遍及各处的

pervasive [pər'veɪsɪv] *adj.* 普遍的；遍布的；充斥各处的；弥漫的

Notes

1. An embedded system can be broadly defined as a device that contains tightly coupled hardware and software components to perform a single function, forms part of a larger system, is not intended to be independently programmable by the user, and is expected to work with minimal or no human interaction.

广义地讲，嵌入式系统可以定义为一种执行单一功能的设备，它包含紧密耦合的硬件和软件，是构成更大系统的一部分，不是要让用户独立编程，而是要在尽量不需要人机交互的情况下工作。

2. Most embedded systems interact directly with processes or the environment, making decisions on the fly, based on their inputs.

大多数嵌入式系统直接与进程或环境交互，根据它们的输入，动态地做出决策。

3. Besides, these systems operate in constrained environments where memory, computing power, and power supply are limited.

此外，这些系统在内存、计算能力和供电受限的环境中运行。

4. One of the earliest electronic computing devices credited with the term "embedded system" and closer to our present conception of such was the Apollo Guidance Computer (AGC).

阿波罗制导计算机（AGC）是最早被称为“嵌入式系统”的电子计算设备之一，与我们目前的概念更接近。

5. The system software of the Apollo Guidance Computer was written in AGC assembly language and supported a non-preemptive real-time operating system that could simultaneously run up to eight prioritized jobs.

阿波罗制导计算机的系统软件是用 AGC 汇编语言编写的，支持一个非抢占式实时操作系统，可以同时运行多达 8 个优先作业。

6. This chip was a 4-bit CPU that incorporated in the same die 1 KB of ROM and 256 bits of RAM to offer a complete computer functionality in a single-chip, making it the first microcomputer-on-a-chip (a.k.a. microcontroller).

该芯片是一个 4 位 CPU，单片集成了 1 KB 的 ROM 和 256 位的 RAM，在一个芯片中实现了完整的计算机功能，使其成为第一个片上微型计算机（又称微控制器）。

7. Despite being omnipresent in virtually every aspect of our modern lives, embedded systems are ubiquitous devices almost invisible to the user, working in a pervasive way to make possible the "intelligent" operation of machines and appliances around us.

尽管嵌入式系统实际上普遍存在于现代生活的方方面面，但用户几乎看不见它，因为嵌入式系统以一种无处不在的方式融入我们周围的机器和设备，使得“智能”操作成为可能。

2.2 Structure of an Embedded System

Regardless of the function performed by an embedded system, the broadest view of its structure reveals two major, tightly coupled sets of components: a set of hardware components that include a central processing unit, typically in the form of a microcontroller; and a set of software components, typically included as firmware that gives functionality to the hardware. **Firmware is a computer program typically stored in a non-volatile memory embedded in a hardware device. It is tightly coupled to the hardware where it resides and although it can be upgradeable in some applications, it is not intended to be changed by users.**

Figure 2.5 depicts this general view, denoting these two major sets of components and their interrelation. **Typical inputs in an embedded system are process variables and parameters that arrive via sensors and input/output (I/O) ports. The outputs are in the form of control actions on system actuators or processed information for users or other subsystems within the application.** In some instances, the exchange of input-output information occurs with users via

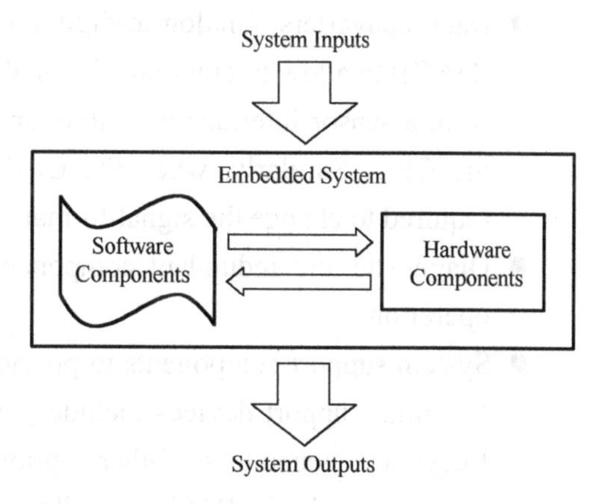

Figure 2.5 General view of an embedded system

some sort of user interface that might include keys and buttons, sensors, light emitting diodes (LEDs), liquid crystal displays (LCDs), and other types of display devices, depending on the application.

2.2.1 Hardware Components

When viewed from a general perspective, the hardware components of an embedded system include all the electronics necessary for the system to perform the function it was designed for. Therefore, the specific structure of a particular system could substantially differ from another, based on the application itself. **Despite these dissimilarities, three core hardware components are essential in an embedded system: the central processing unit (CPU), the system memory, and a set of input/output ports.** The CPU executes software instructions to process the system inputs and to make the decisions that guide the system operation. Memory stores programs and data necessary for system operation. Most systems differentiate between program and data memories. Program memory stores the software programs executed by the CPU. Data memory stores the data processed by the system. The I/O ports allow conveying signals between the CPU and the world external to it. Beyond this point, a number of other supporting and I/O devices needed for system functionality might be present, depending on the application. These include:

- Communication ports for serial and/or parallel information exchanges with other devices or systems. USB ports, printer ports, wireless RF and infrared ports, are some representative examples of I/O communication ports.
- User interfaces to interact with humans. Keypads, switches, buzzers and audio, lights, numeric, alphanumeric, and graphic displays, are examples of I/O user interfaces.
- Sensors and electromechanical actuators to interact with the environment external to the system. Sensors provide inputs related to physical parameters such as temperature, pressure, displacement, acceleration, rotation, etc. Motor speed controllers, stepper motor controllers, relays, and power drivers are some examples of actuators to receive outputs from the system I/O ports. These are just a few of the many devices that allow interaction with processes and the environment.
- Data converters (analog-to-digital converter (ADC) and/or digital-to-analog converter (DAC)) to allow interaction with analog sensors and actuators. When the signal coming out from a sensor interface is analog, an ADC converts it to the digital format understood by the CPU. Similarly, when the CPU needs to command an analog actuator, a DAC is required to change the signal format.
- Diagnostic and redundant components to verify and provide for robust, reliable system operation.
- System support components to provide essential services that allow the system to operate. Essential support devices include power supply and management components, and clock frequency generators. Other optional support components include timers, interrupt management logic, DMA controllers, etc.
- Other subsystems to enable functionality, which might include application specific

integrated circuits (ASICs), field programmable gate arrays (FPGAs), and other dedicated units, according to the complexity of the application.

Figure 2.6 illustrates how these hardware components are integrated to provide the desired system functionality.

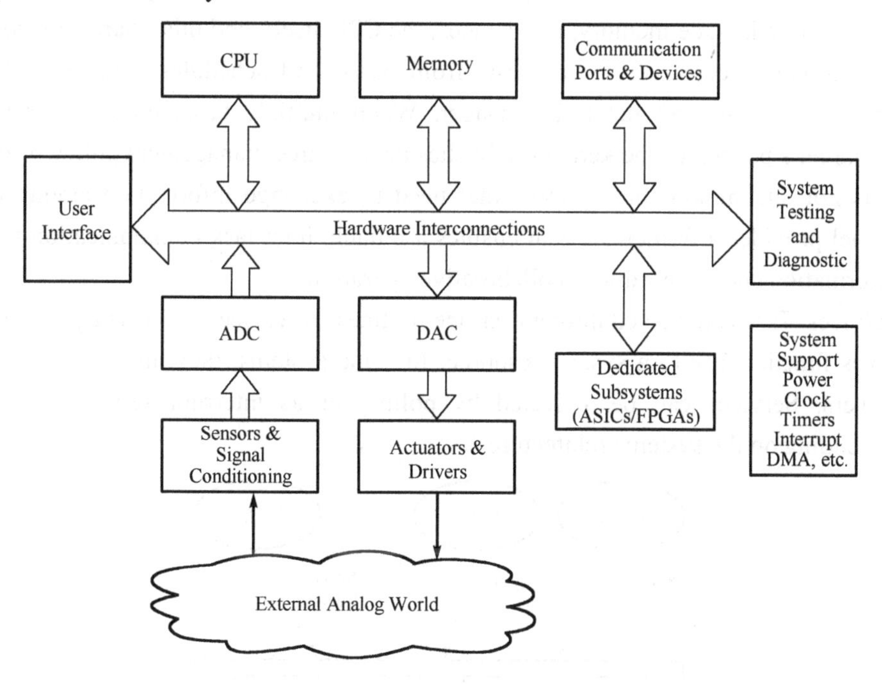

Figure 2.6 Hardware components in an embedded system

2.2.2 Software Components

The software components of an embedded system include all the programs necessary to give functionality to the system hardware. These programs, frequently referred to as the system firmware, are stored in some sort of non-volatile memory. Firmware is not meant to be modifiable by users, although some systems could provide means of performing upgrades. System programs are organized around some form of operating system and application routines. **The operating systems can be simple and informal in small applications, but as the application complexity grows, the operating system requires more structure and formality.** In some of these cases, designs are developed around real-time operating systems (RTOS). Figure 2.7 illustrates the software structure in an embedded system.

The major components identified in a system software include:

- System tasks. The application software in an embedded system is divided into a set of smaller programs called tasks. Each task handles a distinct action in the system and requires the use of specific system resources. Tasks submit service requests to the kernel in order to perform their designated actions. In our microwave oven example, the system operation can be decomposed into a set of tasks that include reading the keypad to determine user selections, presenting information on the oven display, turning on the magnetron at a certain power level for a certain amount of time, just to mention a few. Service requests can be placed via registers or interrupts.

- System kernel. The software component that handles the system resources in an embedded application is called the kernel. System resources are all those components needed to serve tasks. These include memory, I/O devices, the CPU itself, and other hardware components. The kernel receives service requests from tasks, and schedules them according to the priorities dictated by the task manager. When multiple tasks contend for a common resource, a portion of the kernel establishes the resource management policy of the system. It is not uncommon finding tasks that need to exchange information among them. The kernel provides a framework that enables a reliable inter-task communication to exchange information and to coordinate collaborative operation.
- Services. Tasks are served through service routines. A service routine is a piece of code that gives functionality to a system resource. In some systems, they are referred to as device drivers. Services can be activated by polling or as interrupt service routines (ISR), depending on the system architecture.

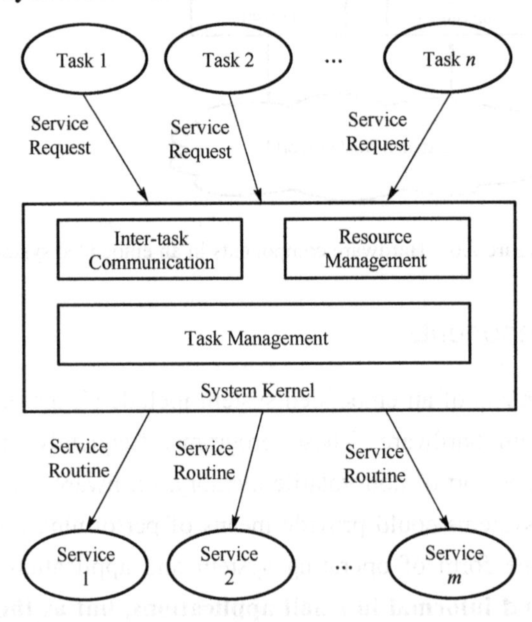

Figure 2.7 Software structure in an embedded system

Words and Expressions

firmware ['fɜːrmwer] *n.* 固件
non-volatile memory 非易失性存储器
upgradeable ['ʌp'greɪdəbl] *adj.* 可升级的
interrelation [ˌɪntərɪ'leɪʃn] *n.* 相互关系
variable ['veriəbl] *n.* 变量；可变情况；可变因素
parameter [pə'ræmɪtər] *n.* 参数；决定因素；规范；范围

sensor ['sensər]	n. 传感器；（探测光、热、压力等的）敏感元件；探测设备
actuator ['æktjuˌeɪtər]	n. 致动（促动，激励，调节）器；执行机构（元件）；传动（装置，机构）
light emitting diode (LED)	发光二极管
liquid crystal display (LCD)	液晶显示器
differentiate [ˌdɪfə'renʃieɪt]	v. 区分；辨别；表明……间的差别；求导
convey [kən'veɪ]	vt. 传达；输送；传送；运送
USB：universal serial bus	通用串行总线
RF：radio frequency	射频；无线电频率
infrared [ˌɪnfrə'red]	adj. 红外线的；使用红外线的 n. 红外线；红外区
electromechanical [ɪˌlektroʊmə'kænɪkəl]	adj. 电机的；机电的
stepper motor controller	步进电动机控制器
relay ['riːleɪ]	vt. 转发；转播（电视或广播信号） n. 中继设备
analog-to-digital converter (ADC)	模数转换器
digital-to-analog converter (DAC)	数模转换器
diagnostic [ˌdaɪəg'nɑːstɪk]	adj. 诊断的；判断的 n. 诊断；诊断法
redundant [rɪ'dʌndənt]	adj. 冗余的；多余的
robust [roʊ'bʌst]	adj. 结实的；耐用的；坚固的
clock frequency generator	时钟频率发生器
DMA：direct memory access	直接存储器访问
application specific integrated circuit (ASIC)	专用集成电路
field programmable gate array (FPGA)	现场可编程门阵列
dedicated ['dedɪkeɪtɪd]	adj. 专用的
real-time operating system (RTOS)	实时操作系统
magnetron ['mægnəˌtrɑn]	n. 磁控管；磁（控）电（子）管
collaborative [kə'læbəreɪtɪv]	adj. 合作的；协同的
service routine	服务例程

Notes

1. Firmware is a computer program typically stored in a non-volatile memory embedded in a hardware device. It is tightly coupled to the hardware where it resides and although it can be upgradeable in some applications, it is not intended to be changed by users.

固件是一种计算机程序，通常存储于嵌入硬件设备中的非易失性存储器中。它与所在的硬件紧密耦合，虽然在某些应用中可以升级，但用户不可对其进行更改。

2. Typical inputs in an embedded system are process variables and parameters that arrive via sensors and input/output (I/O) ports. The outputs are in the form of control actions on system

actuators or processed information for users or other subsystems within the application.

嵌入式系统中的典型输入是通过传感器和输入/输出（I/O）端口获得的过程变量和参数。输出则是系统执行器上的控制动作，或者是用户或应用中的其他子系统所需的处理信息。

3. Despite these dissimilarities, three core hardware components are essential in an embedded system: the central processing unit (CPU), the system memory, and a set of input/output ports.

尽管存在以上差异，但嵌入式系统中有 3 个核心硬件组件是必不可少的：中央处理器（CPU）、系统内存和一组输入/输出端口。

4. Figure 2.6 illustrates how these hardware components are integrated to provide the desired system functionality.

图 2.6 说明了如何将这些硬件组件集成起来，以实现所需的系统功能。

5. The operating systems can be simple and informal in small applications, but as the application complexity grows, the operating system requires more structure and formality.

在小型应用当中，操作系统可以简单而非正式，但随着应用的复杂性增强，操作系统需要更丰富的结构和形式。

2.3 Microprocessors Versus Microcontrollers

The minimal set of components required to establish a computing system is denominated as a microcomputer. The basic structural description of an embedded system introduced in the previous section shows us the integration between hardware and software components. **Before we delve any deeper into the structure of the different components of a microcomputer system, let's first establish the fundamental difference between microprocessors and microcontrollers.**

2.3.1 Microprocessor Units

A Microprocessor unit, commonly abbreviated MPU, fundamentally contains a general purpose CPU in its die. To develop a basic system using an MPU, all components depicted in Figure 2.6 other than the CPU, i.e., the buses, memory, and I/O interfaces, are implemented externally (hence the name peripherals). **Other characteristics of MPUs include an optimized architecture to move code and data from external memory into the chip such as queues and caches, and the inclusion of architectural elements to accelerate processing such as multiple functional units, ability to issue multiple instructions at once, and other features such as branch prediction units and numeric co-processors.**

The most common examples of systems designed around MPUs are personal computers and mainframes. But these are not the only ones. There are many other systems developed around traditional MPUs. Manufacturers of MPUs include Intel, Freescale, Zilog, Fujitsu, Siemens, and many others. Microprocessor design has advanced from the initial models in the early 1970s to present day technology. Intel's initial 4004 in 1971 was built using 10 μm technology, ran at 400 kHz and contained 2,250 transistors. Intel's Xeon E7 MPU, released in 2011, was built using 32 nm technology, ran at 2 GHz and contained 2.6×10^9 transistors. MPUs indeed, represent the most powerful type of processing components available to implement a microcomputer.

Most small embedded systems, however, do not need the large computational and processing power that characterize microprocessors, and hence the orientation to microcontrollers for these tasks.

2.3.2 Microcontroller Units

A microcontroller unit, abbreviated MCU, is developed using a microprocessor core or central processing unit (CPU), usually less complex than that of an MPU. This basic CPU is then surrounded with memory of both types (program and data) and several types of peripherals, all of them embedded into a single integrated circuit, or chip. This blending of CPU, memory, and I/O within a single chip is what we call a microcontroller.

The assortment of components embedded into a microcontroller allows for implementing complete applications requiring only a minimal number of external components or in many cases solely using the MCU chip. Peripheral timers, input/output (I/O) ports, interrupt handlers, and data converters are among those commonly found in most microcontrollers. The provision of such an assortment of resources inside the same chip is what has gained them the denomination of computers-on-a-chip. Figure 2.8 shows the structure of a typical microcontroller.

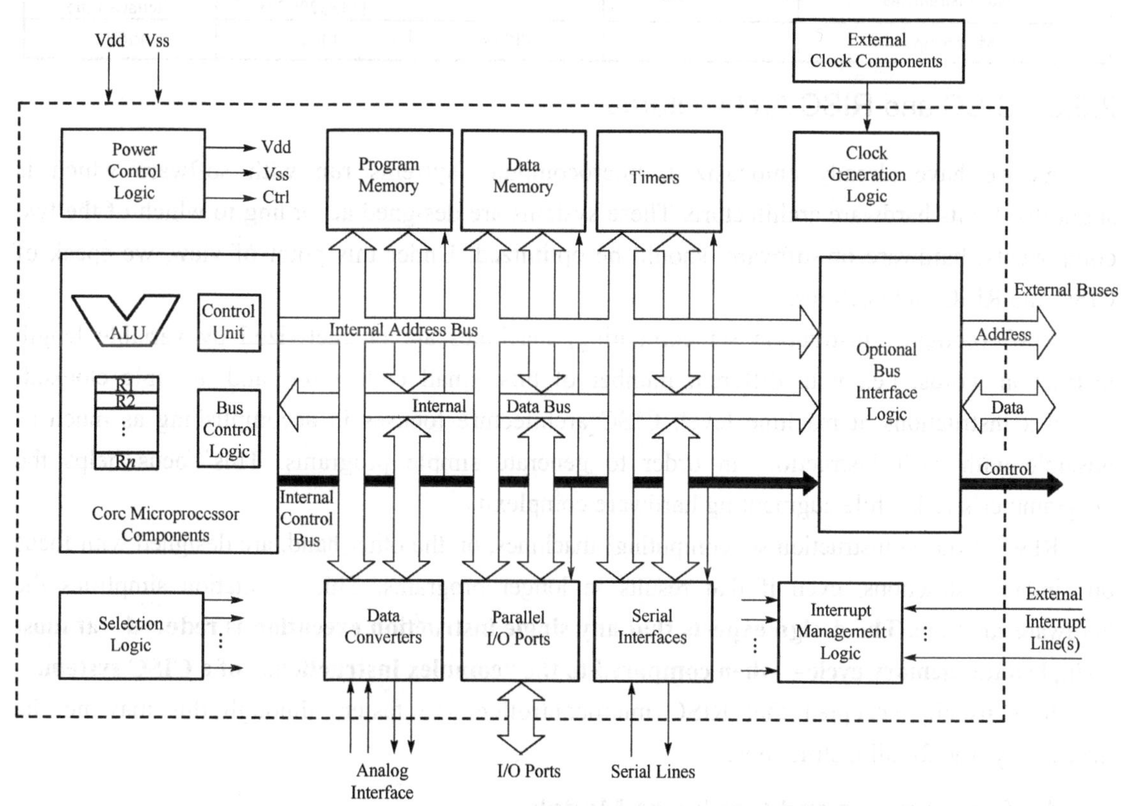

Figure 2.8 Structure of a typical microcontroller

Microcontrollers share a number of characteristics with general purpose microprocessors. Yet, the core architectural components in a typical MCU are less complex and more application oriented than those in a general purpose microprocessor.

Microcontrollers are usually marketed as family members. Each family is developed around a basic architecture which defines the common characteristics of all members. These include, among others, the data and program path widths, architectural style, register structure, base instruction set, and addressing modes. **Features differentiating family members include the amount of on-chip data and program memories and the assortment of on-chip peripherals.**

There are literally hundreds, perhaps thousands, of microprocessors and microcontrollers on the market. Table 2.1 shows a very small sample of microcontroller family models of different sizes from six companies.

Table 2.1 A sample of microcontroller families/series

Company	4-bit	8-bit	16-bit	32-bit
EM Microelectronics	EM6807	EM6819		
Samsung	S3P7xx	S3F9xxx	S3FCxx	S3FN23BXZZ
Freescale		68HC11	68HC12	
Toshiba		TLCS-870	TLCS-900/L1	TLCS-900/H1
Texas Instruments			MSP430 TMS320C24x	TMS320C28x Stellaris Line
Microchip		PIC1x	PIC2x	PIC32

2.3.3 CISC and RISC Architectures

As we have already emphasized, microcomputer systems run with software which is supported by its hardware architecture. These systems are designed according to which of the two components, hardware or software, should be optimized. Under this point of view, we speak of CISC and RISC architectures.

CISC (complex instruction set computing) machines are characterized by variable length instruction words, i.e., with different number of bits, small code sizes, and multiple clocked-complex instructions at machine level. CISC architecture focuses in accomplishing as much as possible with each instruction, in order to generate simple programs. This focus helps the programmer's task while augmenting hardware complexity.

RISC (reduced instruction set computing) machines, on the other hand, are designed with focus on simple instructions, even if that results in longer programs. This orientation simplifies the hardware structure. **The design expects that any single instruction execution is reduced—at most a single data memory cycle—when compared to the "complex instructions" of a CISC system.**

It is usually accepted that RISC microcontrollers are faster, although this may not be necessarily true for all instructions.

2.3.4 Programmer and Hardware Models

Most readers might already have programming experience with some high level language such as Java, C, or some other language. Most probably, the experience did not require knowledge of the hardware system supporting the execution of the program.

Embedded systems programmers need to go one step forward and consider both the hardware and software issues. Hence, they need to look at the system both from a hardware point of view, the hardware model, as well as a software point of view, the programmer's model.

In the hardware model, the user focuses on the hardware characteristics and subsystems that support the instructions and the interactions with the outer world. This knowledge is indispensable from the beginning especially because of the intimate relationship with the programming possibilities. In this model we speak of hardware subsystems, characteristics of peripherals, interfacing with memory, peripherals and outer world, timing, and so on. The hardware supports the programmer's model.

In the programmer's model, we focus on the instruction set and syntaxes, addressing modes, the memory map, transfers and execution time, and so on. Very often, when a microcontroller is designed from scratch, the process starts with the desired instruction set.

Words and Expressions

denominate [dɪ'nɑːmɪneɪt] *vt.* 将……命名为；以（某种货币）为单位；称……为

cache [kæʃ] *n.* 高速缓冲存储器，高速缓存

co-processor['koʊˌproʊsesə] *n.* 协处理器

mainframe ['meɪnfreɪm] *n.*（大型电脑的）主机；中央处理机

assortment [ə'sɔːrtmənt] *n.* 各种各样

complex instruction set computing(CISC) 复杂指令集计算

reduced instruction set computing(RISC) 精简指令集计算

indispensable [ˌɪndɪ'spensəbl] *adj.* 不可或缺的；必不可少的

Notes

1. Before we delve any deeper into the structure of the different components of a microcomputer system, let's first establish the fundamental difference between microprocessors and microcontrollers.

在深入研究微机系统不同组件的结构之前，让我们首先确定微处理器和微控制器之间的基本区别。

2. Other characteristics of MPU include an optimized architecture to move code and data from external memory into the chip such as queues and caches, and the inclusion of architectural elements to accelerate processing such as multiple functional units, ability to issue multiple instructions at once, and other features such as branch prediction units and numeric co-processors.

MPU 的其他特征包括：优化的体系结构，可以将代码和数据从外部存储器移动到芯片中，如队列和高速缓存；提高处理速度的组成要素，如多个功能单元同时发出多条指令的能力；分支预测单元和数字协处理器等其他特征。

3. The assortment of components embedded into a microcontroller allows for implementing complete applications requiring only a minimal number of external components or in many cases solely using the MCU chip.

微控制器包含各种各样的组件，可以实现完整的应用，几乎不需要外部组件。

4. Features differentiating family members include the amount of on-chip data and program memory and the assortment of on-chip peripherals.

根据片上数据和程序存储器的数量及片上外设的种类，可以区分同系列微控制器产品中的不同型号。

5. The design expects that any single instruction execution is reduced—at most a single data memory cycle—when compared to the "complex instructions" of a CISC system.

与 CISC 系统的"复杂指令"相比，RISC 系统的设计期望减少任意单条指令的执行时间，最多为一个数据存储周期。

Exercises

1. Match the terms (1)–(6) with the definitions A–F.

(1) navigation system	A. a device that detects and responds to some type of input from the physical environment. The specific input could be light, heat, motion, ...
(2) sensor	B. a device that has the capability of knowing your current position, and allows you to determine your destination
(3) peripherals	C. a small computer on a single metal-oxide-semiconductor (MOS) integrated circuit (IC) chip
(4) microcontroller	D. an industry standard that establishes specifications for cables, connectors and protocols for connection, communication and power supply (interfacing) between computers, peripherals and other computers
(5) USB	E. an integrated circuit (IC) chip customized for a particular use, rather than intended for general purpose use
(6) ASIC	F. all hardware components that are attached to a computer and are controlled by the computer system, but they are not the core components of the computer

2. Translate into Chinese.

(1) The embedded firmware is responsible for controlling the various peripherals of the embedded hardware and generating response in accordance with the functional requirements of the product.

(2) Designing embedded firmware requires understanding of the particular embedded product hardware, like various component interfacing, memory map details, I/O port details, configuration and register details of various hardware chips used and some programming language (either low level assembly language or high level language like C/C++ or a combination of the two).

(3) Interrupts are mainly triggered by hardware events. Events such as a push-button depression, a threshold reached, and a timer expiration, are few examples of events that might be configured to trigger interrupts.

3. Translate into English.

（1）寄存器用于暂存数据、内存地址和控制信息，以供 CPU 快速访问。

（2）德州仪器公司开发的 MSP430 系列微控制器面向电池供电的便携式嵌入式应用市场。

（3）基于微控制器的应用对电源的要求与任何其他类型的数字系统没有太大区别。

4. Read the following article and write a summary.

The goal of an embedded system development environment is to enable a developer to create the application firmware for a microcontroller unit (MCU) based hardware design. The primary component of a contemporary development system is a software package called an integrated development environment (IDE). The IDE provides a user interface for the application developer to coordinate the development activities.

Application development consists of five steps.

(1) Coding

Applications are developed by writing code using languages such as C or assembly. The code is placed in text files, called source files, on the developer's computer. An embedded application typically consists of multiple source files. The content and format of the source files are determined by the language being used. In order to enable efficient development, the IDE includes a text editor and the ability to create and manage multiple source files.

(2) Building

Once the source files have been written, they need to be converted to a form which is readable by the MCU. The process of converting source files to a format which can be read by an MCU is called building. A building is a two-step process:

a. Each application source file is converted to a machine-readable file called an object file.

b. The object files are then linked together, forming a single application file which can be programmed into the MCU.

Building is performed by a software tool called a compiler. The IDE offers a user interface, allowing the developer to start the building process by invoking the compiler.

(3) Soft Simulation

The improperly written application can cause damage to the external hardware or the circuity of a development board. Many developers wish to determine if their applications are functioning properly before they program them into MCUs. The process of verifying an application's operation before it is programmed is called software simulation.

Software simulation is a feature offered by the IDE. After a project is built, and before it is programmed into the MCU, the developer can run an interactive software simulation of the code.

(4) Programming

Placing a built application file into an MCU's program memory requires the use of a hardware device called a programmer (or programmer/debugger). A programmer typically connects to the developer's computer through USB and to the MCU through the MCU's specified I/O pins. After a building has been completed, using the IDE, the developer initiates the programming of the built application file into the MCU.

(5) Debugging

Sometimes, after an application has been coded, built, and programmed into an MCU, its performance may differ from what the developer expects. To identify and correct the causes of these anomalies, the developer will need to "see" what the MCU sees as it executes the application code. The process of controlling an MCU's operation and observing its performance is called debugging (or hardware debugging).

Debugging is performed through the IDE and requires the device programmer to have a set of enhanced capabilities. The programmer must have the ability to pass control signals from the IDE to the MCU and return data from the MCU to the IDE. The enhanced programmer is referred to as a programmer/debugger.

5. Language study: Describing purpose

Study these ways of describing the purpose of random access memory:

- RAM is used for the temporary storage of programs and data.
- RAM is used for storing programs and data temporarily.
- RAM is used to store programs and data temporarily.

Identify each of the electronic components or pieces of equipment described below. Compare answers with your partner.

(1) It is used to control data going in and out of the computer.

(2) It is used for permanent storage.

(3) It is used to perform the functions of a computer's central processing unit.

(4) It is used for moving and controlling a mechanism or a system with a control signal and a source of energy.

(5) It is used to supplement the functions of the primary processor (the CPU).

Unit 3 Integrated Circuits

3.1 Introduction

Integrated circuit (IC) (see Figure 3.1), also called microelectronic circuit, microchip, or chip, is an assembly of electronic components, fabricated as a single unit, in which miniaturized active devices (e.g., transistors and diodes) and passive devices (e.g., capacitors and resistors) and their interconnections are built up on a thin substrate of semiconductor material (typically silicon). The resulting circuit is thus a small monolithic "chip," which may be as small as a few square centimeters or only a few square millimeters. The individual circuit components are generally microscopic in size.

Integrated circuits have their origin in the invention of the transistor in 1947 by William B. Shockley and his team at the American Telephone and Telegraph Company's Bell Laboratories. Shockley's team (including John Bardeen and Walter H. Brattain) found that, under the right circumstances, electrons would form a barrier at the surface of a certain crystal, and they learned to control the flow of electricity through the crystal by manipulating this barrier. Controlling electron flow through a crystal allowed the team to create a device that could perform certain electrical operations, such as signal amplification, that were previously done by vacuum tubes. They named this device a transistor, from a combination of the words transfer and resistor. The study of methods of creating electronic devices using solid materials became known as solid-state electronics. **Solid-state devices proved to be much sturdier, easier to work with, more reliable, much smaller, and less expensive than vacuum tubes.** Using the same principles and materials, engineers soon learned to create other electronic components, such as resistors (see Figure 3.2) and capacitors. Now that electrical devices could be made so small, the largest part of a circuit was the awkward wiring between the devices.

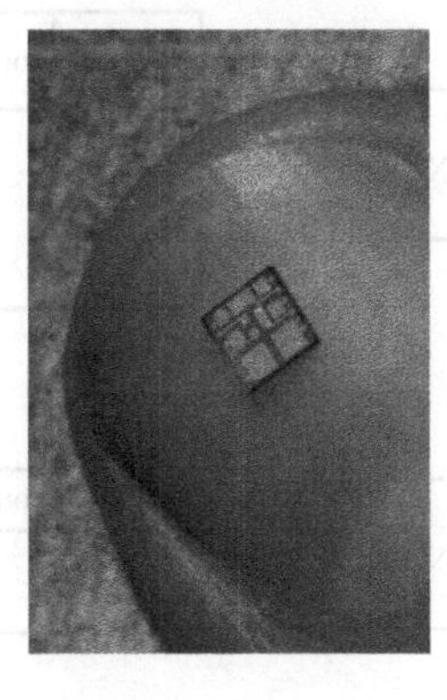

Figure 3.1 IC

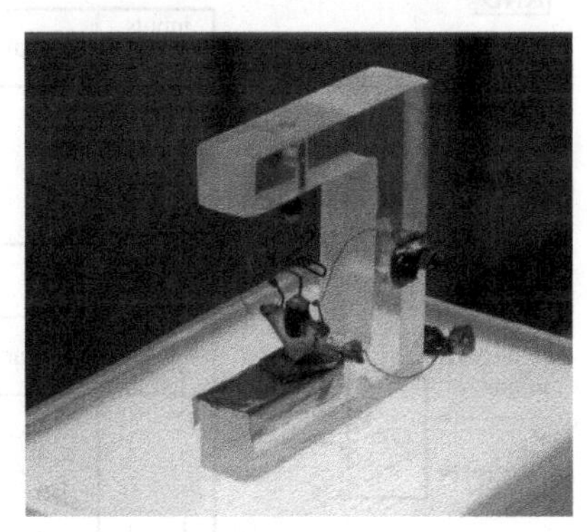

Figure 3.2 The first transistor

In 1958, Jack Kilby of Texas Instruments, Inc., and Robert Noyce of Fairchild Semiconductor Corporation independently thought of a way to reduce circuit size further. They laid very thin paths of metal (usually aluminum or copper) directly on the same piece of material as their devices. These small paths acted as wires. With this technique, an entire circuit could be "integrated" on a single piece of solid material and an integrated circuit (IC) was thus created. An IC can contain hundreds of thousands of individual transistors on a single piece of material the size of a pea. Working with that many vacuum tubes would have been unrealistically awkward and expensive. The invention of the integrated circuit made technologies of the information age feasible. ICs are now used extensively in all walks of life, from cars to toasters to amusement park rides.

There are several basic types of IC:

1. Analog Versus Digital Circuits

Analog, or linear, circuits typically use only a few components and are thus some of the simplest types of ICs. Generally, analog circuits are connected to devices that collect signals from the environment or send signals back to the environment. For example, a microphone converts fluctuating vocal sounds into an electrical signal of varying voltage. An analog circuit then modifies the signal in some useful way—such as amplifying it or filtering it of undesirable noise. Such a signal might then be fed back to a loudspeaker, which would reproduce the tones originally picked up by the microphone. Another typical use for an analog circuit is to control some device in response to continual changes in the environment. **For example, a temperature sensor sends a varying signal to a thermostat, which can be programmed to turn an air conditioner, heater, or oven on and off once the signal has reached a certain value.**

A digital circuit, on the other hand, is designed to accept only voltages of specific given values. A circuit that uses only two states is known as a binary circuit. Circuit design with binary quantities, "on" and "off" representing 1 and 0 (i.e., true and false), uses the logic of Boolean algebra. (Arithmetic is also performed in the binary number system employing Boolean algebra.) Figure 3.3 shows different logic circuits. These basic elements are combined in the design of ICs for digital computers and associated devices to perform the desired functions.

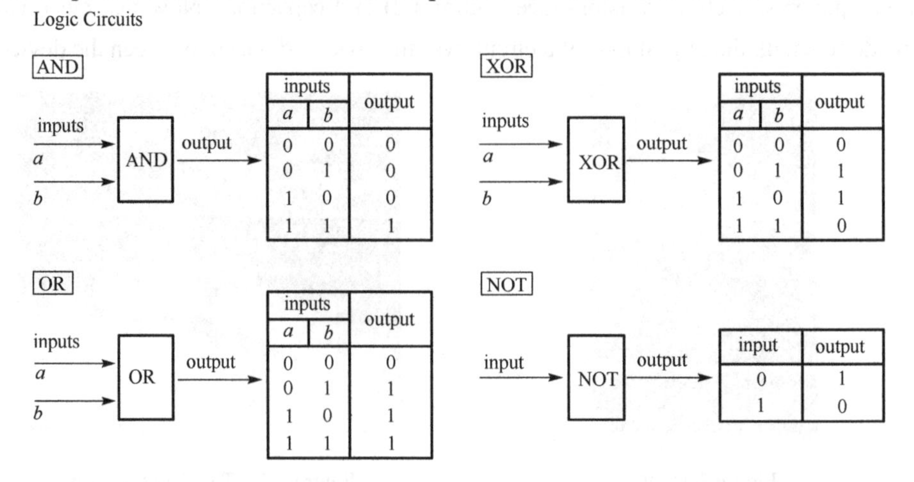

inputs a	inputs b	output
0	0	0
0	1	0
1	0	0
1	1	1

inputs a	inputs b	output
0	0	0
0	1	1
1	0	1
1	1	0

inputs a	inputs b	output
0	0	0
0	1	1
1	0	1
1	1	1

input	output
0	1
1	0

Figure 3.3 Different logic circuits

2. Microprocessor Circuits

Microprocessors are the most complicated ICs. They are composed of billions of transistors that have been configured as thousands of individual digital circuits, each of which performs some specific logic function. A microprocessor is built entirely of these logic circuits synchronized to each other. A microprocessor typically contains the central processing unit (CPU) of a computer.

Just like a marching band, the circuits perform their logic functions only on direction by the bandmaster. The bandmaster in a microprocessor, so to speak, is called the clock. The clock is a signal that quickly alternates between two logic states. Every time the clock changes state, every logic circuit in the microprocessor does something. Calculations can be made very quickly, depending on the speed (clock frequency) of the microprocessor.

Microprocessors contain some circuits, known as registers, which store information. Registers are predetermined memory locations. Each processor has many different types of registers. Permanent registers are used to store the preprogrammed instructions required for various operations (such as addition and multiplication). Temporary registers store numbers that are to be operated on and also the results. **Other examples of registers include the program counter (also called the instruction pointer), which contains the address in memory of the next instruction; the stack pointer (also called the stack register), which contains the address of the last instruction put into an area of memory called the stack; and the memory address register, which contains the address of where the data to be worked on is located or where the data that has been processed will be stored.**

Microprocessors can perform billions of operations per second on data. In addition to computers, microprocessors are common in video game systems, televisions, cameras, and automobiles.

3. Memory Circuits

Microprocessors typically have to store more data than can be held in a few registers. This additional information is relocated to special memory circuits. Memory is composed of dense arrays of parallel circuits that use their voltage states to store information. Memory also stores the temporary sequence of instructions, or program, for the microprocessor.

Manufacturers continually strive to reduce the size of memory circuits—to increase capability without increasing space. In addition, smaller components typically use less power, operate more efficiently, and cost less to manufacture.

4. Digital Signal Processors

A signal is an analog waveform—anything in the environment that can be captured electronically. A digital signal is an analog waveform that has been converted into a series of binary numbers for quick manipulation. As the name implies, a digital signal processor (DSP) processes signals digitally, as patterns of 1 s and 0 s. For instance, using an analog-to-digital converter, commonly called an A-to-D or A/D converter, a recording of someone's voice can be converted into digital 1 s and 0 s. The digital representation of the voice can then be modified by a DSP using complex mathematical formulas. For example, the DSP algorithm in the circuit may be configured to recognize gaps between spoken words as background noise and digitally remove ambient noise from the waveform. Finally, the processed signal can be converted back (by a D/A converter) into an analog signal for listening. **Digital processing can filter out background noise so fast that there is no discernible delay and the signal appears to be heard in "real time".** For instance, such processing enables "live" television broadcasts to focus on a quarterback's signals in an American gridiron football game.

DSPs are also used to produce digital effects on live television. For example, the yellow marker lines displayed during the football game are not really on the field; a DSP adds the lines after the cameras shoot the picture but before it is broadcast. Similarly, some of the advertisements seen on stadium fences and billboards during televised sporting events are not really there.

5. Application Specific ICs

An application specific IC (ASIC) can be either a digital or an analog circuit. As its name implies, an ASIC is not reconfigurable; it performs only one specific function. For example, a speed controller IC for a remote control car is hard-wired to do one job and could never become a microprocessor. An ASIC does not contain any ability to follow alternate instructions.

6. Radio-Frequency ICs

Radio-frequency ICs (RFICs) are widely used in mobile phones and wireless devices. RFICs are analog circuits that usually run in the frequency range of 3 kHz to 2.4 GHz (3,000 hertz to 2.4 billion hertz), and circuits operating at about 1 THz (1 trillion hertz) are under development. They are usually thought of as ASICs even though some may be configurable for several similar applications.

Most semiconductor circuits that operate above 500 MHz (500 million hertz) cause the electronic components and their connecting paths to interfere with each other in unusual ways. Engineers must use special design techniques to deal with the physics of high-frequency microelectronic interactions.

A special type of RFIC is known as a monolithic microwave IC (MMIC; also called microwave monolithic IC). These circuits usually run in the 2- to 100- GHz range, or microwave frequencies, and are used in radar systems, in satellite communications, and as power amplifiers for cellular telephones. **Just as sound travels faster through water than through air, electron velocity is different through each type of semiconductor material.** Silicon offers too much resistance for microwave-frequency circuits, and so the compound gallium arsenide (GaAs) is often used for MMICs. Unfortunately, GaAs is mechanically much less sound than silicon. It breaks easily, so GaAs wafers are usually much more expensive to build than silicon wafers.

Words and Expressions

microelectronic [ˌmaɪkroʊɪˌlek'trɑːnɪk]	*adj*. 微电子（学）的；超小型电子的
miniaturize ['mɪnətʃəraɪz]	*vt*. 使微型化；使成为缩影
substrate ['sʌbstreɪt]	*n*. 基底；底物；底层；基层；衬底
semiconductor ['semikəndʌktər]	*n*. 半导体；半导体装置

microscopic [ˌmaɪkrə'skɑːpɪk]	*adj*. 极小的；微小的；需用显微镜观察的
electron [ɪ'lektrɑːn]	*n*. 电子
barrier ['bæriər]	*n*. 垫垒；障碍；屏障；阻力
crystal ['krɪstl]	*n*. 结晶；晶体；晶振
vacuum tube	真空管
transistor [træn'zɪstər]	*n*. 晶体管
solid-state electronics	固态电子学；固体电子学
thermostat ['θɜːrməstæt]	*n*. 恒温器；温度自动调节器
binary ['baɪnəri]	*adj*. 二进制的（用 0 和 1 记数）；二元的 *n*. 二进制数
Boolean algebra	布尔代数
synchronize ['sɪŋkrənaɪz]	*v*. 使同步；（使）同步，在时间上一致，同速进行
program counter	程序计数器
stack pointer	堆栈指针；堆栈指示器；堆栈指针寄存器
digital signal processor(DSP)	数字信号处理器
capture ['kæptʃər]	*vt*. 俘获；捕获；把……输入计算机
ambient noise	环境噪声；氛围噪声；背景噪声
discernible [dɪ'sɜːrnəbl]	*adj*. 可辨的；看得清的；辨别得出的
digital effect	数字特效
reconfigurable [ˌrikən'fɪgjərəbl]	*adj*. 可重构的；可重配置的
remote control	遥控器；遥控
radio frequency IC(RFIC)	射频集成电路；射频芯片
microwave monolithic IC(MMIC)	微波单片集成电路
gallium arsenide	砷化镓（GaAs）
silicon wafer	硅晶圆；硅晶片；硅片

Notes

1. Integrated circuit (IC), also called microelectronic circuit, microchip, or chip, is an assembly of electronic components, fabricated as a single unit, in which miniaturized active devices (e.g., transistors and diodes) and passive devices (e.g., capacitors and resistors) and their interconnections are built up on a thin substrate of semiconductor material (typically silicon).

集成电路（IC），也被称为微电子电路、微芯片或芯片，集成了多个电子元件，制造为单个单元，其中微型有源器件（如晶体管和二极管）和无源器件（如电容器和电阻器）及其互连都建立在半导体材料（通常为硅）的薄衬底上。

2. Solid-state devices proved to be much sturdier, easier to work with, more reliable, much smaller, and less expensive than vacuum tubes.

事实证明，固态器件比真空管更坚固、更易用、更可靠、体积更小且成本更低。

3. For example, a temperature sensor sends a varying signal to a thermostat, which can be programmed to turn an air conditioner, heater, or oven on and off once the signal has reached a certain value.

例如，温度传感器向恒温器发送一个随温度变化的信号，一旦此信号达到某个特定值，预编程的恒温器就会自动打开或关闭空调、加热器或烤箱。

4. Other examples of registers include the program counter (also called the instruction pointer), which contains the address in memory of the next instruction; the stack pointer (also called the stack register), which contains the address of the last instruction put into an area of memory called the stack; and the memory address register, which contains the address of where the data to be worked on is located or where the data that has been processed will be stored.

寄存器的其他例子包括：程序计数器（也称指令指针），存储下一条指令在内存中的地址；堆栈指针（也称堆栈寄存器），记录放入堆栈的最后一条指令的地址；内存地址寄存器，其中包含要处理的数据所在的地址或已处理的数据将被存储的地址。

5. Manufacturers continually strive to reduce the size of memory circuits—to increase capability without increasing space. In addition, smaller components typically use less power, operate more efficiently, and cost less to manufacture.

制造商不断努力缩小内存电路的尺寸，以在不增加物理空间的情况下增加存储容量。此外，较小的组件通常使用较少的功率，运行效率更高，制造成本更低。

6. Digital processing can filter out background noise so fast that there is no discernible delay and the signal appears to be heard in "real time".

数字处理可以快速滤除背景噪声，因此没有明显的延迟，信号似乎是"实时"听到的。

7. Just as sound travels faster through water than through air, electron velocity is different through each type of semiconductor material.

正如声音在水中的传播速度比在空气中的传播速度快一样，电子在不同半导体材料中的传播速度也不同。

3.2 PN Junction and Diode

Any material can be classified as one of three types: conductor, insulator, or semiconductor. A conductor (such as copper or salt water) can easily conduct electricity because it has an abundance of free electrons. An insulator (such as ceramic or dry air) conducts electricity very poorly because it has few or no free electrons. A semiconductor (such as silicon or gallium arsenide) is somewhere between a conductor and an insulator. It is capable of conducting some electricity, but not much.

Most ICs are made of silicon, which is abundant in ordinary beach sand. Pure crystalline silicon, as with other semiconducting materials, has a very high resistance to electric current at normal room temperature. However, with the addition of certain impurities, known as dopants, the silicon can be made to conduct usable currents. **In particular, the doped silicon can be used as a switch, turning current off and on as desired.** The process of introducing impurities is known as doping or implantation. Depending on a dopant's atomic structure, the result of implantation will be either an N-type (negative) or a P-type (positive) semiconductor. An N-type semiconductor results from implanting donor atoms that have more electrons in their outer (bonding) shell than silicon. The resulting semiconductor crystal contains excess, or free, electrons that are available for conducting current. A P-type semiconductor results from implanting acceptor atoms that have fewer electrons in their outer shell than silicon. **The resulting crystal contains "holes" in its bonding structure where electrons would normally be located. In essence, such holes can move through the crystal conducting positive charges.** Figure 3.4 shows semiconductor bonds.

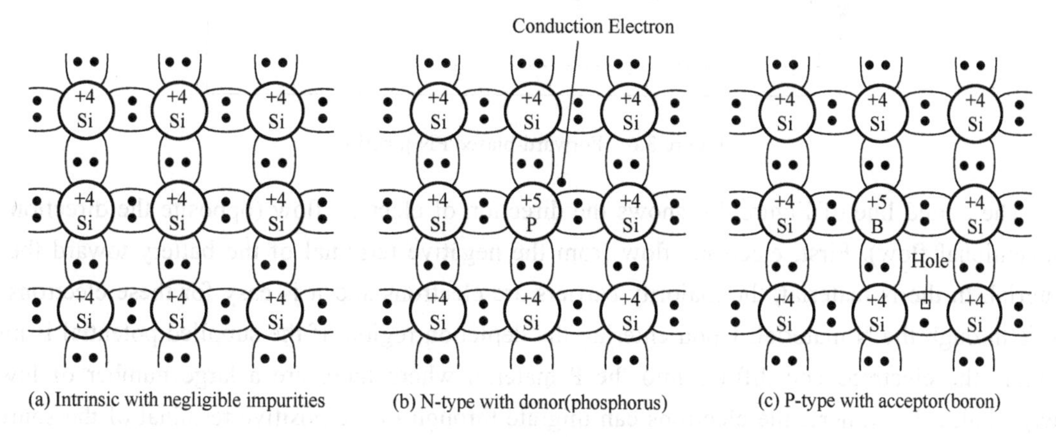

Figure 3.4 Semiconductor bonds

By creating a single zone of N material adjacent to a zone of P material, we wind up with the PN junction. The PN junction is arguably the fundamental building block of solid state semiconductor devices. The PN junctions can be found in a variety of devices including bipolar junction transistors (BJTs) and junction field effect transistors (JFETs). The most basic device built from the PN junction is the diode. Diodes are designed for a wide variety of uses including rectifying, lighting (LEDs) and photodetection (photodiodes).

Assuming the crystal is not at absolute zero, the thermal energy in the system will cause some of the free electrons in the N material to "fall" into the excess holes of the adjoining P material. This will create a region that is devoid of charge carriers (remember, electrons are the majority charge carriers in the N material while holes are the majority charge carriers in the P material). In other words, the area where the N and P materials abut is depleted of available electrons and holes, and thus we refer to it as a depletion region. This is depicted in Figure 3.5.

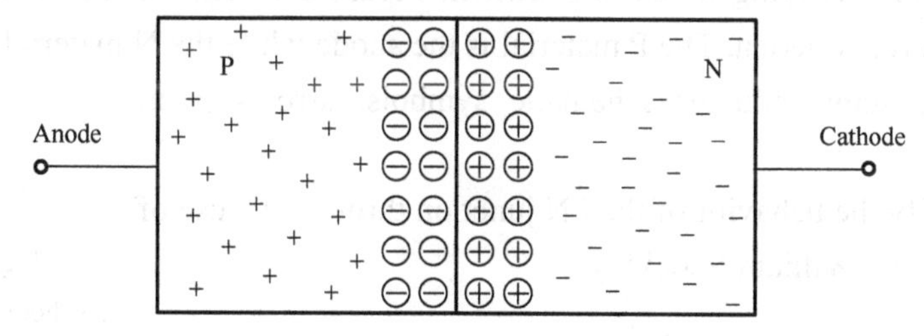

Figure 3.5 The PN junction

Now let's consider what happens if we were to connect this device to an external voltage source as shown in Figure 3.6. Obviously, there are two ways to orient the PN junction with respect to the voltage source. This version is termed forward-bias.

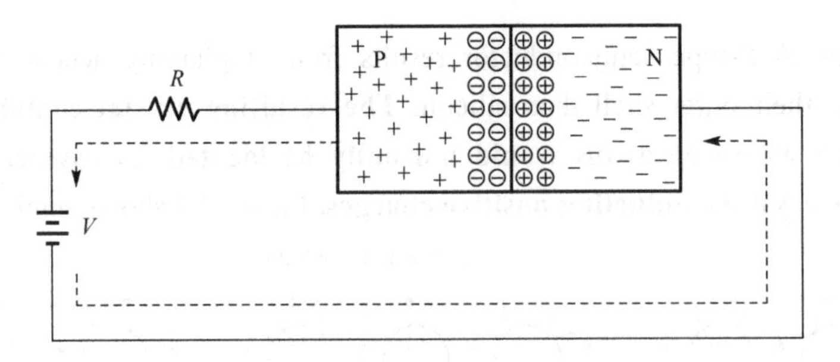

Figure 3.6 Forward-biased PN junction

The dotted line of Figure 3.6 shows the direction of electron flow (opposite the direction of conventional flow). First, electrons flow from the negative terminal of the battery toward the N material. In the N material, the majority carriers are electrons and it is easy for these electrons to move through the N material. Upon entering the depletion region, if the supplied potential is high enough, the electrons can diffuse into the P material where there are a large number of lower energy holes. From here, the electrons can migrate through to the positive terminal of the source, completing the circuit (the resistor has been added to limit maximum current flow). The "trick" here is to assure that the supplied potential is large enough to overcome the effect of the depletion region. That is, a certain voltage will be dropped across the depletion region in order to achieve current flow. This required potential is called the barrier potential or forward voltage drop. The precise value depends on the material used. **For silicon devices, the barrier potential is usually estimated at around 0.7 volts. For germanium devices, it is closer to 0.3 volts while LEDs may exhibit barrier potentials in the vicinity of 1.5 to 3 volts, partly depending on the color.**

If the voltage source polarity is reversed in Figure 3.6, the electrons in the N material will be drawn toward the positive terminal of the source while the P material holes will be drawn toward the negative terminal, creating a small, short-lived current. This has the effect of widening the depletion region and once it reaches the supplied potential, the flow of current ceases.

In its basic form, a diode is just a PN junction. It is a device that will allow current to pass easily in one direction but prevent current flow in the opposite direction. The schematic symbol for a basic switching or rectifying diode is shown in Figure 3.7. This is the ANSI standard which predominates in North America. The P material is the anode while the N material is the cathode. As a general rule for semiconductor schematic symbols, arrows point toward N material.

Anode Cathode

Figure 3.7 Diode schematic symbol (ANSI)

We can quantify the behavior of the PN junction through the use of an equation derived by William Shockley:

$$I_D = I_S(e^{\frac{V_D q}{nkT}} - 1)$$

where

I_D is the diode current.

I_S is the reverse saturation current.

V_D is the voltage across the diode.

q is the charge on an electron, 1.6E-19 coulombs.

n is the quality factor (typically between 1 and 2).

k is the Boltzmann constant, 1.38E-23 joules/kelvin.

T is the temperature in kelvin.

At 300 kelvin, q/kT is approximately 38.6. Consequently, for even very small forward (positive) voltages, the "−1" term can be ignored. **Also, I_S is not a constant. It increases with temperature, approximately doubling for each 10℃ rise.** For negative voltages (reverse-bias) the Shockley equation predicts negligible diode current. This is true up to a point. The equation does not model the effects of breakdown. When the reverse voltage is large enough, the diode will start to conduct. This is shown in Figure 3.8.

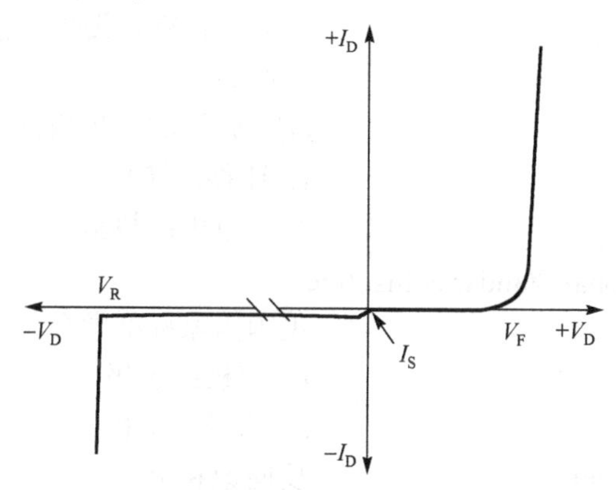

Figure 3.8 Simplified forward and reverse I_D-V_D curve for diode

V_F is the forward "knee" voltage (roughly 0.7 volts for silicon). I_S is the reverse saturation current (ideally zero but in reality a very small amount of current will flow). V_R is the reverse breakdown voltage. Note that the current increases rapidly once this reverse voltage is reached.

Words and Expressions

PN junction	PN 结
diode ['daɪoʊd]	n. 二极管
ceramic [sə'ræmɪk]	n. 陶瓷器；陶瓷制品
crystalline silicon	晶体硅
room temperature	室温
impurity [ɪm'pjʊrəti]	n. 杂质
dopant ['doʊpənt]	n. 掺杂物；掺杂剂
implantation [ˌɪmplæn'teɪʃn]	n. 离子注入；植入
donor atom	施主原子；供电子原子
outer shell	外电子层
acceptor atom	受主原子；接受体原子
hole [hoʊl]	n. 空穴
intrinsic semiconductor	本征半导体

bipolar junction transistor(BJT)	双极晶体管；双极结晶体管；双极结型晶体管
junction field effect transistor(JFET)	结型场效晶体管；结型场效应晶体管
rectify ['rektɪfaɪ]	*vt.* 整流；矫正；纠正
photodiode ['foʊtoʊˌdaɪoʊd]	*n.* 光电二极管
devoid [dɪ'vɔɪd]	*adj.* 缺乏；完全没有
majority charge carrier	多数载流子
depletion region	耗尽区
forward-bias ['fɔːrwərd'baɪəs]	*n.* 正向偏置
reverse-bias [rɪ'vɜːrs'baɪəs]	*n.* 反向偏置
diffuse [dɪ'fjuːs]	*v.* （使气体或液体）扩散，弥漫，渗透；（使光）模糊，漫射，漫散；传播；使分散；散布 *adj.* 弥漫的；扩散的；漫射的
migrate ['maɪgreɪt]	*v.* 迁移；转移
barrier potential	势垒电压；势垒电位
ANSI: American National Standards Institute	美国国家标准学会
anode ['ænoʊd]	*n.* 阳极；正极
cathode ['kæθoʊd]	*n.* 阴极；负极
reverse saturation current	反向饱和电流
quality factor	品质因数
Boltzmann constant	玻耳兹曼常数
reverse breakdown voltage	反向击穿电压

Notes

1. Any material can be classified as one of three types: conductor, insulator, or semiconductor.

任何材料都可以归为以下 3 种类型之一：导体、绝缘体或半导体。

2. In particular, the doped silicon can be used as a switch, turning current off and on as desired.

特别是，掺杂硅可以用作电流开关，根据需要关闭和打开。

3. The resulting crystal contains “holes” in its bonding structure where electrons would normally be located. In essence, such holes can move through the crystal conducting positive charges.

结果是，晶体的键合结构中原本是电子的位置被“空穴”取而代之。本质上，这样的空穴可以在晶体中移动，相当于传输了正电荷。

4. Assuming the crystal is not at absolute zero, the thermal energy in the system will cause some of the free electrons in the N material to “fall” into the excess holes of the adjoining P material.

假设晶体不是绝对零度，系统中的热能将导致 N 型材料中的一些自由电子“落入”相邻 P 型材料的空穴。

5. For silicon devices, the barrier potential is usually estimated at around 0.7 volts. For germanium devices, it is closer to 0.3 volts while LEDs may exhibit barrier potentials in the vicinity of 1.5 to 3 volts, partly depending on the color.

对于硅器件，势垒电压通常估计在 0.7 V 左右。对于锗器件，势垒电压接近 0.3 V，而 LED 的势垒电压可能在 1.5 V 至 3 V 附近，部分取决于 LED 的发光颜色。

6. Also, I_S is not a constant. It increases with temperature, approximately doubling for each 10℃ rise.

并且，I_S 不是常数。它随着温度的升高而增大，每升高 10℃，大约会增大一倍。

3.3 Transistor

A transistor is a semiconductor device used to amplify or switch electronic signals and electrical power. Transistors are one of the basic building blocks of modern electronics. It is composed of semiconductor material usually with at least three terminals for connection to an external circuit. A voltage or current applied to one pair of the transistor's terminals controls the current through another pair of terminals. Because the controlled (output) power can be higher than the controlling (input) power, a transistor can amplify a signal. Today, some transistors are packaged individually, but many more are found embedded in integrated circuits.

There are two broad classifications of transistors: bipolar junction transistors (BJT) and field effect transistors (FET).

3.3.1 BJT

A bipolar junction transistor consists of a three-layer "sandwich" of doped (extrinsic) semiconductor materials, either PNP in the Figure 3.9 (b) or NPN in the Figure 3.9 (d). Each layer forming the transistor has a specific name, and each layer is provided with a wire contact for connection to a circuit. The schematic symbols are shown in the Figure 3.9 (a) and (c).

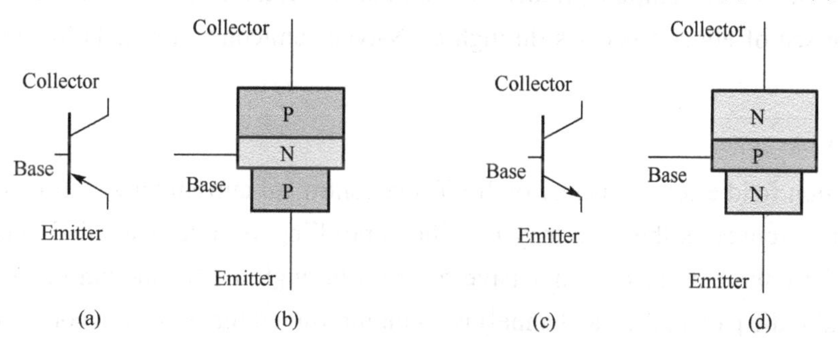

Figure 3.9 BJT: (a) PNP schematic symbol; (b) PNP physical layout; (c) NPN schematic symbol; (d) NPN physical layout

The functional difference between a PNP transistor and an NPN transistor is the proper biasing (polarity) of the junctions when operating. For any given state of operation, the current directions and voltage polarities for each kind of transistor are exactly opposite each other.

Bipolar junction transistors work as current-controlled current regulators. In other words, transistors restrict the amount of current passed according to a smaller, controlling current. The main current that is controlled goes from collector to emitter, or from emitter to collector, depending on the type of transistor it is (PNP or NPN, respectively). The small current that controls the main current goes from base to emitter, or from emitter to base, once again depending on the kind of transistor it is (PNP or NPN, respectively). This is shown in Figure 3.10.

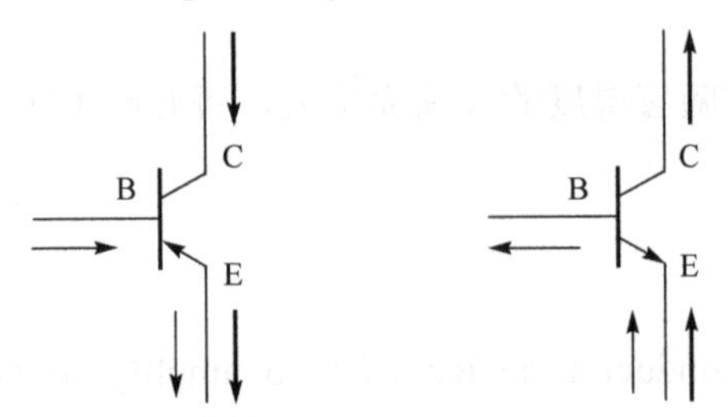

Figure 3.10 Small base-emitter current controls large collector-emitter current

Generally there are three different configurations of transistors and they are common base (CB) configuration, common collector (CC) configuration and common emitter (CE) configuration. The behaviors of these three different configurations of transistors with respect to gain are given below.

- Common base (CB) configuration: no current gain but voltage gain.
- Common collector (CC) configuration: current gain but no voltage gain.
- Common emitter (CE) configuration: current gain and voltage gain.

3.3.2 FET

A field effect transistor(FET) is a device utilizing a small voltage to control current, including the junction field effect transistor(JFET) and the insulated gate field effect transistor(IGFET). All field effect transistors are unipolar rather than bipolar devices. That is, the main current through them is comprised of either electrons through an N-type semiconductor or holes through a P-type semiconductor.

1. JFET

In a junction field effect transistor or JFET, the controlled current passes from source to drain, or from drain to source as the case may be. The controlling voltage is applied between gate and source. Note how the current does not have to cross through a PN junction on its way between source and drain: the path (called a channel) is an uninterrupted block of semiconductor material.

With no voltage applied between gate and source, the channel is a wide-open path for electrons to flow. **However, if a voltage V_{GS} is applied between gate and source of such polarity that it reverse-biases the PN junction, the flow between source and drain connection becomes limited or regulated.** Maximum gate-source voltage "pinches off" all current through source and drain, thus forcing the JFET into cutoff mode. **This behavior is due to the depletion region of the PN junction expanding under the influence of a reverse-bias voltage, eventually occupying**

the entire width of the channel if the voltage is great enough. Figure 3.11 shows the constructions and symbols for both configurations of JFETs.

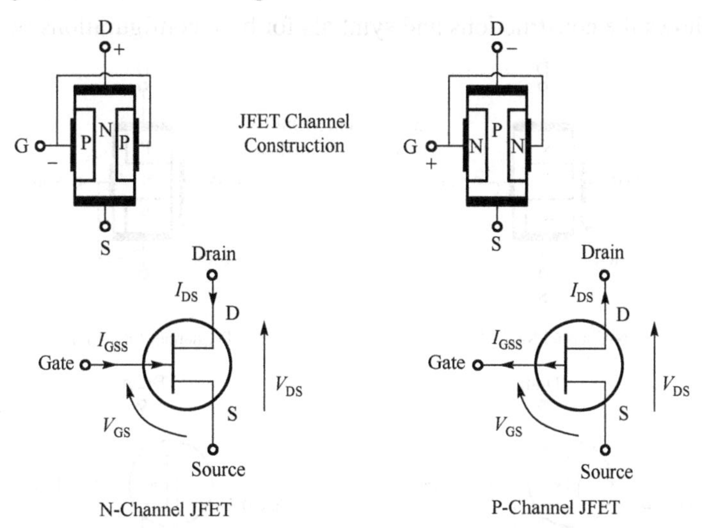

Figure 3.11 The constructions and symbols for both configurations of JFETs

Note how this operational behavior is exactly opposite of the bipolar junction transistor. Bipolar junction transistors are normally-off devices: no current through the base, no current through the collector or the emitter. JFETs, on the other hand, are normally-on devices: no voltage applied to the gate allows maximum current through the source and the drain. Also, take note that the amount of current allowed through a JFET is determined by a voltage signal rather than a current signal as with bipolar junction transistors. In fact, with the gate-source PN junction reverse-biased, there should be nearly zero current through the gate connection. For this reason, we classify the JFET as a voltage-controlled device and the bipolar junction transistor as a current-controlled device.

2. IGFET

Another type of field effect device—the insulated gate field effect transistor, or IGFET—exploits a similar principle of a depletion region controlling conductivity through a semiconductor channel, but it differs primarily from the JFET in that there is no direct connection between the gate lead and the semiconductor material itself. Rather, the gate lead is insulated from the transistor body by a thin barrier, hence the term insulated gate. **This insulating barrier acts like the dielectric layer of a capacitor and allows gate-source voltage to influence the depletion region electrostatically rather than by direct connection.**

The most common type of IGFET which is used in many different types of electronic circuits is called the metal-oxide-semiconductor field effect transistor, MOSFET for short, due to its metal (gate)-oxide (barrier)-semiconductor (channel) construction. In addition to a choice of N-channel versus P-channel design, MOSFETs come in two major types: depletion and enhancement.

- Depletion type: the transistor requires the gate-source voltage, V_{GS}, to switch the device “off ”. The depletion mode MOSFET is equivalent to a “normally closed” switch.

- Enhancement type: the transistor requires a gate-source voltage, V_{GS}, to switch the device "on". The enhancement mode MOSFET is equivalent to a "normally open" switch.

Figure 3.12 shows the constructions and symbols for both configurations of MOSFETs.

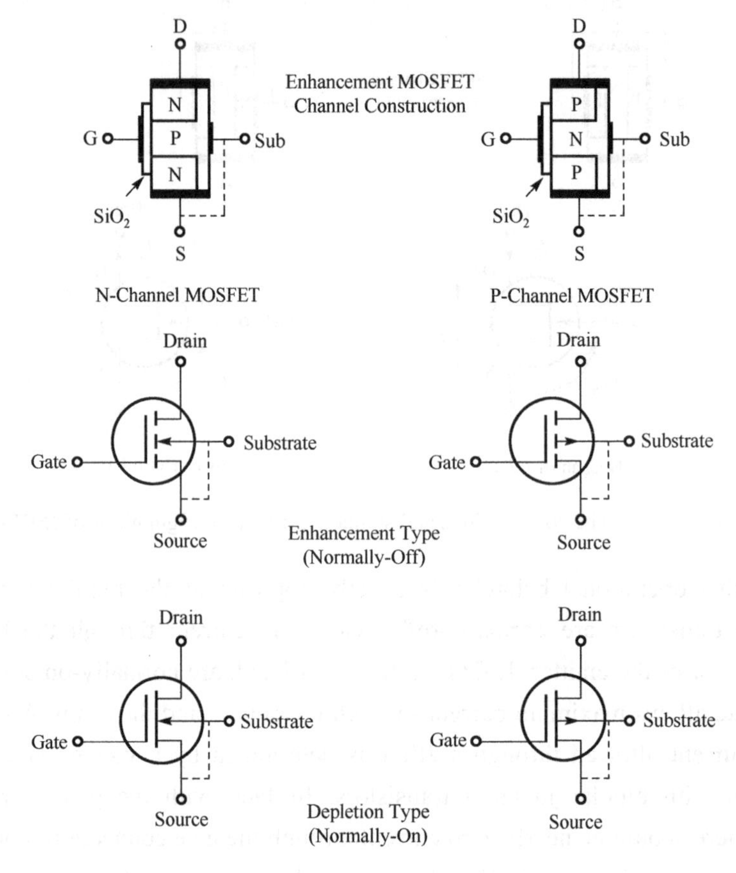

Figure 3.12 The constructions and symbols for both configurations of MOSFETs

The threshold voltage of a MOSFET is usually defined as the gate voltage where an inversion layer forms at the interface between the insulating layer (oxide) and the substrate (body) of the transistor. **As the gate terminal is electrically isolated from the main current carrying channel between drain and source, "No current flows into the gate" and the input resistance of the MOSFET is extremely high way up in the megohm (MΩ) region.**

MOSFETs are ideal for use as electronic switches or as common-source amplifiers as their power consumption is very small. Typical applications for metal-oxide-semiconductor field effect transistors are in microprocessors, memories, calculators, logic CMOS (complementary MOS) gates, etc.

 Words and Expressions

amplify ['æmplɪfaɪ] *v.* 放大；增强（声音等）

package ['pækɪdʒ] *vt.* 封装；将……包装好

extrinsic [eks'trɪnzɪk] *adj.* 非本征的；非固有的；外来的

schematic symbol 电路符号；原理图符号
collector [kə'lektər] *n.* 集电极
base [beɪs] *n.* 基极
emitter [ɪ'mɪtər] *n.* 发射极
regulator ['regjuleɪtər] *n.* 调节器；调整器；校准器；监管者
insulated gate field effect transistor(IGFET) 绝缘栅场效[应]晶体管；绝缘栅场效[应]管
source [sɔːrs] *n.* 源极
drain [dreɪn] *n.* 漏极
gate [geɪt] *n.* 栅极
channel ['tʃænl] *n.* 沟道
cutoff ['kʌˌtɔf] *n.* 截止；切断；截止点；界限
dielectric [ˌdaɪɪ'lektrɪk] *n.* 电介质；绝缘体
electrostatically [ɪˌlektroʊ'stætɪkəli] *adv.* 静电地
metal-oxide-semiconductor field effect transistor(MOSFET) 金属-氧化物-半导体场效应晶体管
enhancement [ɪn'hænsmənt] *n.* 提高；增加；增强
threshold ['θreʃhoʊld] *n.* 阈值；阈；门限；门槛；界；起始点
inversion layer 反型层
complementary MOS(CMOS) 互补金属氧化物半导体

Notes

1. The functional difference between a PNP transistor and an NPN transistor is the proper biasing (polarity) of the junctions when operating.

PNP 晶体管和 NPN 晶体管之间的功能差异在于工作时的偏置电压极性不同。

2. Generally there are three different configurations of transistors and they are common base (CB) configuration, common collector (CC) configuration and common emitter (CE) configuration.

通常晶体管有 3 种不同的配置，它们是共基极（CB）配置、共集电极（CC）配置和共发射极（CE）配置。

3. However, if a voltage V_{GS} is applied between gate and source of such polarity that it reverse-biases the PN junction, the flow between source and drain connection becomes limited or regulated.

然而，如果栅极和源极之间施加 PN 结反向偏置电压 V_{GS}，则源极到漏极之间的电流将受到限制，或者说可以被调节。

4. This behavior is due to the depletion region of the PN junction expanding under the influence of a reverse-bias voltage, eventually occupying the entire width of the channel if the voltage is great enough.

这种行为是由于 PN 结的耗尽区在反向偏置电压的影响下扩展，如果电压足够大，最终耗尽区会占据沟道的整个宽度。

5. This insulating barrier acts like the dielectric layer of a capacitor and allows gate-source

voltage to influence the depletion region electrostatically rather than by direct connection.

该绝缘屏蔽层的作用类似于电容器的介电层，允许栅极-源极电压以静电方式影响耗尽区，而不是通过直接连接。

6. As the gate terminal is electrically isolated from the main current carrying channel between drain and source, "No current flows into the gate" and the input resistance of the MOSFET is extremely high way up in the megohm (MΩ) region.

由于栅极与漏源之间的主载流通道是电隔离的，因此，"没有电流流入栅极"，并且MOSFET的输入电阻高达兆欧（MΩ）量级。

3.4 FPGA

FPGA stands for field programmable gate array. At its core, an FPGA is an array of interconnected digital subcircuits that implement common functions while also offering very high levels of flexibility. But getting a full picture of what an FPGA is requires more nuance. **The following introduces the concepts behind FPGAs and briefly discusses what makes an FPGA different from a microcontroller in design, what logic gates are, and how to program an FPGA.**

3.4.1 FPGA Versus Microcontroller

Microcontrollers have become dominant components in modern electronic design. They're inexpensive and highly versatile, and nowadays they often serve as a person's first introduction to the world of electronics. As microcontrollers become increasingly powerful, there is less and less need to consider alternative solutions to our design challenges. Nonetheless, a microcontroller is built around a processor and processors come with fundamental limitations that need to be recognized and, in some cases, overcome.

So when would an engineer reach for an FPGA over a microcontroller? The answer comes down to software vs hardware. A processor accomplishes its tasks by executing instructions in a sequential fashion. **This means that the processor's operations are inherently constrained: the desired functionality must be adapted to the available instructions and, in most cases, it is not possible to accomplish multiple processing tasks simultaneously.**

The alternative is a hardware-based approach. **It would be extremely convenient if every new design could be built around a digital IC that implements the exact functionality required by the system: no need to write software, no instruction-set constraints, no processing delays, just a single IC that has input pins, output pins, and digital circuitry corresponding precisely to the necessary operations.** This methodology is impractical beyond description because it would involve designing an ASIC (application specific integrated circuit) for every board. However, we can approximate this methodology using FPGAs.

3.4.2 What Is a Programmable Gate Array?

An FPGA is an array of logic gates, and this array can be programmed (actually, "configured" is probably a better word) in the field, i.e., by the user of the device as opposed to the people who designed it.

Logic gates (AND, OR, XOR, etc.) are the basic building blocks of digital circuitry. However, an FPGA is not a vast collection of individual Boolean gates. It's an array of carefully designed and interconnected digital subcircuits that efficiently implement common functions while also offering very high levels of flexibility. The digital subcircuits are called configurable logic blocks (CLB), and they form the core of the FPGA's programmable-logic capabilities.

The CLBs include look-up tables, storage elements (flip-flops or registers), and multiplexers that allow the CLBs to perform Boolean, data-storage, and arithmetic operations. The CLBs need to interact with one another and with external circuitry. For these purposes, the FPGA uses a matrix of programmable interconnects and input/output (I/O) blocks. The FPGA's "program" is stored in SRAM cells that influence the functionality of the CLBs and control the switches that establish the connection pathways. An I/O block consists of various components that facilitate communication between the CLBs and other components on the board. These include pull-up/pull-down resistors, buffers, and inverters. Figure 3.13 shows field programmable gate array.

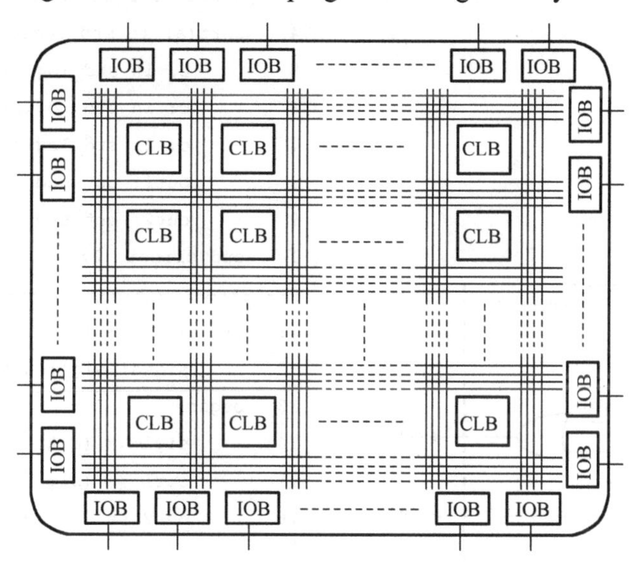

Figure 3.13 Field programmable gate array

3.4.3 How Do You Program an FPGA?

How do we go about turning an array of CLBs into a digital circuit that does precisely what we want it to? At first glance, it seems like a rather complicated task. Indeed, FPGA implementation is generally considered more difficult than programming a microcontroller. However, FPGA development does not require thorough knowledge of CLB functionality or painstaking arrangement of internal interconnects, just as microcontroller development does not require thorough knowledge of a processor's assembly-language instructions or internal control signals.

Actually, it is somewhat misleading to present an FPGA as a standalone component. **FPGAs are always supported by development software that carries out the complicated process of converting a hardware design into the programming bits that determine the behavior of interconnects and CLBs.**

People have created languages that allow us to "describe" hardware. They're called (very appropriately) hardware description languages (HDL), and the two most common are VHDL and Verilog. Despite the apparent similarity between HDL code and code written in a high-level software programming language, the two are fundamentally different. **Software code specifies a sequence of operations, whereas HDL code is more like a schematic that uses text to introduce components and create interconnections.**

Figure 3.14 shows a digital circuit example and Figure 3.15 is the corresponding VHDL code.

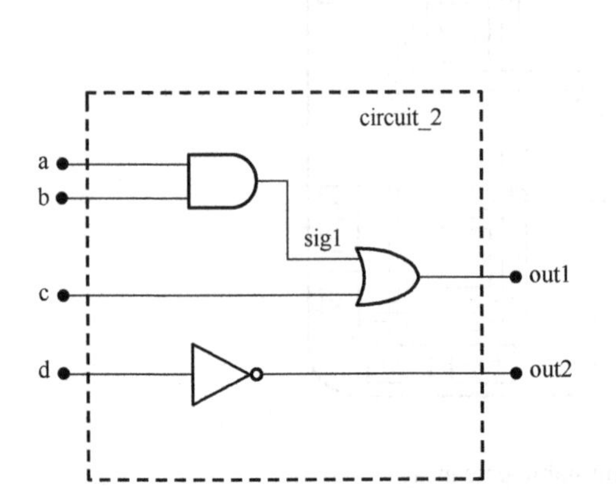

Figure 3.14 A digital circuit example

```
1    library ieee;
2    use ieee.std_logic_1164.all;
------------------------------------
3    entity circuit_2 is
4    port(a: in std_logic;
5         b: in std_logic;
6         c: in std_logic;
7         d: in std_logic;
8         out1: out std_logic;
9         out2: out std_logic);
10   end circuit_2;
------------------------------------
11   architecture Behavioral of circuit_2 is
12        signal sig1: std_logic;
13   begin
14        sig1 <= (a and b);
15        out1 <= (sig1 or c);
16        out2 <= (not d);

17   end Behavioral;
```

Figure 3.15 VHDL code of the example circuit

Lines 1 and 2: These lines add the required library and package to the code. Since the "std_logic" data type is used, we have to add the "std_logic_1164" package.

Lines 3-10: These lines specify the name of the module along with its input/output ports.

Lines 11-17: This part of the code describes the operation of the circuit. As you may have noticed, there is one internal node in Figure 3.14; it is labeled "sig1". We use the "port" statement from "entity" to define the input/output ports, but how can we define the internal nodes of a circuit? For this, we use the "signal" keyword.

In line 12 of the above code, the "signal" keyword tells the synthesis software that there is a node in the circuit labeled “sig1”. Similar to the definition of the ports, we use the keyword “std_logic” after the colon to specify the required data type. Now we can assign a value to this node (line 14) or use its value (line 15).

Note: The “std_logic” data type is a commonly used data type in VHDL. It can be used to describe a one-bit digital signal which can actually take on nine values.

‘U’: Uninitialized

‘1’ : The usual indicator for a logic high, also known as ‘Forcing high’

‘0’: The usual indicator for a logic low, also known as ‘Forcing low’

‘Z’: High impedance

‘-’: Don’t care

‘W’: Weak unknown

‘X’: Forcing unknown

‘H’: Weak high

‘L’: Weak low

Among these values, we commonly use ‘0’, ‘1’, ‘Z’, and ‘-’.

In 2016, long-time industry rivals Xilinx (now part of AMD) and Altera (now an Intel subsidiary) were the FPGA market leaders. They controlled nearly 90 percent of the market. Both Xilinx (now AMD) and Altera (now Intel) provide proprietary electronic design automation software for Windows and Linux (ISE/Vivado and Quartus) which enables engineers to design, analyze, simulate, and synthesize (compile) their designs. **Modern FPGAs are sophisticated, high-performance devices that can be somewhat intimidating for those who are accustomed to using microcontrollers for gathering data, controlling ASICs, and performing mathematical operations.** You might find, however, that in some applications the improved performance and versatility are worth the additional design effort.

Words and Expressions

nuance ['nuːɑːns]	*n.* 细微的差别
versatile ['vɜːrsətl]	*adj.* 多功能的；多用途的；多才多艺的
sequential [sɪ'kwenʃl]	*adj.* 按次序的；顺序的；序列的
approach [ə'proutʃ]	*n.*（处理问题、完成任务的）方法
methodology [ˌmeθə'dɑːlədʒi]	*n.* 方法论；（从事某一活动的）方法，原则
XOR: exclusive OR	异或
look-up table	查找表
flip-flop ['flɪp flɑːp]	*n.* 触发器
multiplexer ['mʌltiˌpleksər]	*n.* 多路复用器
SRAM: static random access memory	静态随机存储器
buffer ['bʌfər]	*n.* 缓冲器；缓存区；缓冲存储器
	vt. 缓存；缓冲；存储
inverter [ɪn'vɜːrtər]	*n.* 反相器；逆变器；变换电路；转换开关；

非门

hardware description language(HDL) 硬件描述语言

synthesis ['sɪnθəsɪs] *n.* 综合；合成；综合体

indicator ['ɪndɪkeɪtər] *n.* 标志；指示器；指针

impedance [ɪm'piːdns] *n.* 阻抗

proprietary [prə'prɑɪəteri] *adj.* 专有的；专利的

Notes

1. The following introduces the concepts behind FPGAs and briefly discusses what makes an FPGA different from a microcontroller in design, what logic gates are, and how to program an FPGA.

下面介绍 FPGA 的相关概念，并简要讨论 FPGA 与微控制器在设计上的区别、什么是逻辑门，以及如何对 FPGA 进行编程。

2. This means that the processor's operations are inherently constrained: the desired functionality must be adapted to the available instructions and, in most cases, it is not possible to accomplish multiple processing tasks simultaneously.

这意味着处理器的操作本身受到限制：所需的功能必须适应可用的指令，并且在大多数情况下，不可能同时完成多个处理任务。

3. It would be extremely convenient if every new design could be built around a digital IC that implements the exact functionality required by the system: no need to write software, no instruction-set constraints, no processing delays, just a single IC that has input pins, output pins, and digital circuitry corresponding precisely to the necessary operations.

如果每种新的系统设计都能基于一个数字集成电路来精确实现所需功能，设计者不需要编写软件，没有指令集约束，也没有处理延迟，只需要一个具有输入引脚、输出引脚并能精确完成所需操作的数字集成电路，那么将会是非常方便的。

4. FPGAs are always supported by development software that carries out the complicated process of converting a hardware design into the programming bits that determine the behavior of interconnects and CLBs.

FPGA 总是由开发软件支持，该开发软件可以将硬件设计转换为编程数据，以定义互连和 CLB 的行为。

5. Software code specifies a sequence of operations, whereas HDL code is more like a schematic that uses text to introduce components and create interconnections.

软件代码指定一系列操作，而 HDL 代码更像是一个原理图，它使用文本来引入电路组件并创建互连。

6. Modern FPGAs are sophisticated, high-performance devices that can be somewhat intimidating for those who are accustomed to using microcontrollers for gathering data, controlling ASICs, and performing mathematical operations.

现代 FPGA 是一种复杂的高性能设备，对那些习惯于使用微控制器收集数据、控制 ASIC 和执行数学运算的人来说，这可能有点惊人。

Exercises

1. Match the terms (1)–(6) with the definitions A–F.

(1) depletion region	A. a two-terminal electronic component that conducts current primarily in one direction
(2) VHDL	B. a three-terminal electronic device used to control the flow of current by the voltage applied to its gate terminal
(3) FET	C. a language that describes the behavior of electronic circuits, most commonly digital circuits
(4) diode	D. it is formed from a conducting region by removal of all free charge carriers, leaving none to carry a current
(5) breakdown voltage	E. an empty spot with a positive charge in the crystal lattice
(6) hole	F. the threshold voltage, beyond which an insulator starts behaving as a conductor and conducts electricity

2. Translate into Chinese.

(1) In April 2019, two of the world's largest semiconductor foundries—Taiwan Semiconductor Manufacturing Company Limited (TSMC) and Samsung Foundry—announced their success in reaching the 5 nm technology node, propelling the miniaturization of transistors one step further to a new age.

(2) Since digital circuits involve millions of times as many components as analog circuits, much of the design work is done by copying and reusing the same circuit functions, especially by using digital design software that contains libraries of prestructured circuit components.

(3) In general, MOSFET amplifiers tend to have good high frequency performance, offer low noise and exhibit low distortion with modestly sized input signals. Compared to BJTs, their voltage gain magnitude is lower.

3. Translate into English.

（1）发光二极管通电时可以发光，而光电二极管在光照下产生电流。

（2）MOSFET 的尺寸是用栅极长度来衡量的，也就是俗称的特征尺寸或特征长度，用符号 *L* 来表示。

（3）随着晶体管尺寸的缩小，它们的工作速度也随之提高。

4. Read the following article and write a summary.

The debate of whether Moore's law is "dying" (or already "dead") has been going on for years. It has been discussed by pretty much everyone. But before we can give an aswer to that, let's first clarify the meaning of Moore's law.

Moore's law stems from the observation of Gordon Moore, co-founder and chairman emeritus of Intel, made in 1965. At the time, he said that the number of transistors in a dense integrated circuit had doubled roughly every year and would continue to do so for the next 10 years. In 1975, he revised his observation to say that this would occur every two years indefinitely. Moore's observation became the driving force behind the semiconductor technology revolution that led to the proliferation of computers and other electronic devices.

Over time, the details of Moore's law were amended to reflect the true growth of transistor density. First, the doubling interval was increased to two years and then decreased to around 18 months. The exponential nature of Moore's law continued and created decades of opportunity for the semiconductor industry and the electronics that use them.

The issue for Moore's law is the inherent complexity of semiconductor process technology, and these complexities have been growing. Transistors are now three-dimensional, and the small feature size of today's advanced process technologies has required multiple exposures to reproduce these features on silicon wafers. This has added extreme complexity to the design process and has "slowed down" Moore's law.

This slowing down has led many to ask, "Is Moore's law dead?" The simple answer to this is no, Moore's law is not dead. While it's true that chip densities are no longer doubling every two years (thus, Moore's law isn't happening anymore by its strictest definition), Moore's law is still delivering exponential improvements, albeit at a slower pace. The trend is very much still here.

Intel's CEO Pat Gelsinger believes that Moore's law is far from obsolete. As a goal for the next 10 years, he announced in 2021 not only to uphold Moore's law, but to outpace it. There are many industry veterans who agree with this. Mario Morales, a program vice president at IDC, said he believes Moore's law is still relevant in theory in an interview with TechRepublic.

"If you look at what Moore's law has enabled, we're seeing an explosion of more computing across the entire landscape," he said, "It used to be computing was centered around mainframes and then it became clients and now edge and endpoints, but they're getting more intelligent, and now they're doing AI inferencing, and you need computing to do that. So, Moore's law has been able to continue to really push computing to the outer edge."

While the consensus is that Moore's law is slowing down and that it might soon be augmented, it is still driving improvements in processing technology and the amount of progress that follows these improvements. If it were dead, it simply couldn't do this.

5. Language study: Describing components

Two questions we may need to answer when we describe components are:

(1) What is it called?

(2) What does it do?

In other words, we need to be able to:

(1) label components.

(2) describe their functions.

We can use these ways of labeling components:

(1) It is called a photodiode.

(2) It is known as a Schottky transistor.

We can describe the functions of components like these:

(1) An FPGA **provides** a way to implement digital logic circuits in a short amount of time.

(2) Batteries **convert** chemical energy into electrical energy.

Unit 4 Signals and Systems

4.1 Introduction to Signals

Signals are represented mathematically as functions of one or more independent variables. For example, a speech signal can be represented mathematically by acoustic pressure as a function of time, and a picture can be represented by brightness as a function of two spatial variables. There are two basic types of signals, continuous-time signals and discrete-time signals. In the case of continuous-time signals, the independent variable is continuous, and thus these signals are defined for a continuum of values of the independent variable. **On the other hand, discrete-time signals are defined only at discrete times, and consequently, for these signals, the independent variable takes on only a discrete set of values.** To distinguish between continuous-time and discrete-time signals, we will use the symbol t to denote the continuous-time independent variable and n to denote the discrete-time independent variable. It is important to note that the discrete-time signal is defined only for integer values of the independent variable. A speech signal as a function of time and atmospheric pressure as a function of altitude are examples of continuous-time signals. The weekly Dow-Jones stock market index, as illustrated in Figure 4.1, is an example of a discrete-time signal.

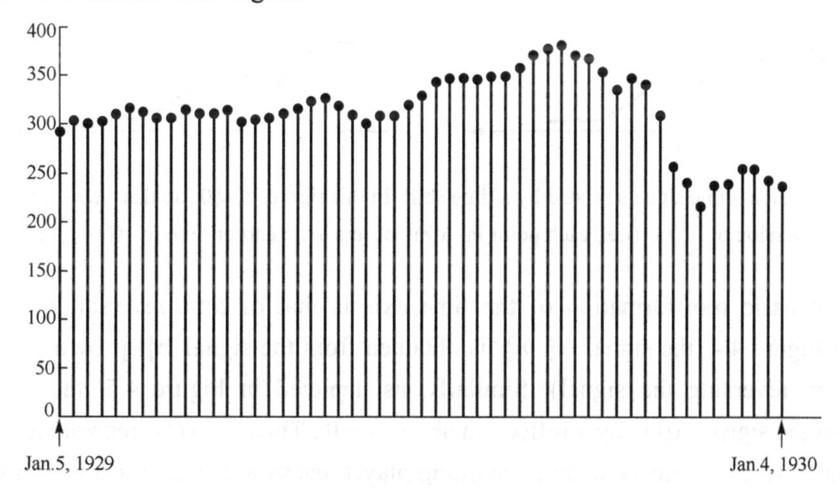

Figure 4.1 An example of a discrete-time signal: The weekly Dow-Jones stock market index from January 5, 1929, to January 4, 1930

A central concept in signal and system analysis is that of the transformation of a signal. In this section, we focus on a very limited but important class of elementary signal transformations that involve simple modification of the independent variable, i.e., the time axis. **A simple and very important example of transforming the independent variable of a signal is a time shift.** A time shift in discrete time is illustrated in Figure 4.2, in which we have two signals $x[n]$ and $x[n-n_0]$ that are identical in shape, but that are displaced or shifted relative to each other. We will also encounter time shifts in continuous time, as illustrated in Figure 4.3, in which $x(t-t_0)$ represents a delayed (if t_0 is positive) or advanced (if t_0 is negative) version of $x(t)$. Signals that are related in this fashion arise in applications such as radar, sonar, and seismic signal processing, in which several receivers at different locations observe a signal being transmitted through a medium (water, rock, air, etc.). **In this case, the difference in propagation time from the point of origin of the transmitted signal to any two receivers results in a time shift between the signals at the two receivers.**

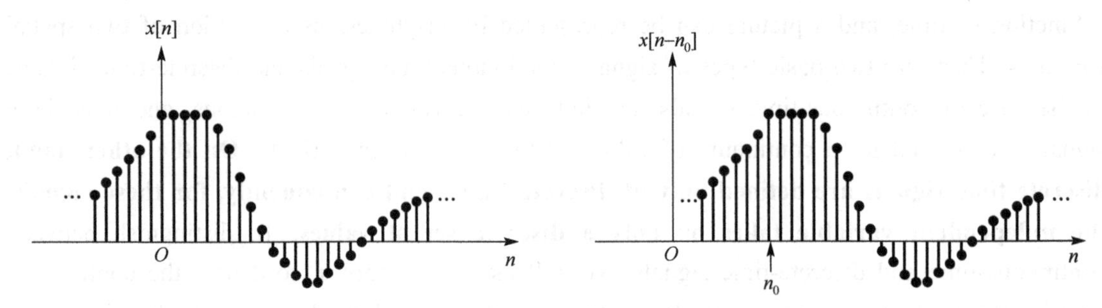

Figure 4.2 Discrete-time signals related by a time shift. In this figure $n_0 > 0$, so that $x[n-n_0]$ is a delayed version of $x[n]$ (i.e., each point in $x[n]$ occurs later in $x[n-n_0]$)

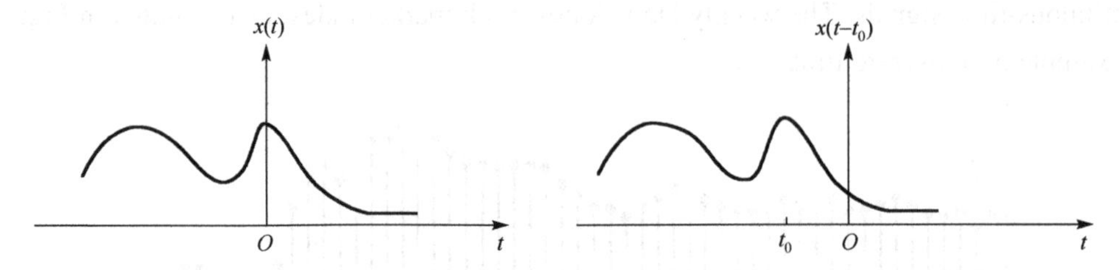

Figure 4.3 Continuous-time signals related by a time shift. In this figure $t_0 < 0$, so that $x(t-t_0)$ is an advanced version of $x(t)$ (i.e., each point in $x(t)$ occurs at an earlier time in $x(t-t_0)$)

A second basic transformation of the time axis is that of time reversal. For example, as illustrated in Figure 4.4, the signal $x[-n]$ is obtained from the signal $x[n]$ by a reflection about $n=0$ (i.e., by reversing the signal). Similarly, as depicted in Figure 4.5, the signal $x(-t)$ is obtained from the signal $x(t)$ by a reflection about $t=0$. Thus, if $x(t)$ represents an audio tape recording, then $x(-t)$ is the same tape recording played backward. Another transformation is that of time scaling. In Figure 4.6, we have illustrated three signals, $x(t)$, $x(2t)$, $x(t/2)$, that are replaced by linear scale changes in the independent variable. If we again think of the example of $x(t)$ as a tape recording, then $x(2t)$ is that recording played at twice the speed, and $x(t/2)$ is the recording played at half-speed.

It is often of interest to determine the effect of transforming the independent variable of

a given signal $x(t)$ to obtain a signal of the form $x(\alpha t+\beta)$, where α and β are given numbers. Such a transformation of the independent variable preserves the shape of $x(t)$, except that the resulting signal may be linearly stretched if $|\alpha|<1$, linearly compressed if $|\alpha|>1$, reversed in time if $\alpha<0$, and shifted in time if β is nonzero.

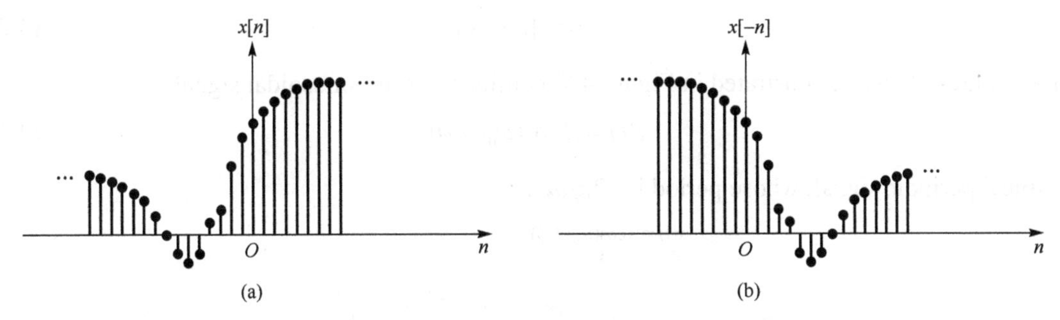

Figure 4.4 (a) A discrete-time signal $x[n]$; (b) its reflection $x[-n]$ about $n=0$

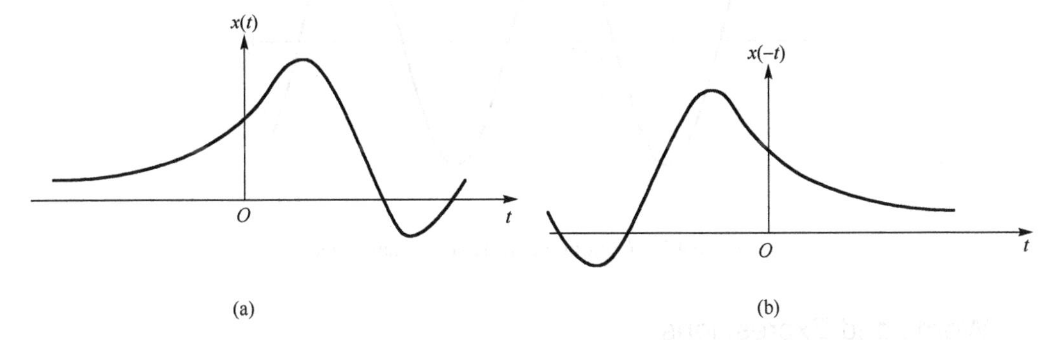

Figure 4.5 (a) A continuous-time signal $x(t)$; (b) its reflection $x(-t)$ about $t=0$

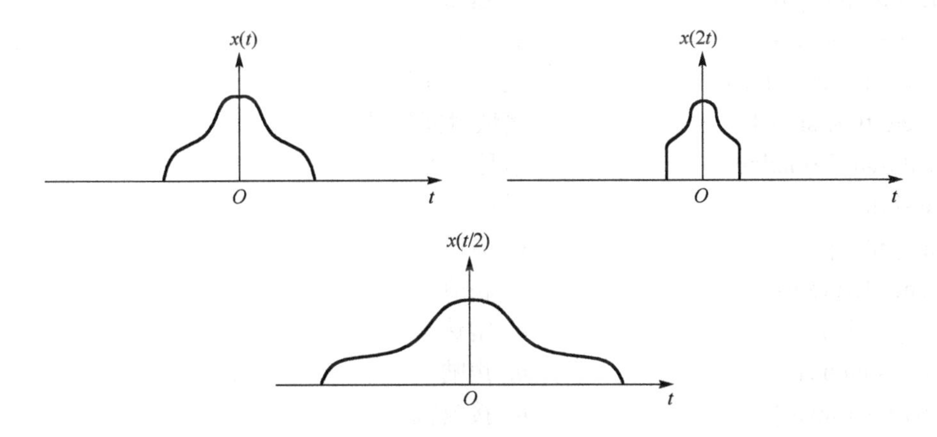

Figure 4.6 Continuous-time signals related by time scaling

An important class of signals that we will encounter frequently is the class of periodic signals. A periodic continuous-time signal $x(t)$ has the property that there is a positive value of T for which

$$x(t)=x(t+T) \tag{4.1}$$

for all values of t. In other words, a periodic signal has the property that it is unchanged by a time shift of T. In this case, we say that $x(t)$ is periodic with period T. Periodic signals are defined analogously in discrete time. Specially, a discrete-time signal $x[n]$ is periodic with period N, where N is a positive integer, if it is unchanged by a time shift of N, i.e., if

$$x[n] = x[n+N] \tag{4.2}$$

for all values of n. As illustrated in Figure 4.7, continuous-time sinusoidal signal

$$x(t) = A\cos(\omega_0 t + \phi) \tag{4.3}$$

is typical periodic signal, whose period is $2\pi/\omega_0$.

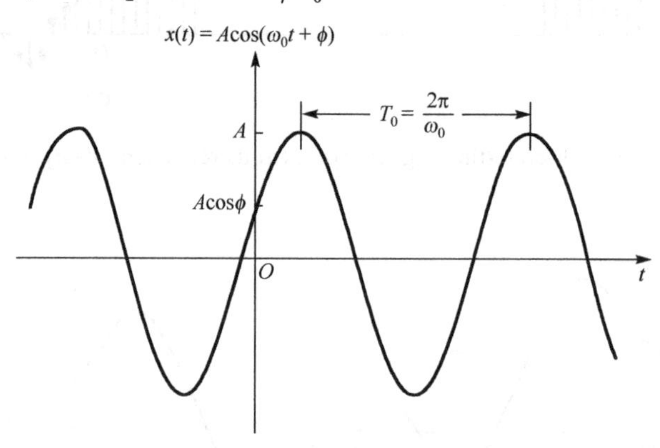

Figure 4.7 Continuous-time sinusoidal signal

Words and Expressions

function ['fʌŋkʃn]	*n.* 函数
variable ['veriəbl]	*n.* 变量
continuous-time signal	连续时间信号
discrete-time signal	离散时间信号
signal transformation	信号变换
time shift	时移
delay [dɪ'leɪ]	*v.* 滞后
advance [əd'væns]	*v.* 超前
radar ['reɪdɑːr]	*n.* 雷达
sonar ['soʊnɑːr]	*n.* 声呐
receiver [rɪ'siːvər]	*n.* 接收机
medium ['miːdiəm]	*n.* 媒介
propagation [ˌprɑːpə'geɪʃn]	*n.* 传播
time reversal	时间变量的翻转
time scaling	时间变量的尺度变换
periodic [ˌpɪri'ɑːdɪk]	*adj.* 周期的
period ['pɪriəd]	*n.* 周期

sinusoidal signal 正弦信号

Notes

1. Signals are represented mathematically as functions of one or more independent variables.

数学上信号表示为一个或多个自变量的函数。

2. On the other hand, discrete-time signals are defined only at discrete times, and consequently, for these signals, the independent variable takes on only a discrete set of values.

另一方面，离散时间信号仅在离散时间定义，因此，对这些信号，自变量仅具有一组离散值。

3. A simple and very important example of transforming the independent variable of a signal is a time shift.

时移是信号自变量变换的一个简单且重要的例子。

4. In this case, the difference in propagation time from the point of origin of the transmitted signal to any two receivers results in a time shift between the signals at the two receivers.

在此情况下，从发射信号原始位置到任意两个接收机之间的传输时间差异会导致两个接收机中信号的时移。

5. It is often of interest to determine the effect of transforming the independent variable of a given signal $x(t)$ to obtain a signal of the form $x(\alpha t+\beta)$, where α and β are given numbers.

通常感兴趣的是确定对给定信号 $x(t)$ 的自变量进行变换以获得 $x(\alpha t+\beta)$ 形式信号的效果，其中 α 和 β 是给定的数字。

4.2 Frequency Response of LTI Systems

The frequency-domain characterization of an LTI (linear time-invariant) system in terms of its frequency response represents an alternative to the time-domain characterization through convolution. In analyzing LTI systems, it is often particularly convenient to utilize the frequency domain because differential and difference equations and convolution operations in the time domain become algebraic operations in the frequency domain. Moreover, concepts such as frequency-selective filtering are readily and simply visualized in the frequency domain. However, in system design, there are typically both time-domain and frequency-domain considerations.

The Fourier transform is in general complex valued and can be represented in terms of its real and imaginary components or in terms of magnitude and phase. The magnitude-phase representation of the continuous-time Fourier transform $X(\mathrm{j}\omega)$ is

$$X(\mathrm{j}\omega)=\left|X(\mathrm{j}\omega)\right|\mathrm{e}^{\mathrm{j}\measuredangle X(\mathrm{j}\omega)} \tag{4.4}$$

Similarly, the magnitude-phase representation of the discrete-time Fourier transform $X(\mathrm{e}^{\mathrm{j}\omega})$ is

$$X(\mathrm{e}^{\mathrm{j}\omega})=\left|X(\mathrm{e}^{\mathrm{j}\omega})\right|\mathrm{e}^{\mathrm{j}\measuredangle X(\mathrm{e}^{\mathrm{j}\omega})} \tag{4.5}$$

From the convolution property for continuous-time Fourier transforms, the transform

$Y(\mathrm{j}\omega)$ **of the output of an LTI system is related to the transform** $X(\mathrm{j}\omega)$ **of the input to the system by the equation**

$$Y(\mathrm{j}\omega)=H(\mathrm{j}\omega)X(\mathrm{j}\omega) \tag{4.6}$$

where $H(\mathrm{j}\omega)$ **is the frequency response of the system, i.e., the Fourier transform of the system's impulse response.** Similarly, in discrete time, the Fourier transforms of the input $X(\mathrm{e}^{\mathrm{j}\omega})$ and output $Y(\mathrm{e}^{\mathrm{j}\omega})$ of an LTI system with frequency response $H(\mathrm{e}^{\mathrm{j}\omega})$ are related by

$$Y(\mathrm{e}^{\mathrm{j}\omega})=H(\mathrm{e}^{\mathrm{j}\omega})X(\mathrm{e}^{\mathrm{j}\omega}) \tag{4.7}$$

Thus, the effect that an LTI system has on the input is to change the complex amplitude of each of the frequency components of the signal. By looking at this effect in terms of the magnitude-phase representation, we can understand the nature of the effect in more detail. Specifically, in continuous time,

$$|Y(\mathrm{j}\omega)|=|H(\mathrm{j}\omega)||X(\mathrm{j}\omega)| \tag{4.8}$$

and

$$\measuredangle Y(\mathrm{j}\omega)=\measuredangle H(\mathrm{j}\omega)+\measuredangle X(\mathrm{j}\omega) \tag{4.9}$$

and exactly analogous relationships hold in the discrete-time case. From Equation (4.8), we see that the effect an LTI system has on the magnitude of the Fourier transform of the signal is to scale it by the magnitude of the frequency response. For this reason, $|H(\mathrm{j}\omega)|$ (or $|H(\mathrm{e}^{\mathrm{j}\omega})|$) is commonly referred to as the gain of the system. Also, from Equation (4.9), we see that the phase of the input $\measuredangle X(\mathrm{j}\omega)$ is modified by the LTI system by adding the phase $\measuredangle H(\mathrm{j}\omega)$ to it, and $\measuredangle H(\mathrm{j}\omega)$ is typically referred to as the phase shift of the system. The phase shift of the system can change the relative phase relationships among the components of the input, possibly resulting in significant modifications to the time-domain characteristics of the input even when the gain of the system is constant for all frequencies. **The changes in the magnitude and phase that result from the application of an input to an LTI system may be either desirable, if the input signal is modified in a useful way, or undesirable, if the input is changed in an unwanted manner.** In the latter case, the effects in Equations (4.8) and (4.9) are commonly referred to as magnitude and phase distortions.

When the phase shift at the frequency ω is a linear function of ω, there is a particularly straightforward interpretation of the effect in the time domain. Consider the continuous-time LTI system with frequency response

$$H(\mathrm{j}\omega)=\mathrm{e}^{-\mathrm{j}\omega t_0} \tag{4.10}$$

so that the system has unit gain and linear phase, i.e.,

$$|H(\mathrm{j}\omega)|=1,\ \measuredangle H(\mathrm{j}\omega)=-\omega t_0 \tag{4.11}$$

The system with this frequency response characteristic produces an output that is simply a time shift of the input, i.e., $y(t)=x(t-t_0)$.

In the discrete-time case, the effect of linear phase is similar to that in the continuous-time case when the slope of the linear phase is an integer. Specially, we know that the LTI system with frequency response $e^{-j\omega n_0}$ with linear phase function $-\omega n_0$ produces an output that is a simple shift of the input, i.e., $y[n]=x[n-n_0]$. Thus, a linear phase shift with an integer slope corresponds to a shift of $x[n]$ by an integer number of samples.

In graphically displaying continuous-time or discrete-time Fourier transforms and system frequency responses in polar form, it is often convenient to use a logarithmic scale for the magnitude of the Fourier transform. One of the principal reasons for doing this can be seen from Equations (4.8) and (4.9), which relate the magnitude and phase of the output of an LTI system to those of the input and frequency response. Note that the phase relationship is additive, while the magnitude relationship involves the product of $|H(j\omega)|$ and $|X(j\omega)|$. Thus, if the magnitudes of the Fourier transform are displayed on a logarithmic amplitude scale, Equation (4.8) takes the form of an additive relationship, namely, $\log|Y(j\omega)|=\log|H(j\omega)|+\log|X(j\omega)|$. There is an exactly analogous expression in discrete time. **Consequently, if we have a graph of the log magnitude and phase of the Fourier transform of the input and the frequency response of an LTI system, the Fourier transform of the output is obtained by adding the log-magnitude plots and by adding the phase plots.** In a similar fashion, since the frequency response of the cascade of LTI systems is the product of the individual frequency responses, we can obtain plots of the log magnitude and phase of the overall frequency response of cascade systems by adding the corresponding plots for each of the component systems. In addition, plotting the magnitude of the Fourier transform on a logarithmic scale allows detail to be displayed over a wider dynamic range. For continuous-time system, it is also common and useful to use a logarithmic frequency scale. Plots of $20\log_{10}|H(j\omega)|$ and $\measuredangle H(j\omega)$ versus $\log_{10}\omega$ are referred to as Bode plots. A typical Bode plot is illustrated in Figure 4.8.

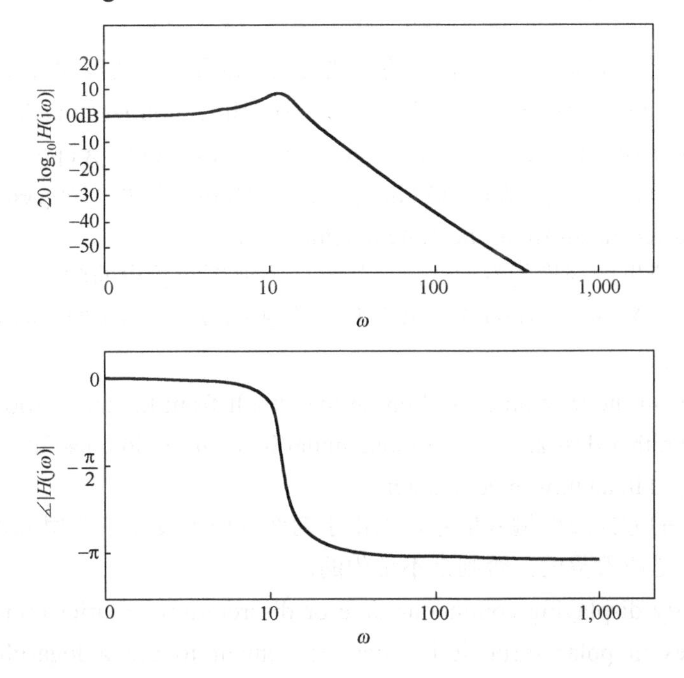

Figure 4.8 A typical Bode plot (Note that ω is plotted using a logarithmic scale.)

Words and Expressions

linear time-invariant (LTI) system 线性时不变系统
time-domain characterization 时域表征
convolution [ˌkɑːnvəˈluːʃn] *n.* 卷积
differential and difference equations 微分与差分方程
filtering [ˈfɪltərɪŋ] *n.* 滤波
real and imaginary components 实部和虚部分量
magnitude [ˈmæɡnɪtuːd] *n.* 幅度
phase [feɪz] *n.* 相位
Fourier transform 傅里叶变换
frequency response 频率响应
impulse response 冲激(脉冲)响应
gain [geɪn] *n.* 增益
phase shift 相移
magnitude and phase distortion 幅度和相位失真
slope [sloʊp] *n.* 斜率
cascade [kæˈskeɪd] *n.* 级联
Bode plot 伯德图

Notes

1. The frequency-domain characterization of an LTI (linear time-invariant) system in terms of its frequency response represents an alternative to the time-domain characterization through convolution.

线性时不变系统的以频率响应为代表的频域表征是以卷积为代表的时域表征的替代方法。

2. From the convolution property for continuous-time Fourier transforms, the transform $Y(\mathrm{j}\omega)$ of the output of an LTI system is related to the transform $X(\mathrm{j}\omega)$ of the input to the system by the equation $Y(\mathrm{j}\omega)=H(\mathrm{j}\omega)X(\mathrm{j}\omega)$, where $H(\mathrm{j}\omega)$ is the frequency response of the system, i.e., the Fourier transform of the system's impulse response.

根据连续时间傅里叶变换的卷积性质，线性时不变系统输出的频谱$Y(\mathrm{j}\omega)$与系统激励的频谱$X(\mathrm{j}\omega)$通过公式$Y(\mathrm{j}\omega)=H(\mathrm{j}\omega)X(\mathrm{j}\omega)$关联，其中$H(\mathrm{j}\omega)$是系统的频率响应，即系统冲激响应的傅里叶变换。

3. The changes in the magnitude and phase that result from the application of an input to an LTI system may be either desirable, if the input signal is modified in a useful way, or undesirable, if the input is changed in an unwanted manner.

如果以有意义的方式改变输入信号，则由于向线性时不变系统应用输入信号而导致的幅度和相位变化可能是所希望的，否则是不希望的。

4. In graphically displaying continuous-time or discrete-time Fourier transforms and system frequency responses in polar form, it is often convenient to use a logarithmic scale for the magnitude of the Fourier transform.

以极坐标形式展示连续时间或离散时间傅里叶变换和系统频率响应时，对傅里叶变换的幅度采用对数刻度往往是方便的。

5. Consequently, if we have a graph of the log magnitude and phase of the Fourier transform of the input and the frequency response of an LTI system, the Fourier transform of the output is obtained by adding the log-magnitude plots and by adding the phase plots.

因此，如果我们已知线性时不变系统输入的傅里叶变换和频率响应的对数幅度和相位图，那么通过将两者的对数幅度图相加和它们的相位图相加可获得输出的傅里叶变换。

4.3 Sampling Theorem

Under certain conditions, a continuous-time signal can be completely represented by and recoverable from knowledge of its values, or samples, at points equally spaced in time. This somewhat surprising property follows from a basic result that is referred to as the sampling theorem. This theorem is extremely important and useful. It is exploited, for example, in moving pictures, which consist of a sequence of individual frames, each of which represents an instantaneous view (i.e., a sample in time) of a continuously changing scene. **When these samples are viewed in sequence at a sufficiently fast rate, we perceive an accurate representation of the original continuously moving scene.** As another example, printed pictures typically consist of a very fine grid of points, each corresponding to a sample of the spatially continuous scene represented in the picture. If the samples are sufficiently close together, the picture appears to be spatially continuous, although under a magnifying glass its representation in terms of samples becomes evident.

In order to develop the sampling theorem, we need a convenient way in which to represent the sampling of a continuous-time signal at regular intervals. **A useful way to do this is through the use of a periodic impulse train multiplied by the continuous-time signal $x(t)$ that we wish to sample.** This mechanism, known as impulse-train sampling, is depicted in Figure 4.9. The periodic impulse train $p(t)$ is referred to as the sampling function, the period T as the sampling period, and the fundamental frequency of $p(t)$, $\omega_s = 2\pi/T$, as the sampling frequency. In the time domain,

$$x_p(t) = x(t)p(t) \tag{4.12}$$

where

$$p(t) = \sum_{n=-\infty}^{+\infty} \delta(t - nT) \tag{4.13}$$

Because of the sampling property of the unit impulse, we know that multiplying $x(t)$ by a unit impulse samples the value of the signal at the point at which the impulse is located; i.e., $x(t)\delta(t-t_0) = x(t_0)\delta(t-t_0)$. Applying this to Equation (4.12), we see, as illustrated in Figure 4.9, that $x_p(t)$ is an impulse train with the amplitudes of the impulses equal to the samples of $x(t)$ at intervals spaced by T; that is

$$x_p(t)=\sum_{n=-\infty}^{+\infty}x(nT)\delta(t-nT) \tag{4.14}$$

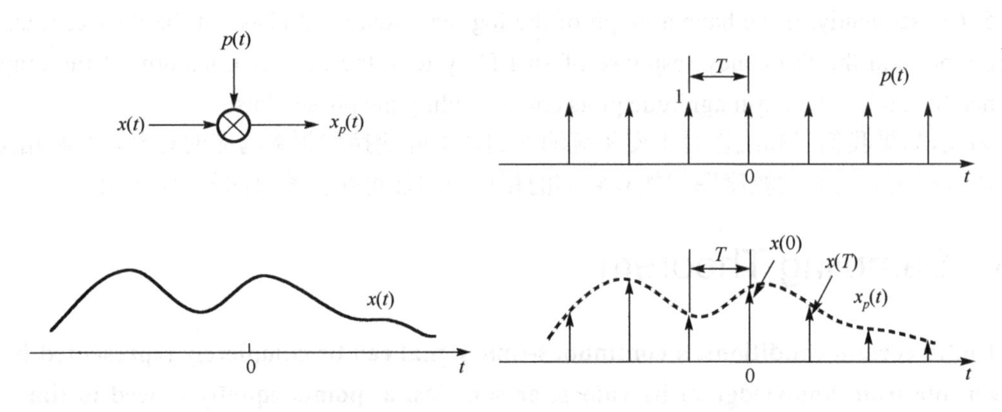

Figure 4.9 Impulse-train sampling

From the multiplication property of Fourier transform, we know that

$$X_p(\mathrm{j}\omega)=\frac{1}{2\pi}[X(\mathrm{j}\omega)*P(\mathrm{j}\omega)] \tag{4.15}$$

and the Fourier transform of impulse train

$$P(\mathrm{j}\omega)=\frac{2\pi}{T}\sum_{n=-\infty}^{+\infty}\delta(\omega-n\omega_{\mathrm{s}}) \tag{4.16}$$

Since convolution with an impulse simply shifts a signal [i.e., $X(\mathrm{j}\omega)*\delta(\omega-\omega_0)=X(\mathrm{j}(\omega-\omega_0))$], it follows that

$$X_p(\mathrm{j}\omega)=\frac{1}{T}\sum_{n=-\infty}^{+\infty}X(\mathrm{j}(\omega-n\omega_{\mathrm{s}})) \tag{4.17}$$

That is, $X_p(\mathrm{j}\omega)$ is a periodic function of ω consisting of a superposition of shifted replicas of $X(\mathrm{j}\omega)$, scaled by $1/T$, as illustrated in Figure 4.10. In Figure 4.10(c), $\omega_{\mathrm{M}}<\omega_{\mathrm{s}}-\omega_{\mathrm{M}}$, or equivalently, $\omega_{\mathrm{s}}>2\omega_{\mathrm{M}}$, and thus there is no overlap between the shifted replicas of $X(\mathrm{j}\omega)$, whereas in Figure 4.10(d), with $\omega_{\mathrm{s}}<2\omega_{\mathrm{M}}$, there is overlap. For the case illustrated in Figure 4.10(c), $X(\mathrm{j}\omega)$ is faithfully reproduced at integer multiples of the sampling frequency. Consequently, if $\omega_{\mathrm{s}}<2\omega_{\mathrm{M}}$, $x(t)$ can be recovered exactly from $x_p(t)$ by means of a low-pass filter with gain T and a cutoff frequency greater than ω_{M} and less than $\omega_{\mathrm{s}}-\omega_{\mathrm{M}}$, as indicated in Figure 4.11. This basic result, referred to as the sampling theorem, can be stated as follows.

Sampling Theorem:

Let $x(t)$ be a band-limited signal with $X(\mathrm{j}\omega)=0$ for $|\omega|>\omega_{\mathrm{M}}$. Then $x(t)$ is uniquely determined by its samples $x(nT)$, $n=0,\pm1,\pm2,\cdots$, if $\omega_{\mathrm{s}}>2\omega_{\mathrm{M}}$, where $\omega_{\mathrm{s}}=\dfrac{2\pi}{T}$. Given these samples, we can reconstruct $x(t)$ by generating a periodic impulse train in

which successive impulses have amplitudes that are successive sample values. This impulse train is then processed through an ideal low-pass filter with gain T and cutoff frequency greater than ω_M and less than $\omega_s - \omega_M$. The resulting output signal will exactly equal $x(t)$.

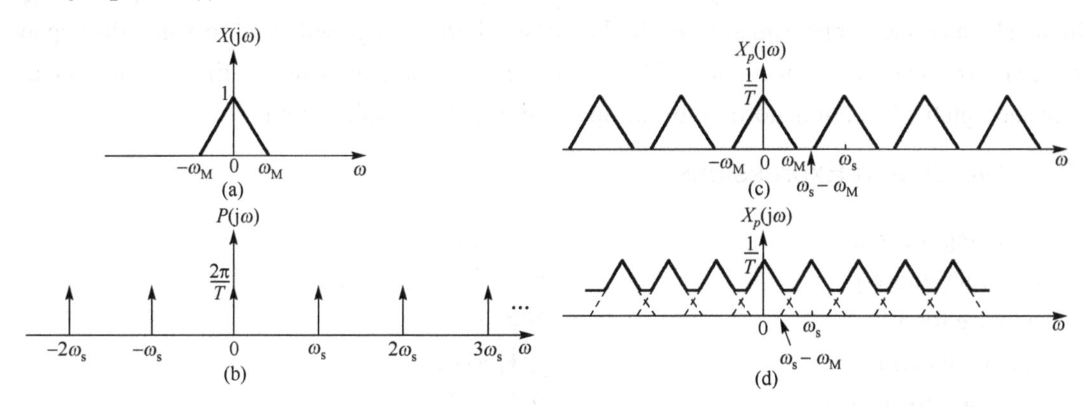

Figure 4.10 Effect in the frequency domain of sampling in the time domain: (a) spectrum of original signal; (b) spectrum of sampling function; (c) spectrum of sampled signal with $\omega_s > 2\omega_M$; (d) spectrum of sampled signal with $\omega_s < 2\omega_M$

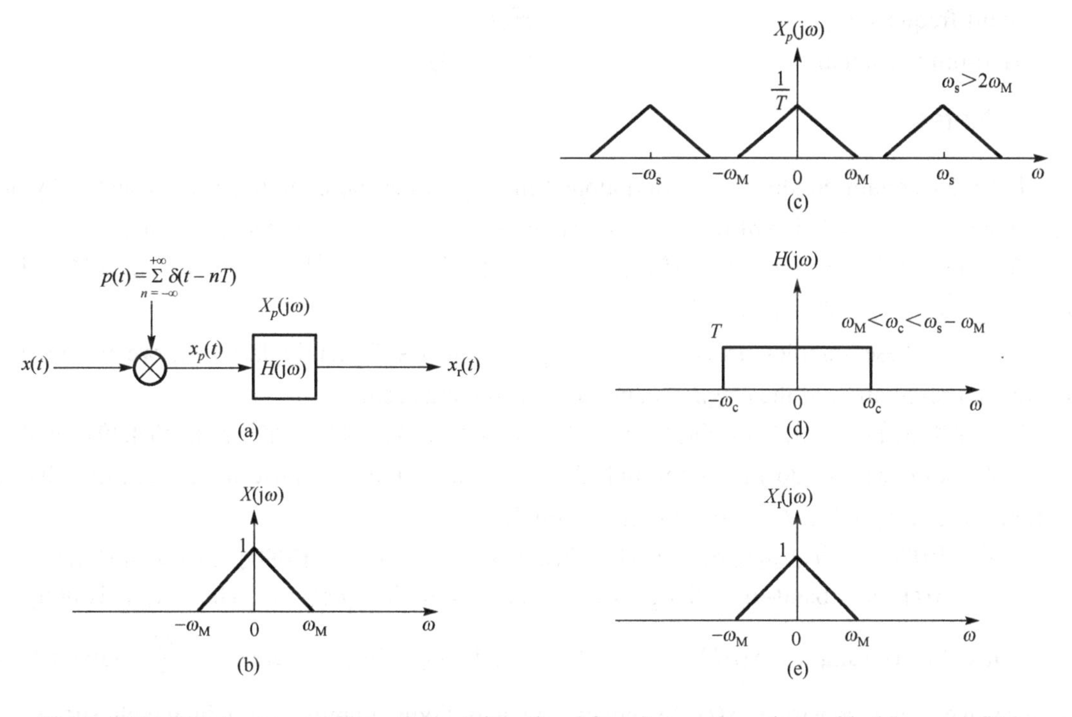

Figure 4.11 Exact recovery of a continuous-time signal from its samples using an ideal low-pass filter: (a) system for sampling and reconstruction; (b) representative spectrum for $x(t)$; (c) corresponding spectrum for $x_p(t)$; (d) ideal low-pass filter to recover $X(j\omega)$ from $X_p(j\omega)$; (e) spectrum of $x_r(t)$

The frequency $2\omega_M$, which, under the sampling theorem, must be exceeded by the sampling frequency, is commonly referred to as the Nyquist rate.

Ideal filters are generally not used in practice for a variety of reasons. **In any practical application, the ideal low-pass filter in Figure 4.11 would be replaced by a nonideal filter $H(\mathrm{j}\omega)$ that approximated the desired frequency characteristic accurately enough for the problem of interest (i.e., $H(\mathrm{j}\omega)\approx 1$ for $|\omega|<\omega_{\mathrm{M}}$, and $H(\mathrm{j}\omega)\approx 0$ for $|\omega|>\omega_{\mathrm{s}}-\omega_{\mathrm{M}}$).** Obviously, any such approximation in the low-pass filtering stage will lead to some discrepancy between $x(t)$ and $x_{\mathrm{r}}(t)$ in Figure 4.11. The particular choice of nonideal filter is then dictated by the acceptable level of distortion for the application under consideration.

Words and Expressions

sampling theorem	采样定理
interval ['ɪntərvl]	*n.* 区间；时间间隔
impulse train	冲激串
sampling period	采样周期
sampling frequency	采样频率
superposition [ˌsuːpərpə'zɪʃn]	*n.* 叠加
overlap [ˌoʊvər'læp]	*n.* 交叠；重叠
low-pass filter	低通滤波器
cutoff frequency	截止频率
band-limited signal	带限信号

Notes

1. Under certain conditions, a continuous-time signal can be completely represented by and recoverable from knowledge of its values, or samples, at points equally spaced in time.

在一定条件下，一个连续时间信号完全可以用该信号在等时间间隔点上的值或样本来表示，并且可以用这些样本值把该信号全部恢复出来。

2. When these samples are viewed in sequence at a sufficiently fast rate, we perceive an accurate representation of the original continuously moving scene.

当以足够快的速度来看这些时序样本时，我们就会感觉到是原来连续活动景象的重现。

3. A useful way to do this is through the use of a periodic impulse train multiplied by the continuous-time signal $x(t)$ that we wish to sample.

一种有用的采样方法是使用一个周期冲激串乘以我们希望采样的连续时间信号 $x(t)$。

4. Let $x(t)$ be a band-limited signal with $X(\mathrm{j}\omega)=0$ for $|\omega|>\omega_{\mathrm{M}}$. Then $x(t)$ is uniquely determined by its samples $x(nT)$, $n=0,\pm1,\pm2,\cdots$, if $\omega_{\mathrm{s}}>2\omega_{\mathrm{M}}$, where $\omega_{\mathrm{s}}=\dfrac{2\pi}{T}$. Given these samples, we can reconstruct $x(t)$ by generating a periodic impulse train in which successive impulses have amplitudes that are successive sample values. This impulse train is then processed through an ideal low-pass filter with gain T and cutoff frequency greater than ω_{M} and less than $\omega_{\mathrm{s}}-\omega_{\mathrm{M}}$. The resulting output signal will exactly equal $x(t)$.

设 $x(t)$ 是某一个带限信号，在 $|\omega|>\omega_{\mathrm{M}}$ 时，$X(\mathrm{j}\omega)=0$。如果 $\omega_{\mathrm{s}}>2\omega_{\mathrm{M}}$，其中 $\omega_{\mathrm{s}}=\dfrac{2\pi}{T}$，

那么 $x(t)$ 就唯一地由其样本 $x(nT)$，$n=0,\pm1,\pm2,\cdots$ 所确定。已知这些样本值，我们能用如下方法重建 $x(t)$：产生一个周期冲激串，其冲激强度就是这些依次而来的样本值；然后将该冲激串通过一个增益为 T、截止频率大于 ω_{M} 而小于 $\omega_{\mathrm{s}}-\omega_{\mathrm{M}}$ 的理想低通滤波器，该滤波器的输出就是 $x(t)$。

5. In any practical application, the ideal low-pass filter in Figure 4.11 would be replaced by a nonideal filter $H(\mathrm{j}\omega)$ that approximated the desired frequency characteristic accurately enough for the problem of interest (i.e., $H(\mathrm{j}\omega)\approx1$ for $|\omega|<\omega_{\mathrm{M}}$, and $H(\mathrm{j}\omega)\approx0$ for $|\omega|>\omega_{\mathrm{s}}-\omega_{\mathrm{M}}$).

在任何实际应用中，图 4.11 中的理想低通滤波器都用一个非理想滤波器 $H(\mathrm{j}\omega)$ 所代替，该滤波器对所关心的问题来说已足够准确地近似于所要求的频率特性（即 $H(\mathrm{j}\omega)\approx1$，$|\omega|<\omega_{\mathrm{M}}$ 和 $H(\mathrm{j}\omega)\approx0$，$|\omega|>\omega_{\mathrm{s}}-\omega_{\mathrm{M}}$）。

Exercises

1. Match the terms (1)–(6) with the definitions A–F.

(1) signals	A. system which has properties of linearity and time-invariance
(2) linear time-invariant system	B. the integral of the product of the two functions after one is reversed and shifted
(3) convolution	C. functions of one or more independent variables
(4) frequency response	D. two times of the highest frequency (bandwidth) of a continuous-time signal
(5) Nyquist rate	E. a graph of the frequency response of a system in decibels
(6) Bode plot	F. the quantitative measure of the magnitude and phase of the output as a function of input frequency

2. Translate into Chinese.

(1) An extremely important class of continuous-time systems is that for which the input and output are related through a linear constant-coefficient differential equation. For example, the response of the RC circuit can be described through linear constant-coefficient differential equation.

(2) Such a representation, referred to as the convolution sum in discrete-time case and the convolution integral in continuous time, provides considerable analytical convenience in dealing with LTI systems.

(3) A signal for which the system output is a (possibly complex) constant times the input is referred to as an eigenfunction of the system, and the amplitude factor is referred to as the system's eigenvalue.

3. Translate into English.

（1）离散时间单位脉冲序列与单位阶跃序列之间有着密切的关系。特别地，离散时间单位脉冲序列是单位阶跃序列的一阶后向差分。相反地，离散时间单位阶跃序列是单位脉冲序列的求和。

（2）对数幅度刻度通常是有用且重要的。然而，在许多情况下，使用线性幅度刻度是很方便的。例如，在讨论频率响应幅度在某些频带上为非零常数，在其他频带上为零的理想滤波器时，线性幅度刻度更合适。

（3）采样定理最容易用冲激串采样来解释，这表明带限信号可由其样本唯一表示。然而在实践中，近似于冲激信号的大幅度窄脉冲也相对难以生成和传输，而以零阶保持的形式生成采样信号通常更为方便。

4. Read the following article and write a summary.

In signal processing, a signal is a function that conveys information about a phenomenon. In electronics and telecommunications, it refers to any time-varying voltage, current, or electromagnetic wave that carries information. A signal may also be defined as an observable change in a quality such as quantity.

Any quality, such as physical quantity that exhibits variation in space or time， can be used as a signal to share messages between observers. According to the *IEEE Transactions on Signal Processing*, a signal can be audio, video, speech, image, sonar, and radar-related and so on. In another effort to define signal, anything that is only a function of space, such as an image, is excluded from the category of signals. Also, it is stated that a signal may or may not contain any information.

In nature, signals can be actions done by an organism to alert other organisms, ranging from the release of plant chemicals to warn nearby plants of a predator, to sounds or motions made by animals to alert other animals of food. Signaling occurs in all organisms even at cellular levels, with cell signaling. Signaling theory, in evolutionary biology, proposes that a substantial driver for evolution is the ability for animals to communicate with each other by developing ways of signaling. In human engineering, signals are typically provided by the sensors, and often the original form of a signal is converted to another form of energy using a transducer. For example, a microphone converts an acoustic signal to a voltage waveform, and a speaker does the reverse.

Information theory serves as the formal study of signals and their content, and the information of a signal is often accompanied by noise. The term "noise" refers to unwanted signal modifications but is often extended to include unwanted signals conflicting with desired signals (crosstalk). The reduction of noise is covered in part under the heading of signal integrity. The separation of desired signals from background noise is the field of signal recovery, one branch of which is estimation theory, a probabilistic approach to suppressing random disturbances.

Engineering disciplines such as electrical engineering have led the way in the design, study, and implementation of systems involving transmission, storage, and manipulation of information. In the latter half of the 20th century, electrical engineering itself separated into several disciplines: electronic engineering and computer engineering developed to specialize in the design and analysis of systems that manipulate physical signals, while design engineering developed to address the functional design of signals in user-machine interfaces.

Signal processing is the manipulation of signals. A common example is signal transmission between different locations. The embodiment of a signal in electrical form is made by a transducer that converts the signal from its original form to a waveform expressed as a current or a voltage, or electromagnetic radiation, for example, an optical signal or radio transmission. Once expressed as an electronic signal, the signal is available for further processing by electrical devices such as electronic amplifiers and filters, and can be transmitted to a remote location by a transmitter and received using radio receivers.

5. Language study: Comparison and contrast 1

We can describe similarities like these:

(1) **Both** courses are provided by Bankhead College.

(2) **Like** Course 1, Course 2 deals with electronics.

(3) Course 2 **is similar to** Course 1 in that it deals with electronics.

We can describe differences like these:

(1) Course 2 is much **longer than** Course 1.

(2) Course 2 is day release **but** Course 1 is an evening course.

(3) Course 1 is for complete beginners, **whereas** Course 2 is for technicians.

Study the course descriptions below of two higher level qualifications.

Complete this table of differences between the courses.

	BTech	HND
Duration	3 years	
Award		Diploma
Institutes	Strachclyde University and Bell College	
Main Subjects (unique)	Mathematics; Foreign Language; Engineering Management; Signals and Systems	Quality Management
Options (unique)	Optoelectronics; Signal Processing	

Bachelor of Technology (BTech) in Electrical and Electronic Engineering **Duration** Three years full-time.	**The Course** The degree resulting from joint course between Strachclyde University and Bell College is awarded by Strachclyde University. Over the three years, students spend about half of the course in each institution. The BTech is a balance of theory and practical skills. It will enable graduates to attain the status of Incorporated Engineer after a period of industrial training and experience. It bridges the gap between HND and BEng Honours courses and there are transfer routes possible between all these courses.	**Subjects** First year: Mathematics; Electrotechnology; Digital and Computer Systems; Analog Electronics; Software Engineering; Engineering Applications. Second year: Mathematics; Digital and Analog Electronics; Microprocessor Applications; System Principles; Circuit Analysis; Electronic Design and Production; Foreign Language. Third year: Electrotechnology; Engineering Management; Signals and Systems; Software Development; Measurement and Control; Data Communications Project. *Students will also choose from a range of options including CAD, Optoelectronics, Materials, Power Plant, Signal Processing, and others.*
Higher National Diploma (HND) in Electronic and Electrical Engineering **Duration** Two years full-time.	**The Course** This is a new HND course, planned after market research among employers and former students. This research identified the kinds of jobs, equipment, and management skills which holders of an HND must have in addition to their technological abilities. From this information, we are able to plan the most appropriate course content. All students will study a broad range of subjects before choosing the options which will best suit their intended career. The diploma is taught and awarded by Bell College.	**Subjects** First year: There will be a range of introductory subjects to help everyone become familiar with new subject areas. These will be followed by: Electrotechnology; Electronics; Computer Programming and Applications; Mathematics; Complementary Studies. Second year: Electrotechnology; Computer Programming; Quality Management; Computer Aided Design; Complementary Studies; Project; and a range of options covering Electronics, Power and Machines, Data Communications, Control Systems, and Electronic Production.

Unit 5 Digital Signal Processing

5.1 Introduction to Digital Signal Processing

Signal processing applications span an immense set of disciplines that include entertainment, communications, space exploration, medicine, archeology, geophysics, just to name a few. Signal processing algorithms and hardware are prevalent in a wide range of systems, from highly specialized military systems and industrial applications to low-cost, high-volume consumer electronics. The growing number of applications and demand for increasingly sophisticated algorithms go hand-in-hand with the rapid development of device technology for implementing signal processing systems. **By some estimates, even with impending limitations on Moore's law, the processing capability of both special-purpose signal processing microprocessors and personal computers is likely to increase by several orders of magnitude.**

Signal processing deals with the representation, transformation, and manipulation of signals and the information the signals contain. For example, we may wish to separate two or more signals that have been combined by some operation, such as addition, multiplication, or convolution, or we may want to enhance some signal component or estimate some parameter of a signal model. **In communication system, it is generally necessary to do preprocessing such as modulation, signal conditioning, and compression prior to transmission over a communication channel, and then to carry out postprocessing at the receiver to recover a facsimile of the original signal.** Prior to the 1960s, the technology for such signal processing was almost exclusively continuous-time analog technology. A continual and major shift to digital technologies has resulted from the rapid evolution of digital computers and microprocessors and low-cost chips for analog-to-digital (A/D) and digital-to-analog (D/A) conversion. These developments in technology have been reinforced by many important theoretical developments, such as the fast Fourier transform (FFT) algorithm, parametric signal modeling, multirate techniques, polyphase filter implementation, and new ways of representing signals, such as wavelet expansions. As just an example of this shift, analog radio communication systems are evolving into reconfigurable "software radios" that are implemented almost exclusively with digital computation.

Discrete-time signal processing is based on processing of numeric sequences indexed on integer variables rather than functions of a continuous independent variable. In digital signal processing (DSP), signals are represented by sequences of finite-precision numbers, and processing is implemented using digital computation. The more general term discrete-time signal processing includes digital signal processing as a special case but also includes the possibility that sequences of samples (sampled data) could be processed with other discrete-time technologies. Often the distinction between the terms discrete-time signal processing and digital signal processing is of minor importance, since both are concerned with discrete-time signals. This is particularly true when high-precision computation is employed. Although there are many examples in which signals to be processed are inherently discrete-time sequences, most applications involve the use of discrete-time technology for processing signals that originate as continuous-time signals. In this case, a continuous-time signal is typically converted into a sequence of samples, i.e., a discrete-time signal. Indeed, one of the most important spurs to widespread application of digital signal processing was the development of low-cost A/D, D/A conversion chips based on differential quantization with noise shaping. After discrete-time processing, the output sequence is converted back to a continuous-time signal. Real-time operation is often required or desirable for such systems.

As computer speeds have increased, discrete-time processing of continuous-time signals in real time has become commonplace in communication systems, radar and sonar, speech and video coding and enhancement, biomedical engineering, and many other areas of applications. Non-real-time applications are also common, such as MP3 player. Financial engineering represents another field which incorporates many signal processing concepts and techniques. Effective modeling, prediction and filtering of economic data can result in significant gains in economic performance and stability. Another important area of signal processing is signal interpretation. In such contexts, the objective of the processing is to obtain a characterization of the input signal. For example, in a speech recognition or understanding system, the objective is to interpret the input signal or extract information from it. Typically, such a system will apply digital preprocessing (filtering, parameter estimation, and so on) followed by a pattern recognition system to produce a symbolic representation, such as a phonemic transcription of the speech.

Signal processing problems are not confined, of course, to one-dimensional signals. Although there are some fundamental differences in the theories for one-dimensional and multidimensional signal processing, much of the material that we discuss has a direct counterpart in multidimensional systems. Many image processing applications require the use of two-dimensional signal processing techniques. This is the case in such areas as video coding, medical imaging, enhancement and analysis of aerial photographs, analysis of satellite weather photos, and enhancement of video transmissions from lunar and deep-space probes.

Discrete-time signal processing techniques have already promoted revolutionary advances in some fields of application. **A notable example is in the area of telecommunications, where discrete-time signal processing techniques, microelectronic technology, and fiber-optic transmission have combined to change the nature of communication systems in truly revolutionary ways.** A similar impact can be expected in many other areas. Indeed, signal processing has always been, and will always be, a field that thrives on new applications. The needs of a new filed of application can sometimes be filled by knowledge adapted from other applications, but frequently, new application needs stimulate new algorithms and new hardware systems to implement those algorithms.

Words and Expressions

Moore's law	摩尔定律
microprocessor [ˌmaɪkroʊ'proʊsesər]	*n.* 微处理器
transformation [ˌtrænsfər'meɪʃn]	*n.* 变换
convolution [ˌkɑːnvə'luːʃn]	*n.* 卷积
parameter [pə'ræmɪtər]	*n.* 参数
preprocessing [prep'roʊsesɪŋ]	*n.* 预处理
modulation [ˌmɑːdʒə'leɪʃn]	*n.* 调制
signal conditioning	信号调理
compression [kəm'preʃn]	*n.* 压缩
transmission [trænz'mɪʃn]	*n.* 传输
communication channel	通信信道
postprocessing [poʊstproʊ'sesɪŋ]	*n.* 后处理
receiver [rɪ'siːvər]	*n.* 接收机
facsimile [fæk'sɪməli]	*n.* 传真；复制本
analog ['ænəlɔːg]	*adj.* 模拟的
chip [tʃɪp]	*n.* 芯片
analog-to-digital (A/D) conversion	模数转换
digital-to-analog (D/A) conversion	数模转换
fast Fourier transform (FFT)	快速傅里叶变换
multirate technique	多采样率技术
polyphase filter implementation	多相位滤波器实现
wavelet ['weɪvlət]	*n.* 小波
expansion [ɪk'spænʃn]	*n.* 展开
sequence ['siːkwəns]	*n.* 序列
quantization [ˌkwɑːntɪ'zeɪʃn]	*n.* 量化
biomedical engineering	生物医学工程
one-dimensional [wʌn daɪ'menʃənl]	*adj.* 一维的
multidimensional [ˌmʌltidaɪ'menʃənl]	*adj.* 多维的
telecommunications [ˌtelikəˌmjuːnɪ'keɪʃnz]	*n.* 电信
microelectronic technology	微电子技术

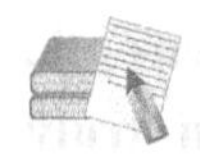

Notes

1. By some estimates, even with impending limitations on Moore's law, the processing capability of both special-purpose signal processing microprocessors and personal computers is likely to increase by several orders of magnitude.

据估计，即使摩尔定律即将受到限制，专用信号处理微处理器和个人计算机的处理能力也可能增加几个数量级。

2. Signal processing deals with the representation, transformation, and manipulation of signals

and the information the signals contain.

信号处理关注的是信号及其所包含信息的表示、变换和处理。

3. In communication system, it is generally necessary to do preprocessing such as modulation, signal conditioning, and compression prior to transmission over a communication channel, and then to carry out postprocessing at the receiver to recover a facsimile of the original signal.

在通信系统中，信号在一条信道上传输之前一般要做一些像调制、信号调理和压缩等这样的预处理，然后在接收机处进行后处理以恢复原始信号。

4. As computer speeds have increased, discrete-time processing of continuous-time signals in real time has become commonplace in communication systems, radar and sonar, speech and video coding and enhancement, biomedical engineering, and many other areas of applications.

随着计算机速度的提高，在通信系统、雷达和声呐、语音和视频编码与增强、生物医学工程及许多其他应用领域，对连续时间信号进行实时离散时间处理已变得司空见惯。

5. A notable example is in the area of telecommunications, where discrete-time signal processing techniques, microelectronic technology, and fiber-optic transmission have combined to change the nature of communication systems in truly revolutionary ways.

电信领域就是一个明显的例子，在该领域中，离散时间信号处理技术、微电子技术和光纤传输技术的结合，正以真正革命性的方式改变着通信系统的面貌。

5.2 Structures for Discrete-Time Systems

An LTI system with a rational system function has the property that the input and output sequences satisfy a linear constant-coefficient difference equation. Since the system function is the *z*-transform of the impulse response, and since the difference equation satisfied by the input and output can be determined by inspection of the system function, it follows that the difference equation, the impulse response, and the system function are equivalent characterizations of the input-output relation of an LTI discrete-time system. When such systems are implemented with discrete-time analog or digital hardware, the difference equation or the system function representation must be converted to an algorithm or structure that can be realized in the desired technology. **Systems described by linear constant-coefficient difference equations can be represented by structures consisting of an interconnection of the basic operations of addition, multiplication by a constant, and delay, the exact implementation of which is dictated by the technology to be used.**

As an illustration of the computation associated with a difference equation, consider the system described by the system function:

$$H(z)=\frac{b_0+b_1z^{-1}}{1-az^{-1}},|z|>|a| \tag{5.1}$$

The impulse response of the system is

$$h[n]=b_0a^nu[n]+b_1a^{n-1}u[n-1] \tag{5.2}$$

and the 1st-order difference equation that is satisfied by the input and output sequence is

$$y[n]-ay[n-1]=b_0x[n]+b_1x[n-1] \tag{5.3}$$

Equation (5.2) gives a formula for the impulse response for this system. However, since the system impulse response has infinite duration, even if we only wanted to compute the output over a finite interval, it would not be efficient to do so by discrete convolution since the amount of computation required to compute $y[n]$ would grow with *n*. However, rewriting Equation (5.3) in the form

$$y[n]=ay[n-1]+b_0x[n]+b_1x[n-1] \tag{5.4}$$

provides the basis for an algorithm for recursive computation of the output at any time n in terms of the previous output $y[n-1]$, the current input sample $x[n]$, and the previous input sample $x[n-1]$. If we further assume initial-rest conditions (i.e., if $x[n]=0$ for $n<0$, then $y[n]=0$ for $n<0$), and if we assume Equation (5.4) as a recurrence formula for computing the output from past values of the output and present and past values of the input, the system will be linear and time invariant. A similar procedure can be applied to the more general case of an Nth-order difference equation. However, the algorithm suggested by Equation (5.4), and its generalization for higher-order difference equations is not the only computational algorithm for implementing a particular system, and often, it is not the best choice. As we will see, an unlimited variety of computational structures result in the same relation between the input sequence $x[n]$ and the output sequence $y[n]$.

The basic elements required for the implementation of an LTI discrete-time systems are adders, multipliers, and memory for storing delayed sequence values and coefficients. The interconnection of these basic elements is conveniently depicted by block diagrams composed of the basic pictorial symbols shown in Figure 5.1. Figure 5.1(a) represents the addition of two sequences. Figure 5.1(b) depicts multiplication of a sequence by a constant, and Figure 5.1(c) depicts delaying a sequence by one sample.

For general N^{th}-order difference equation

$$y[n]-\sum_{k=1}^{N}a_ky[n-k]=\sum_{k=0}^{M}b_kx[n-k] \tag{5.5}$$

the corresponding system function is

$$H(z)=\frac{\sum_{k=0}^{M}b_kz^{-k}}{1-\sum_{k=1}^{N}a_kz^{-k}} \tag{5.6}$$

The Equation (5.5) can be rewritten as

$$y[n]=\sum_{k=1}^{N}a_ky[n-k]+\sum_{k=0}^{M}b_kx[n-k] \tag{5.7}$$

Thus the block diagram representation for the system is shown in Figure 5.2. A block diagram can be rearranged or modified in a variety of ways without changing the overall system function. Each appropriate rearrangement represents a different computational algorithm for implementing the same system. For example, the block diagram of Figure 5.2 can be viewed as a cascade of two systems, the first representing the computation of $v[n]$ (as shown in Figure 5.2, $v[n]$ is equal to $\sum_{k=0}^{M} b_k x[n-k]$)from $x[n]$ and the second representing the computation of $y[n]$ from $v[n]$. **Since each of the two systems is an LTI system (assuming initial-rest conditions for the delay registers), the order in which the two systems are cascaded can be reversed, as shown in Figure 5.3, without affecting the overall system function.** In Figure 5.3, for convenience, we have assumed that $M=N$. Clearly, there is no loss of generality, since if $M \neq N$, some of the coefficients a_k or b_k in the figure would be zero, and the diagram could be simplified accordingly.

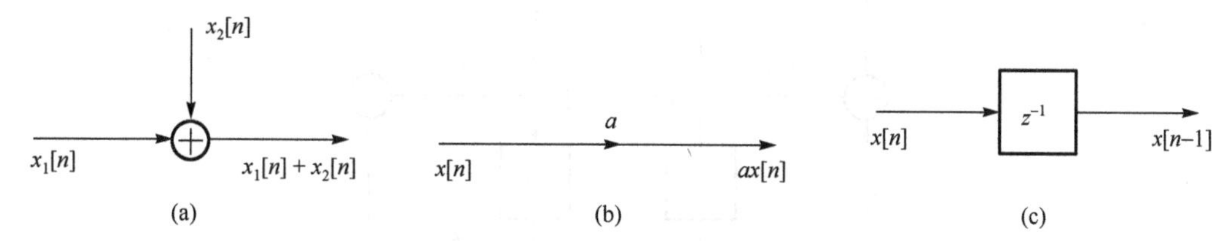

Figure 5.1 Block diagram symbols: (a) addition of two sequences; (b) multiplication of a sequence by a constant; (c) unit delay

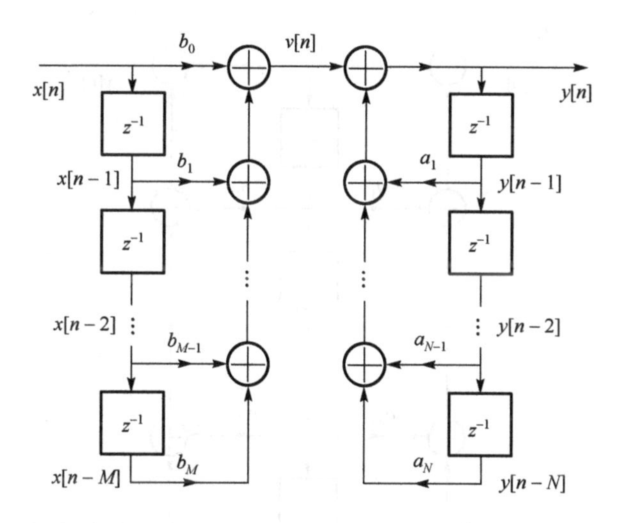

Figure 5.2 Block diagram representation for a general N^{th}-order difference equation

As drawn, the systems in Figure 5.2 and Figure 5.3 each have a total of $N+M$ delay elements. However, the block diagram of Figure 5.3 can be redrawn by noting that exactly the same signal, $w[n]$ (as shown in Figure 5.4, $w[n]$ is equal to $\sum_{k=1}^{N} a_k w[n-k]+x[n]$), is stored in the two chains of delay elements in the figure. Consequently, the two can be collapsed into one chain, as indicated in Figure 5.4. An implementation with the minimum number of delay elements is commonly referred to as a canonic form implementation. The noncanonic block diagram in Figure 5.2 is referred to as the direct form I implementation of the general N^{th}-order system because it is a direct realization of the difference equation satisfied by the input and output. Figure 5.4 is often referred to as the direct form II or canonic direct form implementation.

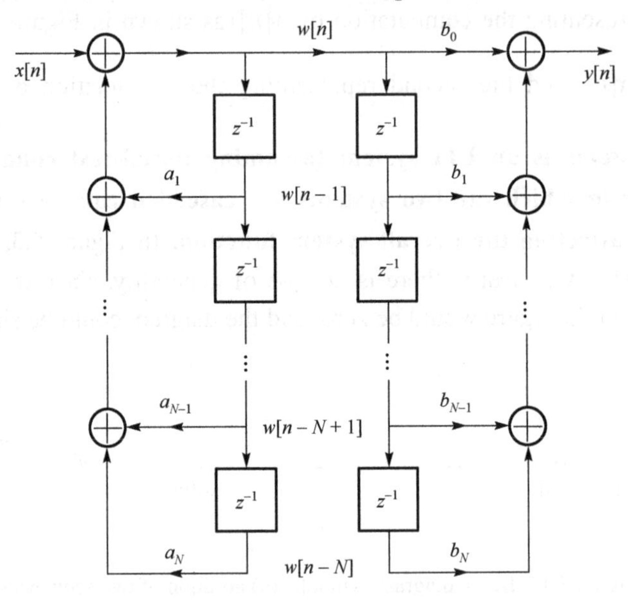

Figure 5.3 Rearrangement of block diagram of Figure 5.2

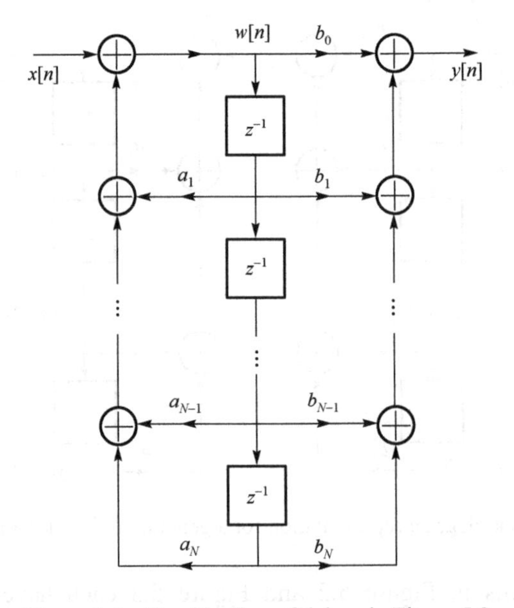

Figure 5.4 Combination of delays in Figure 5.3

Words and Expressions

rational ['ræʃnəl] *adj.* 有理的

system function 系统函数

interconnection [ˌɪntərkə'nekʃn] *n.* 互连

constant ['kɑːnstənt] *n.* 常数

delay [dɪ'leɪ]	*n.* 延迟
formula ['fɔːrmjələ]	*n.* 公式
recursive [rɪ'kɜːrsɪv]	*adj*. 递归的
initial-rest condition	初始松弛条件
recurrence [rɪ'kɜːrəns]	*n.* 重现
block diagram	方框图
cascade [kæ'skeɪd]	*n.* 级联
canonic form implementation	规范形式实现

Notes

1. An LTI system with a rational system function has the property that the input and output sequences satisfy a linear constant-coefficient difference equation.

具有有理系统函数的线性时不变系统有这样的性质，即其输入和输出序列满足线性常系数差分方程。

2. Since the system function is the z-transform of the impulse response, and since the difference equation satisfied by the input and output can be determined by inspection of the system function, it follows that the difference equation, the impulse response, and the system function are equivalent characterizations of the input-output relation of an LTI discrete-time system.

由于系统函数是脉冲响应的 z 变换，并且输入和输出满足的差分方程可以通过系统函数来确定，因此差分方程、脉冲响应和系统函数是线性时不变离散时间系统输入-输出关系的等效表征。

3. Systems described by linear constant-coefficient difference equations can be represented by structures consisting of an interconnection of the basic operations of addition, multiplication by a constant, and delay, the exact implementation of which is dictated by the technology to be used.

由线性常系数差分方程描述的系统能够用由加法、常数乘法和延迟的基本运算互连而成的结构来表示，它的真正实现则取决于所用的技术。

4. The basic elements required for the implementation of an LTI discrete-time systems are adders, multipliers, and memory for storing delayed sequence values and coefficients.

实现线性时不变离散时间系统所需的基本单元是加法器、乘法器和存储延迟序列值和系数的存储器。

5. Since each of the two systems is an LTI system (assuming initial-rest conditions for the delay registers), the order in which the two systems are cascaded can be reversed, as shown in Figure 5.3, without affecting the overall system function.

由于两个系统都是线性时不变系统（假设延迟寄存器初始松弛），那么两个系统在级联中的次序就可以交换成如图 5.3 所示，而不会影响总的系统函数。

5.3 Filter Design Techniques

Filters are a particularly important class of LTI systems. Strictly speaking, the term frequency-selective filter suggests a system that passes certain frequency components of an input signal and totally rejects all others, but in a broader context, any system that modifies certain frequencies relative to others is all called filter. **The design of discrete-time filters corresponds to determining the parameters of a transfer function or difference equation that approximates a desired impulse response or frequency response within specified tolerances. Discrete-time systems implemented with difference equations fall into two basic categories: infinite impulse response (IIR) systems and finite impulse response (FIR) systems.** Designing IIR filters implies obtaining an approximating transfer function that is rational function of z, whereas designing FIR filters implies polynomial approximation.

The commonly used design techniques for these two classes take different forms. When discrete-time filters first came into common use, their design were based on mapping well-formulated and well-understood continuous-time filter designs to discrete-time designs through techniques such as impulse invariance and the bilinear transformation. The basis for impulse invariance is to choose an impulse response for the discrete-time filter that is similar in some sense to the impulse response of the continuous-time filter. **The use of this procedure may be motivated by a desire to maintain the shape of the impulse response or by the knowledge that if the continuous-time filter is bandlimited, consequently the discrete-time filter frequency response will closely approximate the continuous-time frequency response.** When the primary objective is to control some aspect of the time response, such as the impulse response, a natural approach might be to design the discrete-time filter by impulse invariance. However, impulse invariance technique is appropriate only for bandlimited filters. Highpass or bandstop continuous-time filters would require additional bandlimiting to avoid severe aliasing distortion if impulse invariance design is used. **The bilinear transformation avoids the problem of aliasing encountered with the use of impulse invariance, because it maps the entire imaginary axis of the s-plane onto the unit circle in the z-plane.** The price paid for this, however, is the nonlinear compression of the frequency axis. Consequently, the design of discrete-time filters using the bilinear transformation is useful only when this compression can be tolerated or compensated. The most prevalent approaches to designing FIR filters are the use of windowing and the class of iterative algorithms.

The design of filters involves the following stages: the specification of the desired properties of the system, the approximation of the specifications using a causal discrete-time system, and the realization of the system. In a practical setting, the desired filter is generally implemented with digital hardware and often used to filter a signal that is derived from a continuous-time signal by means of periodic sampling followed by A/D conversion. Figure 5.5 depicts the typical representation of the tolerance limits associated with approximating a discrete-time low-pass filter that ideally has unity gain in the passband and zero gain in the stopband. The plot such as Figure 5.5 is referred to as a tolerance scheme. Since the approximation cannot have an abrupt transition from passband to stopband, a transition region from the passband edge frequency ω_{p} to the beginning of the stopband at ω_{s} is allowed, in which the filter gain is unconstrained. Many of the filters using in practice are specified by a tolerance scheme with no constraints on the phase response other than those imposed implicitly by requirements of stability and causality. For example, the poles of the system function for a causal and stable IIR filter must lie inside the unit circle. In designing FIR filters, we often impose the constraint of linear phase. This removes the phase of the signal from consideration in the design process.

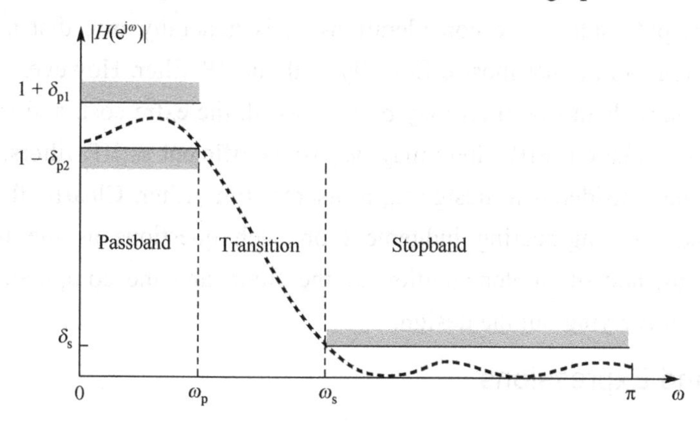

Figure 5.5 Low-pass filter tolerance scheme

Impulse response invariance method and bilinear transformation method are traditional approaches to the design of discrete-time IIR filters, which involve the transformation of a continuous-time filter into a discrete-time filter meeting prescribed specifications. For the design of FIR filters, window function method and optimum approximation method are applied. The choice between an FIR filter and an IIR filter depends on the importance to the design problem of the advantages of each type. IIR filters, for example, have the advantage that a variety of frequency-selective filters can be designed using closed-form design formulas. That is, once the problem has been specified in terms appropriate for a given approximation method (e.g., Butterworth, Chebyshev, or elliptic), then the order of the filter that will meet the specifications can be computed, and the coefficients (or poles and zeros) of the discrete-time filter can be obtained by straightforward substitution into a set of design equations. This kind of simplicity of the design procedure makes it feasible to design IIR filters by manual computation if necessary, and it leads to straightforward noniterative computer programs for IIR filter design. These methods are limited to frequency-selective filters, and they permit only the magnitude response to be specified. If other magnitude shapes are desires, or if it is necessary to approximate a prescribed phase or group-delay response, an algorithm procedure will be required. In contrast, the FIR filter can have a precisely (generalized) linear phase. However, closed-form design equations do not exist for FIR filters. Although the window method is straightforward to apply, some iteration may be necessary to meet a prescribed specification. The Parks-McClellan algorithm leads to lower order filters than the window method does, and both methods can be implemented on a personal computer or a workstation. Also, the window method and most of the algorithmic methods afford the possibility of approximating rather arbitrary frequency-response characteristics with little more difficulty than is encountered in the design of low-pass filters. In addition, the design problem for FIR filters is much more under control than the IIR design problem, because of the existence of an optimality theorem for FIR filters that is meaningful in a wide range of practical situations.

Questions of economics also arise in implementing a discrete-time filter. Economic concerns are usually measured in terms of hardware complexity, chip area, or computational speed. These factors are more or less directly related to the order of the filter required to meet a given specification. If we put aside phase considerations, it is generally true that a given magnitude-response specification can be met most efficiently with an IIR filter. However, in many cases, the linear phase available with an FIR filter may be well worth the extra cost, and in applications such as decimation or interpolation, FIR filters may be just as efficient as IIR filters. Thus, a multitude of trade-offs must be considered in designing a discrete-time filter. Clearly, the final choice will most often be made by engineering judgement on such questions as the formulation of the specifications, the method of implementation of the filter, and the computational facilities and software available for carrying out the design.

Words and Expressions

bilinear transformation	双线性变换法
aliasing ['eɪliəsɪŋ]	*n.* 混叠
passband ['pæsˌbænd]	*n.* 通带
stopband ['stɑːpˌbænd]	*n.* 阻带
transition [træn'zɪʃn]	*n.* 过渡带
tolerance ['tɑːlərəns]	*n.* 容限
causal ['kɔːzl]	*adj.* 因果的

Notes

1. The design of discrete-time filters corresponds to determining the parameters of a transfer function or difference equation that approximates a desired impulse response or frequency response within specified tolerances.

离散时间滤波器的设计即确定其传递函数或差分方程的参数，由此得到近似于规定容限内的期望系统的脉冲响应或频率响应。

2. Discrete-time systems implemented with difference equations fall into two basic categories: infinite impulse response (IIR) systems and finite impulse response (FIR) systems.

用差分方程实现的离散时间系统分为两类：无限脉冲响应（IIR，也译为无限冲激响应）系统和有限脉冲响应（FIR，也译为有限冲激响应）系统。

3. The use of this procedure may be motivated by a desire to maintain the shape of the impulse response or by the knowledge that if the continuous-time filter is bandlimited, consequently the discrete-time filter frequency response will closely approximate the continuous-time frequency response.

使用本方法的出发点可能是希望保持冲激响应的形状，或者若已知连续时间滤波器是带限的，那么离散时间滤波器的频率响应将非常接近连续时间滤波器的频率响应。

4. The bilinear transformation avoids the problem of aliasing encountered with the use of impulse invariance, because it maps the entire imaginary axis of the s-plane onto the unit circle in the z-plane.

双线性变换法避免了使用冲激响应不变法（也译为冲激不变法）时遇到的混叠问题，因为它将 s 平面的整个虚轴映射到 z 平面的单位圆上。

5. The design of filters involves the following stages: the specification of the desired properties of the system, the approximation of the specifications using a causal discrete-time system, and the realization of the system.

滤波器的设计涉及以下步骤：给定系统所要求特性的技术指标、用因果离散时间系统逼近这些技术指标及实现该系统。

Exercises

1. Match the terms (1)–(6) with the definitions A–F.

(1) digital filter	A. the process of varying one or more properties of a periodic waveform, called the carrier signal, with a separate signal that typically contains information to be transmitted
(2) system function	B. the zero-state response when the excitation signal for system is unit impulse signal
(3) modulation	C. the representation of linear time-invariant system using functional elements
(4) impulse response	D. a boundary in a system's frequency response at which energy flowing through the system begins to be reduced (attenuated or reflected) rather than passing through
(5) block diagram	E. a system that performs mathematical operations on a sampled, discrete-time signal to reduce or enhance certain aspects of that signal
(6) cutoff frequency	F. ratio of the zero-state response of the system to the excitation signal in transformation domain

2. Translate into Chinese.

(1) In processing analog signals using discrete-time systems, it is generally desirable to minimize the sampling rate. This is because the amount of arithmetic processing required to implement the system is proportional to the number of samples to be processed.

(2) FFT algorithms are based on the fundamental principle of decomposing the computation of the discrete Fourier transform of a sequence of length N into smaller-length discrete Fourier transforms that are combined to form the N-point transform.

(3) For a Butterworth low-pass filter, the frequency response magnitude decreases monotonically in both the passband and stopband, and all the zeros of the transfer function are at $z=-1$.

3. Translate into English.

（1）在系统的级联连接中，第一个系统的输出是第二个系统的输入，第二个系统的输出是第三个系统的输入等。最后一个系统的输出是总输出。级联的两个线性时不变系统对应一个线性时不变系统，其脉冲响应是两个系统脉冲响应的卷积。

（2）在一些数字系统的实现中，具有最少常数乘法器和最少延迟分支的结构通常是最理想的。这是因为乘法在数字硬件中通常是一种耗时且昂贵的运算，而且每个延迟单元对应一个内存寄存器。因此，常数乘法器数量的减少意味着速度的提高，延迟单元数量的减少意味着内存需求的减少。

（3）FIR 滤波器的设计技术以直接逼近所需离散时间系统的频率响应为基础。此外，逼近 FIR 系统幅度响应的大多数方法均假设有线性相位的限制条件。最简单的 FIR 滤波器设计方法是窗函数法。

4. Read the following article and write a summary.

Digital signal processing (DSP) is the use of digital processing, such as by computers or more specialized digital signal processors, to perform a wide variety of signal processing operations. The digital signal processed in this manner is a sequence of numbers that represent samples of a continuous variable in a domain such as time, space, or frequency. In digital electronics, a digital signal is represented as a pulse train, which is typically generated by the switching of a transistor.

Digital signal processing and analog signal processing are subfields of signal processing. DSP applications include audio and speech processing, sonar, radar and other sensor array processing, spectral density estimation, statistical signal processing, digital image processing, data compression, video coding, audio coding, image compression, signal processing for telecommunications, control systems, biomedical engineering, and seismology, among others.

DSP can involve linear or nonlinear operations. Nonlinear signal processing is closely related to nonlinear system identification and can be implemented in the time, frequency, and spatio-temporal domains.

The application of digital computation to signal processing allows for many advantages over analog processing in many applications, such as error detection and correction in transmission as well as data compression. Digital signal processing is also fundamental to digital technology, such as digital telecommunications and wireless communications. DSP is applicable to both streaming data and static (stored) data.

In DSP, engineers usually study digital signals in one of the following domains: time domain (one-dimensional signals), spatial domain (multidimensional signals), frequency domain, and wavelet domain. They choose the domain in which to process a signal by making an informed assumption (or by trying different possibilities) as to which domain best represents the essential characteristics of the signal and the processing to be applied to it. A sequence of samples from a measuring device produces a temporal or spatial domain representation, whereas a discrete Fourier transform produces the frequency domain representation.

DSP algorithms may be run on general purpose computers and digital signal processors. DSP algorithms are also implemented on purpose-built hardware such as application specific integrated circuit (ASIC). Additional technologies for digital signal processing include more powerful general purpose microprocessors, graphics processing units, field programmable gate arrays (FPGAs), digital signal controllers (mostly for industrial applications such as motor control), and stream processors.

For systems that do not have a real-time computing requirement and whose signal data (either input or output) exists in data files, processing may be done economically with a general purpose computer. This is essentially no different from any other data processing, except DSP mathematical techniques (such as the DCT and FFT) are used, and the sampled data is usually assumed to be uniformly sampled in time or space. An example of such an application is processing digital photographs with software such as Photoshop.

When the application requirement is real-time, DSP is often implemented using specialized or dedicated processors or microprocessors, sometimes using multiple processors or multiple processing cores. These may process data using fixed-point arithmetic or floating point. For more demanding applications, FPGAs may be used. For the most demanding applications or high-volume products, ASICs might be designed specifically for the applications.

5. Language study: Comparison and contrast 2

In this unit, we will examine some other ways to describe differences: to make contrasts.

On the tape, the expert contrasted:

(1) LPs and CDs.

(2) Analog and digital recording.

(3) CDs and newer systems.

Here are some of the things he said:

(1) They (CDs) use laser light **rather than** needles.

(2) LPs are analog recordings **while** CDs are digital.

(3) It (the analog signal) can have any value... **but** the digital signal is either on or off.

(4) You can process a digital signal **with greater accuracy than** a constantly varying signal.

Here are some other expressions used to make contrasts:

differ from **is/are different from** **in contrast to** **whereas** **unlike**

Read the table of differences. Contrast LPs and CDs for each point in the table. Use the expressions from the examples listed above. For example: Unlike LPs, CDs use a digital recording system.

	LPs	CDs
(1) Recording System	analog	digital
(2) Sound Quality	Poorer than the original	like the original
(3) Access	serial	random
(4) Audio Pattern	grooves	pits
(5) Material	vinyl	perspex
(6) Playing Mechanism	mechanical	laser
(7) Durability	easily damaged	does not deteriorate
(8) Size	12 inches	12 cm
(9) Playing Time	45 minutes	74 minutes

Unit 6 Digital Image Processing

6.1 Components of an Image Processing System

As recently as the mid-1980s, numerous models of image processing systems being sold throughout the world were rather substantial peripheral devices that attached to equally substantial host computers. Late in the 1980s and early in the 1990s, the market shifted to image processing hardware in the form of single boards designed to be compatible with industry standard buses and to fit into engineering workstation cabinets and personal computers. **In the late 1990s and early 2000s, a new class of add-on boards, called graphics processing units (GPU), were introduced for work on 3D applications, such as games and other 3D graphics applications.** It was not long before GPUs found their way into image processing applications involving large-scale matrix implementations, such as training deep convolutional networks. In addition to lowering costs, the market shift from substantial peripheral devices to add-on processing boards also served as a catalyst for a significant number of new companies specializing in the development of software written specifically for image processing.

The trend continues toward miniaturizing and blending of general purpose small computers with specialized image processing hardware and software. Figure 6.1 shows the basic components comprising a typical general purpose system used for digital image processing. The function of each component will be discussed in the following paragraphs, starting with image sensing.

Two subsystems are required to acquire digital images. The first is a physical sensor that responds to the energy radiated by the object we wish to image. The second, called a digitizer, is a device for converting the output of the physical sensing device into digital form. For instance, in a digital video camera, the sensors (CCD chips) produce an electrical output proportional to light intensity. The digitizer converts these outputs to digital data.

Specialized image processing hardware usually consists of the digitizer just mentioned, plus hardware that performs other primitive operations, such as an arithmetic logic unit (ALU), that performs arithmetic and logical operations in parallel on entire images. One example of how an ALU is used is in averaging images as quickly as they are digitized, for the purpose of noise reduction. This type of hardware sometimes is called a front-end subsystem, and its most distinguishing characteristic is speed. In other words, this unit performs functions that require fast data throughputs (e.g., digitizing and averaging video images at 30 frames/s) that the typical main computer cannot handle. One or more GPUs (see above) also are common in image processing systems that perform intensive matrix operations.

The computer in an image processing system is a general purpose computer and can range from a PC to a supercomputer. In dedicated applications, sometimes custom computers are used to achieve a required level of performance, but our interest here is on general purpose image processing systems. In these systems, almost any well-equipped PC-type machine is suitable for off-line image processing tasks.

Software for image processing consists of specialized modules that perform specific tasks. A well-designed package also includes the capability for the user to write code that, as a minimum, utilizes the specialized modules. More sophisticated software packages allow the integration of those modules and general purpose software commands from at least one computer language. Commercially available image processing software, such as the well-known MATLAB Image Processing Toolbox, is also common in a well-equipped image processing system.

Mass storage is a must in image processing applications. An image of size 1,024×1,024 pixels, in which the intensity of each pixel is an 8-bit quantity, requires one megabyte of storage space if the image is not compressed. **When dealing with image databases that contain thousands, or even millions, of images, providing adequate storage in an image processing system can be a challenge.** Digital storage for image processing applications falls into three principal categories: (1) short-term storage for use during processing; (2) on-line storage for relatively fast recall; and (3) archival storage, characterized by infrequent access. Storage is measured in bytes (eight bits), kilobytes (2^{10} bytes), megabytes (2^{20} bytes), gigabytes (2^{30} bytes), and terabytes (2^{40} bytes).

One method of providing short-term storage is computer memory. Another is by specialized boards, called frame buffers, that store one or more images and can be accessed rapidly, usually at video rates (e.g., at 30 complete images per second). The latter method allows virtually instantaneous image zoom, as well as scroll (vertical shifts) and pan (horizontal shifts). Frame buffers usually are housed in the specialized image processing hardware unit in Figure 6.1. On-line storage generally takes the form of magnetic disks or optical-media storage. The key factor characterizing on-line storage is frequent access to the stored data. Finally, archival storage is characterized by massive storage requirements but infrequent need for access. Magnetic tapes and optical disks housed in "jukeboxes" are the usual media for archival applications.

Image displays in use today are mainly color, flat screen monitors. Monitors are driven by the outputs of image and graphics display cards that are an integral part of the computer system. Seldom are there requirements for image display applications that cannot be met by display cards and GPUs available commercially as part of the computer system. **In some cases, it is necessary to have stereo displays, and these are implemented in the form of headgear containing two small displays embedded in goggles worn by the user.**

Hardcopy devices for recording images include laser printers, film cameras, heat-sensitive devices, ink-jet units, and digital units, such as optical and CD-ROM disks. Film provides the highest possible resolution, but paper is the obvious medium of choice for written material. For presentations, images are displayed on film transparencies or in a digital medium if image projection equipment is used. The latter approach is gaining acceptance as the standard for image presentations.

Networking and cloud communication are almost default functions in any computer system in use today. **Because of the large amount of data inherent in image processing applications, the key consideration in image transmission is bandwidth.** In dedicated networks, this typically is not a problem, but communications with remote sites via the internet are not always as efficient. Fortunately, transmission bandwidth is improving quickly as a result of optical fiber and other broadband technologies. Image data compression continues to play a major role in the transmission of large amounts of image data.

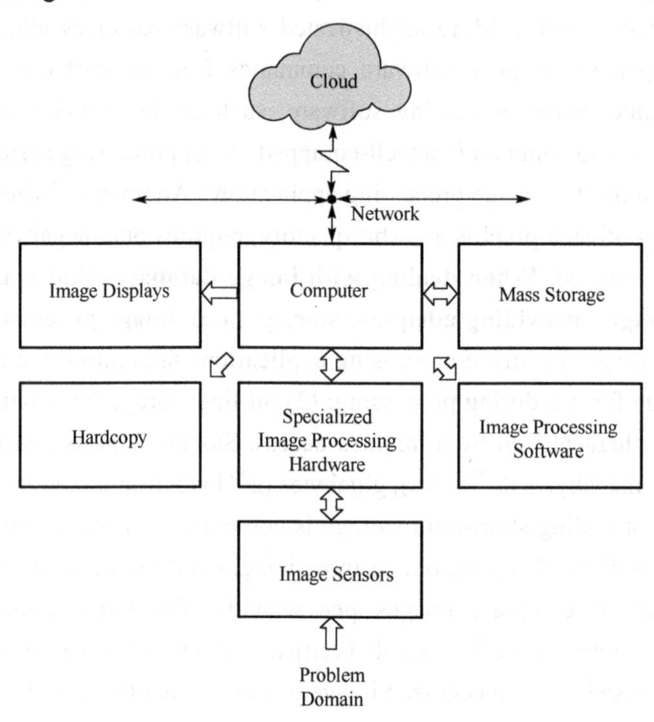

Figure 6.1 Components of a general purpose digital image processing system

 Words and Expressions

peripheral device 外部设备；外围设备
graphics processing unit (GPU) 图形处理单元
deep convolutional network 深度卷积网络
sensor ['sensər] *n*. 传感器
digitizer ['dɪdʒəˌtaɪzər] *n*. 数字化仪；数字化器
arithmetic logic unit (ALU) 算术逻辑单元
frame [freɪm] *n*. 帧
buffer ['bʌfər] *n*. 缓存器；缓冲器；缓存区
magnetic [mæg'netɪk] *adj*. 磁的；有磁性的
stereo ['sterioʊ] *adj*. 立体的；立体声的
resolution [ˌrezə'luːʃn] *n*. 分辨率
projection [prə'dʒekʃn] *n*. 投影
bandwidth ['bændwɪdθ] *n*. 带宽

Notes

1. In the late 1990s and early 2000s, a new class of add-on boards, called graphics processing units (GPU), were introduced for work on 3D applications, such as games and other 3D graphics applications.

在 20 世纪 90 年代末和 21 世纪初，一种称为图形处理单元（GPU）的新型附加板被引入 3D 应用，如游戏和其他 3D 图形应用。

2. Specialized image processing hardware usually consists of the digitizer just mentioned, plus hardware that performs other primitive operations, such as an arithmetic logic unit (ALU), that performs arithmetic and logical operations in parallel on entire images.

专门的图像处理硬件通常包括刚才提到的数字化仪，以及执行其他基本操作的硬件，如算术逻辑单元（ALU），它在整个图像上并行执行算术和逻辑操作。

3. When dealing with image databases that contain thousands, or even millions, of images, providing adequate storage in an image processing system can be a challenge.

在处理包含数千甚至数百万图像的图像数据库时，在图像处理系统中提供足够的存储空间可能是一个挑战。

4. In some cases, it is necessary to have stereo displays, and these are implemented in the form of headgear containing two small displays embedded in goggles worn by the user.

在某些情况下，需要有立体显示器，这些显示器以头盔的形式实现，其中包含嵌入用户佩戴的护目镜中的两个小显示器。

5. Because of the large amount of data inherent in image processing applications, the key consideration in image transmission is bandwidth.

由于图像处理应用中固有的大量数据，图像传输的关键考虑因素是带宽。

6.2 Examples of Fields That Use Digital Image Processing

Today, there is almost no area of technical endeavor that is not impacted in some way by digital image processing. We can cover only a few of these applications in the context and space of the current discussion. However, limited as it is, the material presented in this section will leave no doubt in your mind regarding the breadth and importance of digital image processing. We show in this section numerous areas of application, each of which routinely utilizes the digital image processing techniques.

The areas of application of digital image processing are so varied that some form of organization is desirable in attempting to capture the breadth of this field. **One of the simplest ways to develop a basic understanding of the extent of image processing applications is to categorize images according to their sources (e.g., X-ray, visible, infrared, and so on)**. The principal energy source for images in use today is the electromagnetic(EM) energy spectrum. Other important sources of energy include acoustic, ultrasonic, and electronic (in the form of electron beams used in electron microscopy). Synthetic images, used for modeling and visualization, are generated by computer.

Images based on radiation from the EM spectrum are the most familiar, especially images in the X-ray and visible bands of the spectrum. **Electromagnetic waves can be conceptualized as propagating sinusoidal waves of varying wavelengths, or they can be thought of as a stream of massless particles, each traveling in a wavelike pattern and moving at the speed of light.** Each massless particle contains a certain amount (or bundle) of energy. Each bundle of energy is called a photon. If spectral bands are grouped according to energy per photon, we obtain the spectrum shown in Figure 6.2, ranging from gamma rays (highest energy) at one end to radio waves (lowest energy) at the other. The bands are shown shaded to convey the fact that bands of the EM spectrum are not distinct, but rather transition smoothly from one to the other.

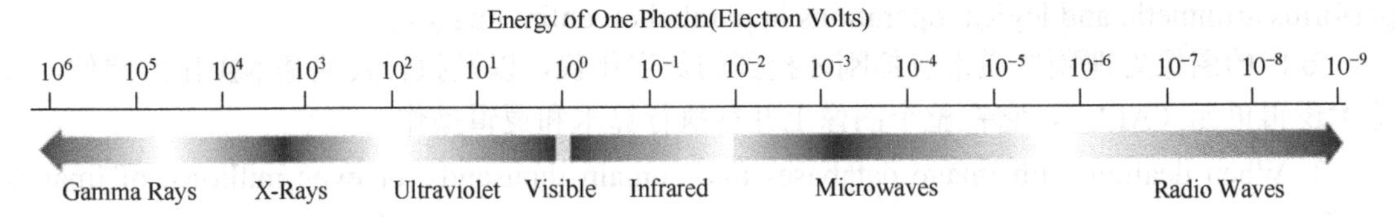

Figure 6.2 The electromagnetic spectrum arranged according to energy per photon

1. Gamma-Ray Imaging

Major uses of imaging based on gamma rays include nuclear medicine and astronomical observations. In nuclear medicine, the approach is to inject a patient with a radioactive isotope that emits gamma rays as it decays. Images are produced from the emissions collected by gamma-ray detectors.

2. X-Ray Imaging

X-rays are among the oldest sources of EM radiation used for imaging. The best-known use of X-rays is medical diagnostics, but they are also used extensively in industry and other areas, such as astronomy.

3. Imaging in the Ultraviolet Band

Applications of ultraviolet "light" are varied. They include lithography, industrial inspection, microscopy, lasers, biological imaging, and astronomical observations.

4. Imaging in the Visible and Infrared Bands

Considering that the visible band of the electromagnetic spectrum is the most familiar in all our activities, it is not surprising that imaging in this band outweighs by far all the others in terms of breadth of application. **The infrared band often is used in conjunction with visual imaging, so we have grouped the visible and infrared bands in this section for the purpose of illustration.** We consider in the following discussion applications in light microscopy, astronomy, remote sensing, industry, and law enforcement. Figure 6.3 shows several examples of images obtained with a light microscope. The examples range from pharmaceuticals and microinspection to materials characterization. Even in microscopy alone, the application areas are too numerous to detail here. It is not difficult to conceptualize the types of processes one might apply to these images, ranging from enhancement to measurements. Another major area of visual processing is remote sensing, which usually includes several bands in the visible and infrared regions of the spectrum.

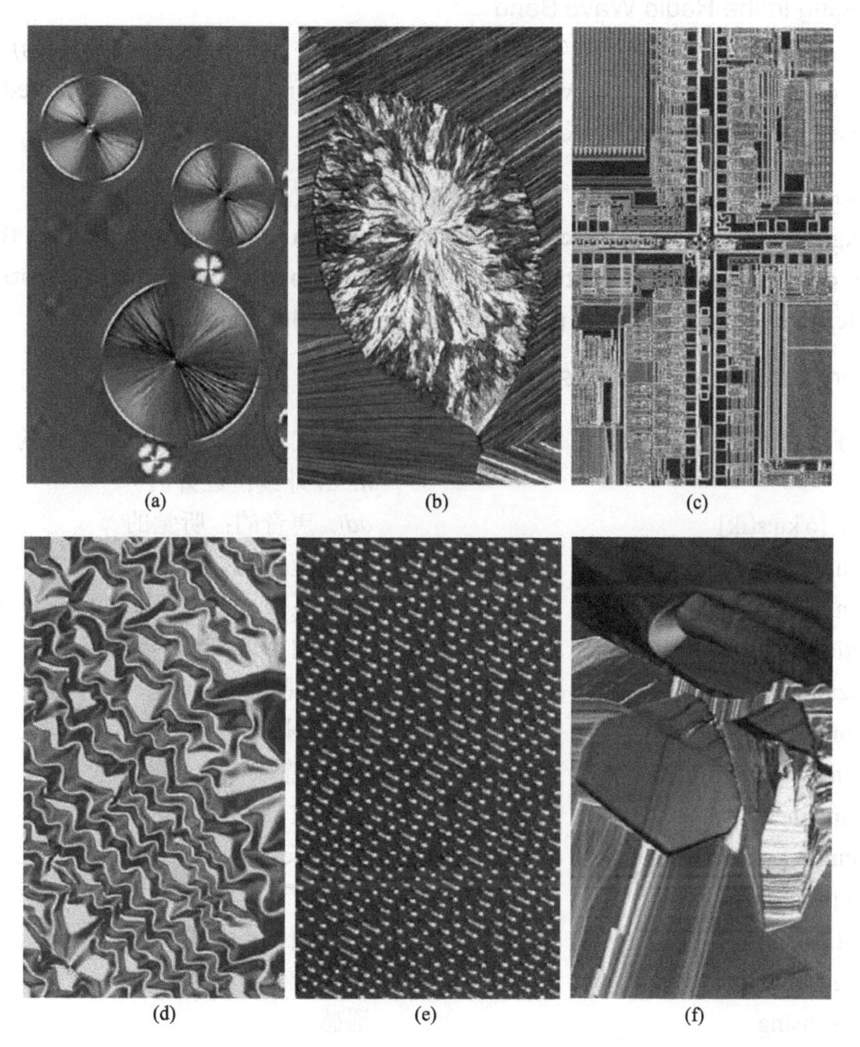

Figure 6.3 Examples of light microscopy images: (a) taxol (anticancer agent), magnified 250×; (b) cholesterol—40×; (c) microprocessor—60×; (d) nickel oxide thin film—600×; (e) surface of audio CD—1750×; (f) organic superconductor— 450 × (Images courtesy of Dr. Michael W. Davidson, Florida State University.)

5. Imaging in the Microwave Band

The principal application of imaging in the microwave band is radar. **The unique feature of imaging radar is its ability to collect data over virtually any region at any time, regardless of weather or ambient lighting conditions.** Some radar waves can penetrate clouds, and under certain conditions, can also see through vegetation, ice, and dry sand. In many cases, radar is the only way to explore inaccessible regions of the Earth's surface. An imaging radar works like a flash camera in that it provides its own illumination (microwave pulses) to illuminate an area on the ground and take a snapshot image. Instead of a camera lens, a radar uses an antenna and digital computer processing to record its images. In a radar image, one can see only the microwave energy that was reflected back toward the radar antenna.

6. Imaging in the Radio Wave Band

As in the case of imaging at the other end of the spectrum (gamma rays), the major applications of imaging in the radio wave band are in medicine and astronomy. In medicine, radio waves are used in magnetic resonance imaging (MRI).

7. Other Imaging Modalities

Although imaging in the electromagnetic spectrum is dominant by far, there are a number of other imaging modalities that are also important, including acoustic imaging, electron microscopy, and synthetic (computer-generated) imaging.

Words and Expressions

infrared [ˌɪnfrə'red]	*adj*. 红外线的；使用红外线的 *n*. 红外线；红外区
acoustic [ə'kuːstɪk]	*adj*. 声音的；听觉的
ultrasonic [ˌʌltrə'sɑːnɪk]	*adj*. 超声的
electron microscopy	电子显微术；电子显微镜学；电子显微法
synthetic image	合成的图像
visualization [ˌvɪʒuələ'zeɪʃn]	*n*. 可视化
radiation [ˌreɪdi'eɪʃn]	*n*. 辐射
spectrum ['spektrəm]	*n*. 频谱
electromagnetic wave	电磁波
wavelength ['weɪvleŋθ]	*n*. 波长
photon ['foʊtɑːn]	*n*. 光子
spectral ['spektrəl]	*adj*. 谱的；光谱的
imaging ['ɪmɪdʒɪŋ]	*n*. 成像
remote sensing	遥感
microwave ['maɪkrəweɪv]	*n*. 微波
snapshot ['snæpʃɑːt]	*n*. 快照
lens [lenz]	*n*. 透镜；镜片
antenna [æn'tenə]	*n*. 天线
magnetic resonance imaging(MRI)	磁共振成像

Notes

1. One of the simplest ways to develop a basic understanding of the extent of image processing applications is to categorize images according to their sources (e.g., X-ray, visible, infrared, and so on).

要基本了解图像处理应用的范围，最简单的方法之一是根据图像的来源（如 X 射线、可见光、红外线等）对图像进行分类。

2. Electromagnetic waves can be conceptualized as propagating sinusoidal waves of varying wavelengths, or they can be thought of as a stream of massless particles, each traveling in a wavelike pattern and moving at the speed of light.

电磁波可以被概念化为传播波长不同的正弦波，也可以被认为是无质量粒子流，每个粒子以波的形式传播，并以光速移动。

3. The infrared band often is used in conjunction with visual imaging, so we have grouped the visible and infrared bands in this section for the purpose of illustration.

红外波段通常与视觉成像结合使用，因此为了便于说明，我们在这部分中将可见光和红外波段合并为一组。

4. The unique feature of imaging radar is its ability to collect data over virtually any region at any time, regardless of weather or ambient lighting conditions.

成像雷达的独特之处在于，它能够在任何时间收集几乎任何地区的数据，而不管天气或环境照明条件如何。

5. Although imaging in the electromagnetic spectrum is dominant by far, there are a number of other imaging modalities that are also important, including acoustic imaging, electron microscopy, and synthetic (computer-generated) imaging.

尽管电磁频谱成像目前占主导地位，但还有许多其他成像方式也很重要，包括声学成像、电子显微镜成像和合成（计算机生成）成像。

6.3 Introduction to Pattern Recognition

Humans possess the most sophisticated pattern recognition capabilities in the known biological world. By contrast, the capabilities of current recognition machines pale in comparison with tasks humans perform routinely, from being able to interpret the meaning of complex images, to our ability for generalizing knowledge stored in our brains. But recognition machines play an important, sometimes even crucial role in everyday life. Imagine what modern life would be like without machines that read barcodes, process bank checks, inspect the quality of manufactured products, read fingerprints, sort mail, and recognize speech.

In image pattern recognition, we think of a pattern as a spatial arrangement of features. A pattern class is a set of patterns that share some common properties. Pattern recognition by machine encompasses techniques for automatically assigning patterns to their respective classes. **That is, given a pattern or sets of patterns whose class is unknown, the job of a pattern recognition system is to assign a class label to each of its input patterns.**

There are four main stages involved in recognition: (1) sensing; (2) preprocessing; (3) feature extraction; and (4) classification. In terms of image processing, sensing is concerned with generating signals in a spatial (2D) or higher-dimensional format. Preprocessing deals with techniques for tasks such as noise reduction, enhancement, restoration, and segmentation. Classification deals with using a set of features as the basis for assigning class labels to unknown input image patterns.

In the following section, we will discuss three basic approaches used for image pattern

classification: (1) classification based on matching unknown patterns against specified prototypes; (2) optimum statistical classifiers; and (3) neural networks. One way to characterize the differences between these approaches is in the level of "engineering" required to transform raw data into formats suitable for computer processing. Ultimately, recognition performance is determined by the discriminative power of the features used.

In classification based on prototypes, the objective is to make the features so unique and easily detectable that classification itself becomes a simple task. A good example of this is bank-check processors, which use stylized font styles to simplify machine processing.

In the second category, classification is cast in decision-theoretic, statistical terms, and the classification approach is based on selecting parameters that can be shown to yield optimum classification performance in a statistical sense. Here, emphasis is placed on both the features used, and the design of the classifier.

In the third category, classification is performed using neural networks. They can operate using engineered features too, but they have the unique ability of being able to generate, on their own, representations (features) suitable for recognition. These systems can accomplish this using raw data, without the need for engineered features.

One characteristic shared by the preceding three approaches is that they are based on parameters that must be either specified or learned from patterns that represent the recognition problem we want to solve. The patterns can be labeled, meaning that we know the class of each pattern, or unlabeled, meaning that the data are known to be patterns, but the class of each pattern is unknown. **A classic example of labeled data is the character recognition problem, in which a set of character samples is collected and the identity of each character is recorded as a label from the group 0 through 9 and a through z.** An example of unlabeled data is when we are seeking clusters in a data set, with the aim of utilizing the resulting cluster centers as being prototypes of the pattern classes contained in the data.

When working with a labeled data, a given data set generally is subdivided into three subsets: a training set, a validation set, and a test set (a typical subdivision might be 50% training, and 25% each for the validation and test sets). The process by which a training set is used to generate classifier parameters is called training. **In this mode, a classifier is given the class label of each pattern, the objective being to make adjustments in the parameters if the classifier makes a mistake in identifying the class of the given pattern.** At this point, we might be working with several candidate designs. At the end of training, we use the validation set to compare the various designs against a performance objective. Typically, several iterations of training/validation are required to establish the design that comes closest to meeting the desired objective. Once a design has been selected, the final step is to determine how it will perform "in the field". For this, we use the test set, which consists of patterns that the system has never "seen" before. If the training and validation sets are truly representative of the data the system will encounter in practice, the results of training/validation should be close to the performance using the test set. If training/validation results are acceptable, but test results are not, we say that training/validation "over fit" the system parameters to the available data, in which case further work on the system architecture is required.

Of course, all this assumes that the given data are truly representative of the problem we want to solve, and that the problem in fact can be solved by available technology.

A system that is designed using training data is said to undergo supervised learning. If we are working with unlabeled data, the system learns the pattern classes itself while in an unsupervised learning mode. **Supervised learning covers a broad range of approaches, from applications in which a system learns parameters of features whose form is fixed by a designer, to systems that utilize deep learning and large sets of raw data sets to learn, on their own, the features required for classification.** These systems accomplish this task without a human designer having to specify the features, a priori.

Words and Expressions

pattern recognition	模式识别
spatial ['speɪʃl]	*adj*. 空间的
feature extraction	特征提取
classification [ˌklæsɪfɪ'keɪʃn]	*n*. 分类
noise reduction	降噪
enhancement [ɪn'hænsmənt]	*n*. 增强
restoration [ˌrestə'reɪʃn]	*n*. 复原
segmentation [ˌsegmen'teɪʃn]	*n*. 分割
label ['leɪbl]	*n*. 标签
prototype ['proʊtətaɪp]	*n*. 原型
optimum statistical classifier	最佳统计分类器
neural network	神经网络
raw data	原始数据
character recognition	字符识别
unlabeled data	未标记的数据
cluster ['klʌstər]	*n*. 群；组
training set	训练集
validation set	验证集
test set	测试集
iteration [ˌɪtə'reɪʃn]	*n*. 迭代
supervised learning	有监督学习
unsupervised learning	无监督学习

Notes

1. That is, given a pattern or sets of patterns whose class is unknown, the job of a pattern recognition system is to assign a class label to each of its input patterns.

也就是说，给定一个或多个类别未知的模式，模式识别系统的工作就是为其每个输入模式分配一个类别标签。

2. In the following section, we will discuss three basic approaches used for image pattern

classification: (1) classification based on matching unknown patterns against specified prototypes, (2) optimum statistical classifiers, and (3) neural networks.

在下面的部分中，我们将讨论用于图像模式分类的 3 种基本方法：（1）根据指定原型匹配未知模式的分类方法；（2）最佳统计分类方法；（3）神经网络方法。

3. A classic example of labeled data is the character recognition problem, in which a set of character samples is collected and the identity of each character is recorded as a label from the group 0 through 9 and a through z.

有标签数据的一个典型示例是字符识别问题，在该问题中，收集一组字符样本，并将每个字符的标识记录为组 0 到 9 及 a 到 z 的标签。

4. In this mode, a classifier is given the class label of each pattern, the objective being to make adjustments in the parameters if the classifier makes a mistake in identifying the class of the given pattern.

在这种模式下，分类器得到每个模式的类别标签，目的是如果分类器在识别给定模式的类型时出错，则对参数进行调整。

5. Supervised learning covers a broad range of approaches, from applications in which a system learns parameters of features whose form is fixed by a designer, to systems that utilize deep learning and large sets of raw data sets to learn, on their own, the features required for classification.

监督学习涵盖了广泛的方法，从系统学习由设计者确定形式的特征参数这类应用，到利用深度学习和大量原始数据集自行学习分类所需特征的系统，这些都属于监督学习的范畴。

Exercises

1. Match the terms (1)–(6) with the definitions A–F.

(1) pixel	A. a measure of the amount of detail in an image
(2) remote sensing	B. the automated recognition of patterns and regularities in data
(3) graphics processing unit	C. a type of data compression applied to digital images, to reduce their cost for storage or transmission
(4) image resolution	D. the smallest addressable element in an image
(5) image compression	E. the use of satellite or aircraft-based sensor technologies to detect and classify objects on Earth
(6) pattern recognition	F. a specialized electronic circuit designed to rapidly manipulate and alter memory to accelerate the creation of images in a frame buffer intended for output to a display device

2. Translate into Chinese.

(1) Image enhancement is more an art than a science, and the definition of a properly enhanced image is highly subjective. Unlike enhancement, restoration techniques for improving images tend to be based on objective, rather than subjective, criteria.

(2) The characteristics generally used to distinguish one color from another are brightness, hue, and saturation. Brightness embodies the chromatic notion of intensity. Hue is an attribute associated with the dominant wavelength in a mixture of light waves. Saturation refers to the

relative purity or the amount of white light mixed with a hue.

(3) Segmentation subdivides an image into its constituent regions or objects. The level to which the subdivision is carried depends on the problem being solved. Segmentation accuracy determines the eventual success or failure of computerized analysis procedures.

3. Translate into English.

（1）总之，具有加性噪声的线性、空间不变退化系统可以在空间域中建模为图像与系统退化函数（点扩散函数）的卷积，然后再加上噪声信号。基于卷积定理，相同的过程可以在频域中表示为图像的傅里叶变换和退化函数的傅里叶变换之积，然后再加上噪声信号的变换。

（2）JPEG 标准是最流行和全面的连续色调静态帧压缩标准之一。它定义了 3 种不同的编码系统：①基于离散余弦变换的有损基线编码系统，适用于大多数压缩应用；②用于更高压缩率、更高精度或渐进式重建应用的扩展编码系统；③用于可逆压缩的无损独立编码系统。

（3）在引入反向传播后的 20 年中，神经网络已成功地应用于广泛的应用领域。其中一些，如语音识别，已经成为日常生活中不可或缺的一部分。当你对着智能手机说话时，神经网络会执行近乎完美的识别。

4. Read the following article and write a summary.

Computer vision is an interdisciplinary scientific field that deals with how computers can gain high-level understanding from digital images or videos. From the perspective of engineering, it seeks to understand and automate tasks that the human visual system can do. Computer vision tasks include methods for acquiring, processing, analyzing and understanding digital images, and extraction of high-dimensional data from the real world in order to produce numerical or symbolic information, e.g., in the forms of decisions. Understanding in this context means the transformation of visual images (the input of the retina) into descriptions of the world that make sense to thought processes and can elicit appropriate action. This image understanding can be seen as the disentangling of symbolic information from image data using models constructed with the aid of geometry, physics, statistics, and learning theory.

The scientific discipline of computer vision is concerned with the theory behind artificial systems that extract information from images. The image data can take many forms, such as video sequences, views from multiple cameras, multidimensional data from a 3D scanner, or medical scanning device. The technological discipline of computer vision seeks to apply its theories and models to the construction of computer vision systems. Sub-domains of computer vision include scene reconstruction, object detection, event detection, video tracking, object recognition, 3D pose estimation, learning, indexing, motion estimation, visual servoing, 3D scene modeling, and image restoration.

The fields most closely related to computer vision are image processing, image analysis and machine vision. There is a significant overlap in the range of techniques and applications. This implies that the basic techniques that are used and developed in these fields are similar, something which can be interpreted as there is only one field with different names. On the other hand, it appears to be necessary for research groups, scientific journals, conferences and companies to present or market themselves as belonging specifically to one of these fields and, hence, various characterizations which distinguish each of the fields from the others have been presented. In image processing, the input is an image and the output is an image as well, whereas in computer vision, an image or a video is taken as an input and the output could be an enhanced image, an understanding of the content of an image or even a behavior of a computer system based on such understanding.

Applications of computer vision range from tasks such as industrial machine vision systems which, say, inspect bottles speeding by on a production line, to research into artificial intelligence and computers or robots that can comprehend the world around them. The computer vision and machine vision fields have significant overlap. Computer vision covers the core technology of automated image analysis which is used in many fields. Machine vision usually refers to a process of combining automated image analysis with other methods and technologies to provide automated inspection and robot guidance in industrial applications. In many computer-vision applications, the computers are pre-programmed to solve a particular task, but methods based on learning are now becoming increasingly common. Examples of applications of computer vision include systems for:

- Automatic inspection, e.g., in manufacturing applications.
- Assisting humans in identification tasks, e.g., a species identification system.
- Controlling processes, e.g., an industrial robot.
- Detecting events, e.g., for visual surveillance or people counting, e.g., in the restaurant industry.
- Interaction, e.g., as the input to a device for computer-human interaction.
- Modeling objects or environments, e.g., medical image analysis or topographical modeling.
- Navigation, e.g., by an autonomous vehicle or mobile robot.
- Organizing information, e.g., for indexing databases of images and image sequences.
- Tracking surfaces or planes in 3D coordinates for allowing augmented reality experiences.

5. Language study: If-sentences

Study this action and its consequence:

Action: A burglar tries to force the alarm open.

Consequence: Sensors trigger the alarm.

We can link action and consequence like these:

(1) **If** a burglar tries to force the alarm open, sensors trigger the alarm.

(2) **If** a burglar tries to force the alarm open, sensors will trigger the alarm.

(3) Sensors will trigger the alarm **if** a burglar tries to force it open.

Complete these sentences with suitable actions or consequences.

(1) If pressure mats are constantly walked on, …

(2) If you fit an exit delay, …

(3) If your system doesn't have an automatic cut-off, …

(4) If a burglar walks in front of a motion sensor, …

(5) Vibration sensors will respond if…

(6) Tamper sensors will trigger the alarm if…

(7) A magnet on the moving part trips a switch if…

(8) The alarm stops after a set time if…

Unit 7 Principles of Communications

7.1 Electrical Communication Systems

A system is a combination of circuits and/or devices that is assembled to accomplish a desired task, such as the transmission of intelligence from one point to another. Many means for the transmission of information have been used down through the ages ranging from the use of sunlight reflected from mirrors by the Romans to our modern era of electrical communications that began with the invention of the telegraph in the 1800s.

A characteristic of electrical communication systems is the presence of uncertainty. **This uncertainty is due in part to the inevitable presence in any system of unwanted signal perturbations, broadly referred to as noise, and in part to the unpredictable nature of information itself.** Systems analysis in the presence of such uncertainty requires the use of probabilistic techniques. Noise has been an ever-present problem since the early days of electrical communication, but it was not until the 1940s that probabilistic systems analysis procedures were used to analyze and optimize communication systems operating in its presence. It is also somewhat surprising that the unpredictable nature of information was not widely recognized until the publication of Claude Shannon's mathematical theory of communications in the late 1940s.

It is an interesting fact that the first electrical communication system, the telegraph, was digital—that is, it conveyed information from point to point by means of a digital code consisting of words composed of dots and dashes. The subsequent invention of the telephone 38 years after the telegraph, wherein voice waves are conveyed by an analog current, swung the pendulum in favor of this more convenient means of word communication for about 75 years.

One may rightly ask, in view of this history, why digital formats are almost completely dominant in today's world. There are several reasons, among which are: (1) signal integrity—a digital format suffers much less deterioration in reproduction than does an analog record; (2) media integration—whether a sound, picture, or naturally digital data such as a word file, all are treated the same when in digital format; (3) flexible interaction—the digital domain is much more convenient for supporting anything from one-on-one to many-to-many interactions; (4) editing—whether text, sound, images, or videos, all are conveniently and easily edited when in digital format.

With this brief introduction and history, we now look in more detail at the various components that make up a typical communication system.

Figure 7.1 shows a commonly used model for a single-link communication system. Although it suggests a system for communication between two remotely located points, this block diagram is also applicable to remote sensing systems, such as radar or sonar, in which the system input and output may be located at the same site. **Regardless of the particular application and configuration, all information transmission systems invariably involve three major subsystems—a transmitter, the channel, and a receiver.** We will now discuss in more detail each functional element shown in Figure 7.1.

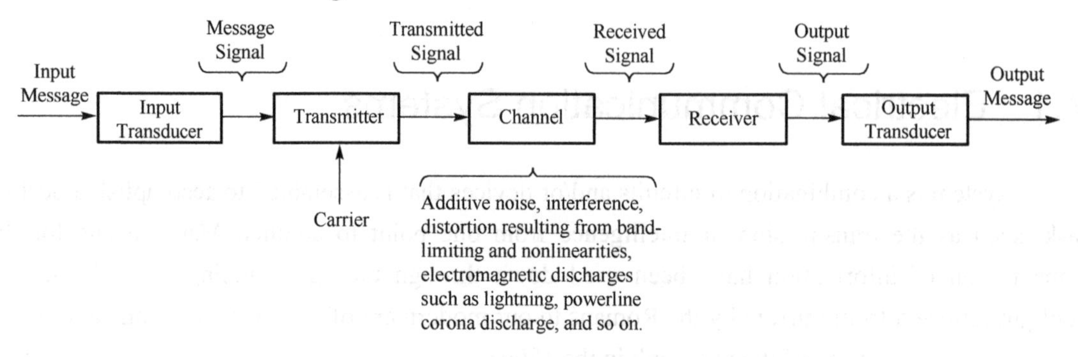

Figure 7.1 The block diagram of a communication system

1. Input Transducer

The wide variety of possible sources of information results in many different forms for messages. Regardless of their exact forms, however, messages may be categorized as analog or digital. The former may be modeled as functions of a continuous-time variable (for example, pressure, temperature, speech, music), whereas the latter consist of discrete symbols (for example, written text or a sampled/quantized analog signal such as speech). **Almost invariably, the message produced by a source must be converted by a transducer to a form suitable for the particular type of communication system employed.** For example, in electrical communications, speech waves are converted by a microphone to voltage variations. Such a converted message is referred to as the message signal. In this book, therefore, a signal can be interpreted as the variation of a quantity, often a voltage or current, with time.

2. Transmitter

The purpose of the transmitter is to couple the message to the channel. Although it is not uncommon to find the input transducer directly coupled to the transmission medium, as for example in some intercom systems, it is often necessary to modulate a carrier wave with the signal from the input transducer. Modulation is the systematic variation of some attribute of the carrier, such as amplitude, phase, or frequency, in accordance with a function of the message signal. There are several reasons for using a carrier and modulating it. Important ones are: (1) for ease of radiation; (2) to reduce noise and interference; (3) for channel assignment; (4) for multiplexing or transmission of several messages over a single channel; and (5) to overcome equipment limitations. Several of these reasons are self-explanatory; others, such as the second, will become more meaningful later.

In addition to modulation, other primary functions performed by the transmitter are filtering, amplification, and coupling the modulated signal to the channel (for example, through an antenna or other appropriate devices).

3. Channel

The channel can have many different forms; the most familiar, perhaps, is the channel that exists between the transmitting antenna of a commercial radio station and the receiving antenna of a radio. In this channel, the transmitted signal propagates through the atmosphere, or free space, to the receiving antenna. However, it is not uncommon to find the transmitter hard-wired to the receiver, as in most local telephone systems. This channel is vastly different from the radio example. However, all channels have one thing in common: the signal undergoes degradation from transmitter to receiver. Although this degradation may occur at any point of the communication system block diagram, it is customarily associated with the channel alone. **This degradation often results from noise and other undesired signals or interference, but also may include other distortion effects as well, such as fading signal levels, multiple transmission paths, and filtering.**

4. Receiver

The receiver's function is to extract the desired message from the received signal at the channel output and to convert it to a form suitable for the output transducer. Although amplification may be one of the first operations performed by the receiver, especially in radio communications, where the received signal may be extremely weak, the main function of the receiver is to demodulate the received signal. **Often it is desired that the receiver output be a scaled, possibly delayed, version of the message signal at the modulator input, although in some cases a more general function of the input message is desired.** However, as a result of the presence of noise and distortion, this operation is less than ideal.

5. Output Transducer

The output transducer completes the communication system. This device converts the electric signal at its input into the form desired by the system user. Perhaps the most common output transducer is a loudspeaker or ear phone.

Finally, it should be noted that Figure 7.1 represents one-way, or simplex (SX), transmission. Two-way communication, of course, requires a transmitter and receiver at each end. A full-duplex (FDX) system has a channel that allows simultaneous transmission in both directions. A half-duplex (HDX) system allows transmission in either direction but not at the same time.

Words and Expressions

communication [kəˌmjuːnɪ'keɪʃn]　　*n.* 通信；交流；传递；交通联系

transmission [trænz'mɪʃn]　　*n.*（电子信号、信息或广播电视节目的）传输；发送

intelligence [ɪn'telɪdʒəns]　　*n.* 情报；智力；才智；智慧

information [ˌɪnfər'meɪʃn]　　*n.* 信息；消息；情报；资料；资讯

uncertainty [ʌn'sɜːrtnti] *n.* 不确定性
perturbation [ˌpɜːrtər'beɪʃn] *n.* 扰动；干扰；微扰；小变异
probabilistic [ˌprɑːbəbɪ'lɪstɪk] *adj.* 基于概率的；或然的
deterioration [dɪˌtɪriə'reɪʃn] *n.* 恶化；变坏；退化
transducer [trænz'duːsər, træns'duːsər] *n.* 传感器；换能器；变换器
transmitter [trænz'mɪtər, træns'mɪtər] *n.*（尤指无线电或电视信号的）发射机；发送端
channel ['tʃænl] *n.* 信道；频道；电视台；波段
receiver [rɪ'siːvər] *n.* 接收机；接收端
additive noise 加性噪声
distortion [dɪ'stɔːrʃn] *n.* 失真；变形；扭曲
powerline corona discharge 电力线电晕放电
remote sensing 遥感；遥测
sonar ['soʊnɑːr] *n.* 声呐；声波定位仪
sample ['sæmpl] *n.* 样品；样本
vt. 采样；抽样
quantize ['kwɑːntaɪz] *v.* 量化
transmission medium 传输介质
intercom system 内部通信系统；对讲机系统
modulate ['mɑːdʒəleɪt] *v.* 调制；调节；调整；控制
carrier ['kæriər] *n.* 载波
attribute [ə'trɪbjuːt] *n.* 属性；特征；性质；定语
amplitude ['æmplɪtuːd] *n.*（声音、无线电波等的）振幅
phase [feɪz] *n.* 相位；阶段；时期
radiation [ˌreɪdi'eɪʃn] *n.* 辐射；放射线；辐射的热（或能量等）
filter ['fɪltər] *n.* 滤波器；筛选（过滤）程序
v. 滤波；（用程序）筛选
antenna [æn'tenə] *n.* 天线
propagate ['prɑːpəgeɪt] *v.* 传播
degradation [ˌdegrə'deɪʃn] *n.* 恶化（过程）
demodulate [diː'mɑːdʒəleɪt] *v.* 解调；使检波
simplex (SX) ['sɪmpleks] *n.* 单工
full-duplex (FDX) [fʊl 'duːpleks] *n.* 全双工
half-duplex (HDX) [hæf 'duːpleks] *n.* 半双工

Notes

1. This uncertainty is due in part to the inevitable presence in any system of unwanted signal perturbations, broadly referred to as noise, and in part to the unpredictable nature of information itself.

这种不确定性，部分是由于在任何系统中不可避免地存在不必要的信号扰动，广义上称为噪声，部分是由于信息本身的不可预测性。

2. Regardless of the particular application and configuration, all information transmission systems invariably involve three major subsystems—a transmitter, the channel, and a receiver.

无论具体应用和配置如何，所有信息传输系统都必须包含 3 个主要子系统：发射机、信道和接收机。

3. Almost invariably, the message produced by a source must be converted by a transducer to a form suitable for the particular type of communication system employed.

几乎无一例外，信源产生的消息必须由传感器转换为适合所采用的特定类型通信系统传输的形式。

4. This degradation often results from noise and other undesired signals or interference, but also may include other distortion effects as well, such as fading signal levels, multiple transmission paths, and filtering.

这种信号恶化通常由噪声和其他不希望的信号或干扰引起，但也可能包括其他失真效应，如衰落信号电平、多路径传输和滤波。

5. Often it is desired that the receiver output be a scaled, possibly delayed, version of the message signal at the modulator input, although in some cases a more general function of the input message is desired.

尽管在某些情况下，希望接收机输出的信号与调制器的输入信号具有更一般的函数关系，一般情况还是希望接收机输出的信号是调制器的输入信号经过缩放、可能时延的版本。

7.2 Modulation and Multiplexing

Modulation and multiplexing are electronic techniques for transmitting information efficiently from one place to another. **Modulation makes the information signal more compatible with the medium, and multiplexing allows more than one signal to be transmitted concurrently over a single medium.** Modulation and multiplexing techniques are basic to electronic communication. Once you have mastered the fundamentals of these techniques, you will easily understand how most modern communication systems work.

7.2.1 Baseband Transmission

Before it can be transmitted, the information or intelligence must be converted to an electronic signal compatible with the medium. For example, a microphone changes voice signals (sound waves) into an analog voltage of varying frequency and amplitude. This signal is then passed over wires to a speaker or headphones. This is the way the telephone system works.

A video camera generates an analog signal that represents the light variations along one scan line of the picture. This analog signal is usually transmitted over a coaxial cable. Binary data is generated by a keyboard attached to a computer. The computer stores the data and processes it in some way. The data is then transmitted on cables to peripherals such as a printer or to other computers over a LAN. Regardless of whether the original information or intelligence signals are analog or digital, they are all referred to as baseband signals.

In a communication system, baseband information signals can be sent directly and unmodified over the medium or can be used to modulate a carrier for transmission over the medium. Putting the original voice, video, or digital signals directly into the medium is referred to as baseband transmission. For example, in many telephone and intercom systems, it is the voice itself that is placed on the wires and transmitted over some distance to the receiver. In most computer networks, the digital signals are applied directly to coaxial or twisted-pair cables for transmission to another computer.

In many instances, baseband signals are incompatible with the medium. Although it is theoretically possible to transmit voice signals directly by radio, realistically it is impractical. As a result, the baseband information signal, be it audio, video, or data, is normally used to modulate a high-frequency signal called a carrier. The higher-frequency carriers radiate into space more efficiently than the baseband signals themselves. Such wireless signals consist of both electric and magnetic fields. **These electromagnetic signals, which are able to travel through space for long distances, are also referred to as radio-frequency (RF) waves, or just radio waves.**

7.2.2 Broadband Transmission

Modulation is the process of having a baseband voice, video, or digital signal modify another, higher-frequency signal, the carrier. The process is illustrated in Figure 7.2. The information or intelligence to be sent is said to be impressed upon the carrier. The carrier is usually a sine wave generated by an oscillator. The carrier is fed to a circuit called a modulator along with the baseband intelligence signal. The intelligence signal changes the carrier in a unique way. The modulated carrier is amplified and sent to the antenna for transmission. This process is called broadband transmission.

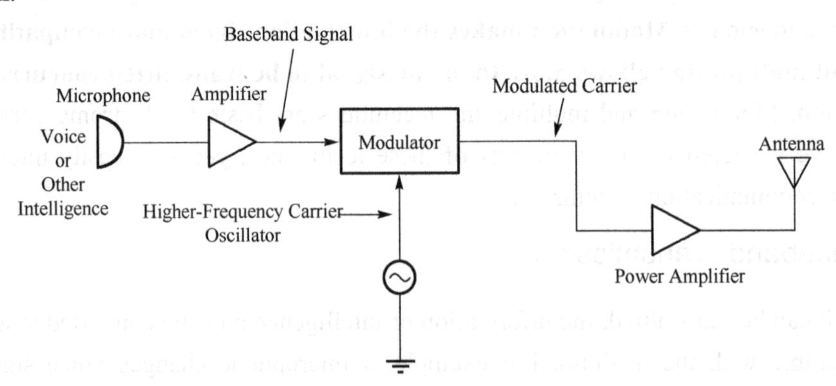

Figure 7.2 Modulation at the transmitter

Consider the common mathematical expression for a sine wave:

$$v = V_p \sin(2\pi ft + \theta) \quad \text{or} \quad v = V_p \sin(\omega t + \theta)$$

where

v = instantaneous value of sine wave voltage

V_p = peak value of sine wave

f = frequency, Hz

ω = angular velocity = $2\pi f$

t = time, s

$\omega t = 2\pi ft$ = angle, rad ($360° = 2\pi$ rad)

θ = phase angle

The three ways to make the baseband signal change the carrier sine wave are to vary its amplitude, vary its frequency, and vary its phase angle. The two most common methods of modulation are amplitude modulation (AM) and frequency modulation (FM). In AM, the baseband information signal called the modulating signal varies the amplitude of the higher-frequency carrier signal, as shown in Figure 7.3(a). It changes the V_p part of the equation. In FM, the information signal varies the frequency of the carrier, as shown in Figure 7.3(b). The carrier amplitude remains constant. FM varies the value of f in the first angle term inside the parentheses. Varying the phase angle produces phase modulation (PM). Here, the second term inside the parentheses (θ) is made to vary by the intelligence signal. Phase modulation produces frequency modulation; therefore, the PM signal is similar in appearance to a frequency-modulated carrier.

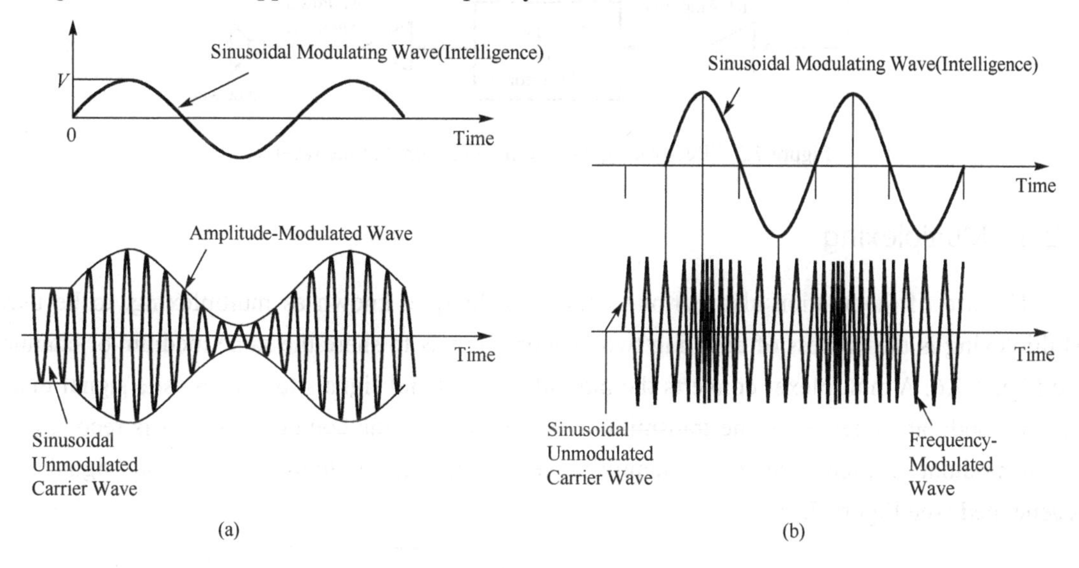

Figure 7.3 Types of modulation: (a) amplitude modulation; (b) frequency modulation

Two common examples of transmitting digital data by modulation are given in Figure 7.4. In Figure 7.4(a), the data is converted to frequency-varying tones. This is called frequency-shift keying (FSK). In Figure 7.4(b), the data introduces a 180° phase shift. This is called phase-shift keying (PSK). Devices called modems (modulator-demodulator) translate the data from digital to analog and back again. Both FM and PM are forms of angle modulation.

At the receiver, the carrier with the intelligence signal is amplified and then demodulated to extract the original baseband signal. Another name for the demodulation process is detection. (See Figure 7.5.)

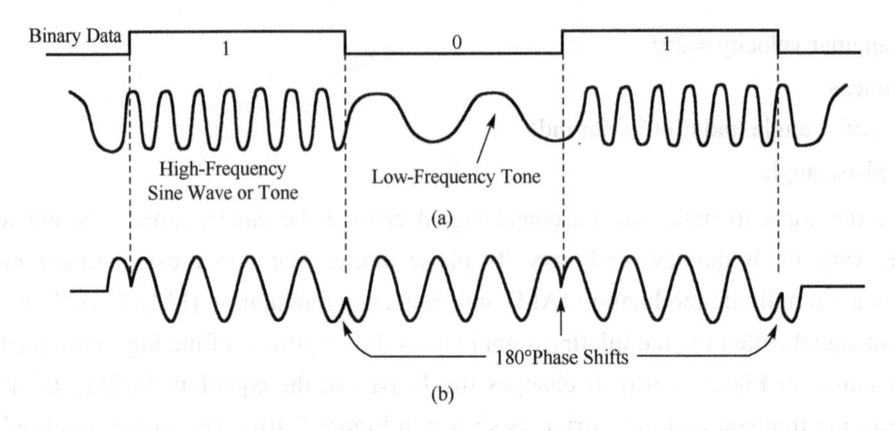

Figure 7.4 Transmitting binary data in analog form: (a) FSK; (b) PSK

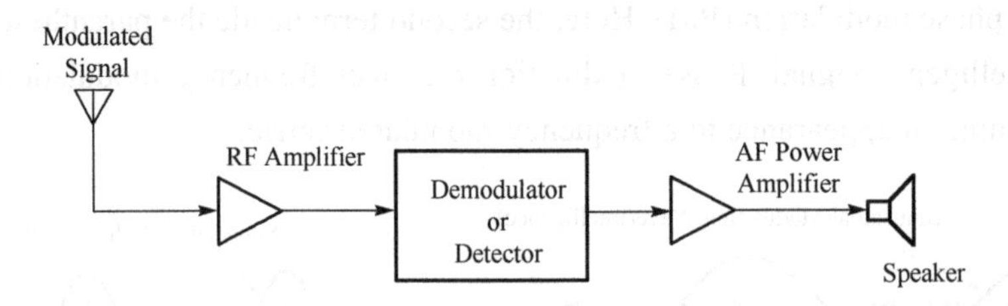

Figure 7.5 Recovering the intelligence signal at the receiver

7.2.3 Multiplexing

The use of modulation also permits another technique, known as multiplexing, to be used. Multiplexing is the process of allowing two or more signals to share the same medium or channel; see Figure 7.6. A multiplexer converts the individual baseband signals to a composite signal that is used to modulate a carrier in the transmitter. At the receiver, the composite signal is recovered at the demodulator, then sent to a demultiplexer where the individual baseband signals are regenerated (see Figure 7.7).

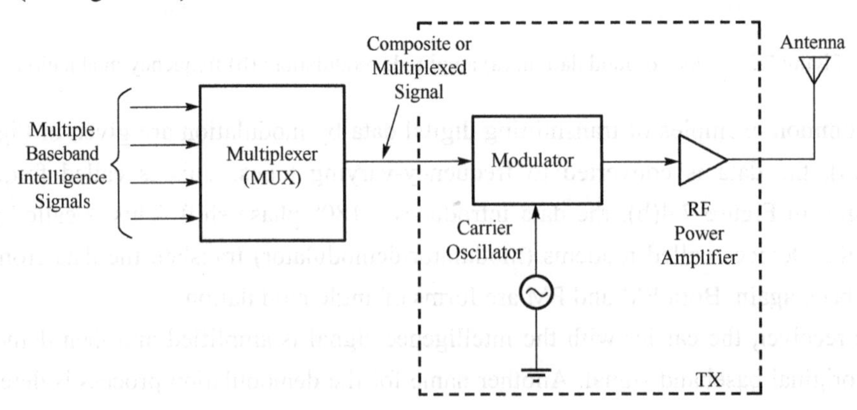

Figure 7.6 Multiplexing at the transmitter

There are three basic types of multiplexing: frequency division multiplexing (FDM), time division multiplexing (TDM), and code division multiplexing (CDM). In frequency division multiplexing, the intelligence signals modulate subcarriers on different frequencies that are then added together, and the composite signal is used to modulate the carrier. In optical networking, wavelength division multiplexing (WDM) is equivalent to frequency division multiplexing for optical signal.

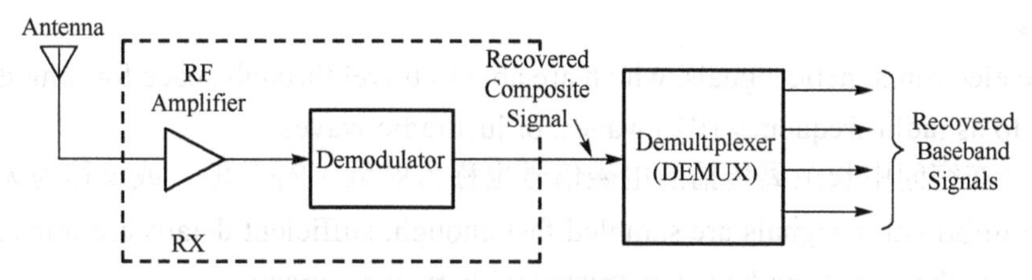

Figure 7.7 Demultiplexing at the receiver

In time division multiplexing, the multiple intelligence signals are sequentially sampled, and a small piece of each is used to modulate the carrier. **If the information signals are sampled fast enough, sufficient details are transmitted that at the receiving end the signal can be reconstructed with great accuracy.**

In code division multiplexing, the signals to be transmitted are converted to digital data that is then uniquely coded with a faster binary code. The signals modulate a carrier on the same frequency. All use the same communication channel simultaneously. The unique coding is used at the receiver to select the desired signal.

Words and Expressions

baseband ['beɪsbænd]	*n.*（无线）基带；基频
coaxial cable	同轴电缆
LAN: local area network	局域网
twisted-pair ['twɪstɪd per]	*n.* 双绞线
broadband ['brɔːdbænd]	*n.* 宽带
oscillator ['ɑːsɪleɪtər]	*n.* 振荡器
amplitude modulation (AM)	调幅
frequency modulation (FM)	调频
phase modulation (PM)	调相
frequency-shift keying (FSK)	频移键控
phase-shift keying (PSK)	相移键控
detector [dɪ'tektər]	*n.* 检波器；探测器
frequency division multiplexing (FDM)	频分复用
time division multiplexing (TDM)	时分复用
code division multiplexing (CDM)	码分复用
wavelength division multiplexing (WDM)	波分复用

Notes

1. Modulation makes the information signal more compatible with the medium, and multiplexing allows more than one signal to be transmitted concurrently over a single medium.

调制技术使得信息信号与传输介质更兼容，而复用技术则允许在单个传输介质上同时传输多路信号。

2. These electromagnetic signals, which are able to travel through space for long distances, are also referred to as radio-frequency (RF) waves, or just radio waves.

这些能够在空间中长距离传播的电磁信号也称为射频（RF）波，或仅称为无线电波。

3. If the information signals are sampled fast enough, sufficient details are transmitted that at the receiving end the signal can be reconstructed with great accuracy.

如果信息信号的采样速度足够快，则会传输足够的细节，以便在接收端可以非常精确地重建信号。

7.3 Digital Communication Techniques

Since the mid-1970s, digital methods of transmitting data have slowly but surely replaced the older, more conventional analog ones. Today, thanks to the availability of fast, low-cost analog-to-digital (A/D) and digital-to-analog (D/A) converters and high-speed digital signal processors, most electronic communications are digital.

7.3.1 Data Conversion

Translating an analog signal to a digital signal is called analog-to-digital (A/D) conversion, digitizing a signal, or encoding. The device used to perform this translation is known as an analog-to-digital (A/D) converter or ADC. A modern A/D converter is usually a single-chip IC that takes an analog signal and generates a parallel or serial binary output. The opposite process is called digital-to-analog (D/A) conversion. The circuit used to perform this is called a digital-to-analog (D/A) converter (or DAC) or a decoder. The input to a D/A converter may be a serial or parallel binary number, and the output is a proportional analog voltage level. Like the A/D converter, a D/A converter is usually a single-chip IC or a part of a large IC.

1. A/D Conversion

An analog signal is a smooth or continuous voltage or current variation. It could be a voice signal, a video waveform, or a voltage representing a variation of some other physical characteristic such as temperature. Through A/D conversion, these continuously variable signals are changed to a series of binary numbers. A/D conversion is a process of sampling or measuring the analog signal at regular time intervals. At the times indicated by the vertical dashed lines in Figure 7.8, the instantaneous value of the analog signal is measured and a proportional binary number is generated to represent that sample. As a result, the continuous analog signal is translated to a series of discrete binary numbers representing samples.

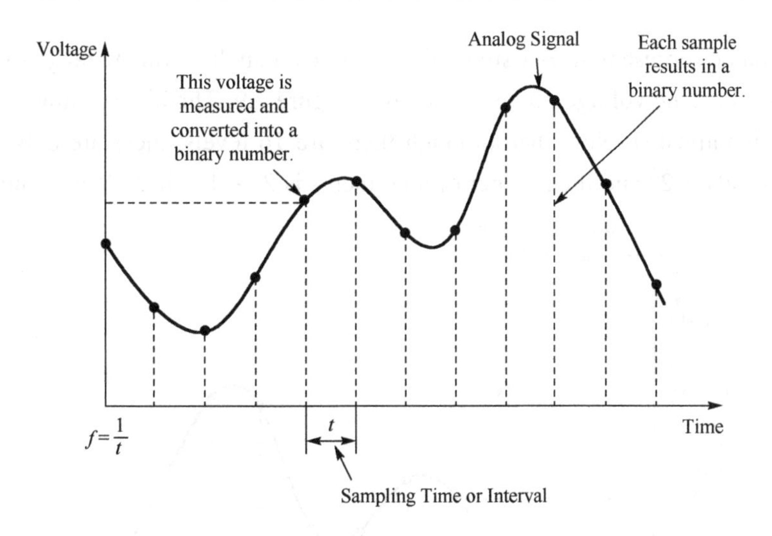

Figure 7.8 Sampling an analog signal

A key factor in the sampling process is the frequency of sampling f, which is the reciprocal of the sampling interval t shown in Figure 7.8. To retain the high frequency information in the analog signal, a sufficient number of samples must be taken so that the waveform is adequately represented. It has been found that the minimum sampling frequency is twice the highest analog frequency content of the signal. For example, if the analog signal contains a maximum frequency variation of 3,000 Hz, the analog wave must be sampled at a rate of at least twice this, or 6,000 Hz. This minimum sampling frequency is known as the Nyquist frequency f_N. (And $f_N = 2f_m$, where f_m is the highest frequency of the input signal.) For bandwidth limited signals with upper and lower limits of f_2 and f_1, the Nyquist sampling rate is just twice the bandwidth or $2(f_2 - f_1)$.

Although theoretically the highest frequency component can be adequately represented by a sampling rate of twice the highest frequency, in practice the sampling rate is much higher than the Nyquist minimum, typically 2.5 to 3 times more. The actual sampling rate depends on the application as well as factors such as cost, complexity, channel bandwidth, and availability of practical circuits.

Another important factor in the conversion process is that, because the analog signal is smooth and continuous, it represents an infinite number of actual voltage values. In a practical A/D converter, it is not possible to convert all analog samples to a precise proportional binary number. Instead, the A/D converter is capable of representing only a finite number of voltage values over a specific range. The samples are converted to a binary number whose value is close to the actual sample value. For example, an 8-bit binary number can represent only 256 states, which may be the converted values from an analog waveform having an infinite number of positive and negative values between +1 V and −1 V.

The physical nature of an A/D converter is such that it divides a voltage range into discrete increments, each of which is then represented by a binary number. The analog voltage measured during the sampling process is assigned to the increment of voltage closest to it. For example, assume that an A/D converter produces 4 output bits. With 4 bits, 2^4 or 16 voltage levels can be represented. For simplicity, assume an analog voltage range of 0 to 15 V. The A/D converter divides the voltage range as shown in Figure 7.9. The binary number represented by each increment is indicated. Note that although there are 16 levels, there are only 15 increments. The number of levels is 2^N and the number of increments is $2^N - 1$, where N is the number of bits.

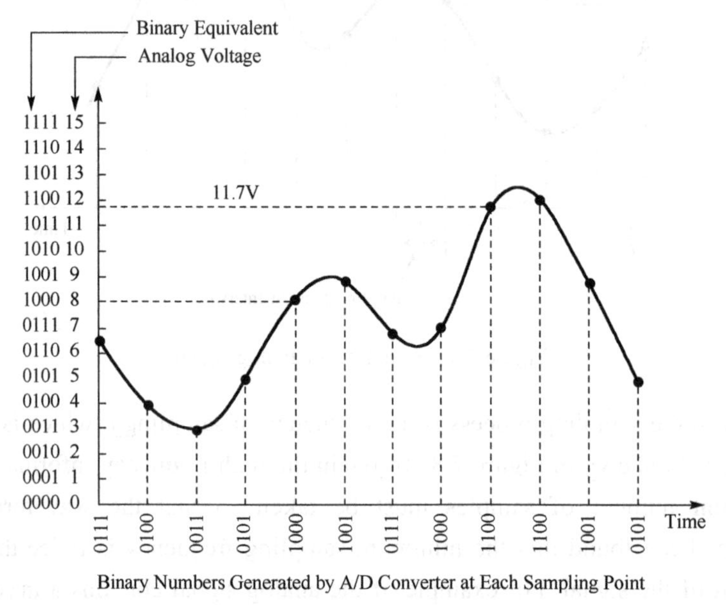

Figure 7.9 The A/D converter divides the input voltage range into discrete voltage increments

Now assume that the A/D converter samples the analog input and measures a voltage of 0 V. The A/D converter will produce a binary number as close as possible to this value, in this case 0000. If the analog input is 8 V, the A/D converter generates the binary number 1000. But what happens if the analog input is 11.7 V, as shown in Figure 7.9? The A/D converter produces the binary number 1011, whose decimal equivalent is 11. In fact, any value of analog voltage between 11 V and 12 V will produce this binary value. As you can see, there is some error associated with the conversion process. This is referred to as quantization error which can be reduced, of course, by simply dividing the analog voltage range into a larger number of smaller voltage increments. To represent more voltage increments, a greater number of bits must be used. In traditional PCM (pulse-code modulation), the analog signal is sampled and converted to a sequence of parallel binary words by an A/D converter. Each time a sample is taken, an 8-bit word is generated by the A/D converter.

2. D/A Conversion

At some point, it is usually desirable to translate the multiple binary numbers back to the equivalent analog voltage. This is the job of the D/A converter, which receives the binary numbers sequentially and produces a proportional analog voltage at the output. Because the input binary numbers represent specific voltage levels, the output of the D/A converter has a stairstep characteristic. Figure 7.10 shows the process of converting the 4-bit binary numbers obtained in the conversion of the waveform in Figure 7.9. If these binary numbers are fed to a D/A converter, the output is a stairstep voltage as shown. Since the steps are very large, the resulting voltage is only an approximation to the actual analog signal. **However, the stairsteps can be filtered out by passing the D/A converter output through a low-pass filter with an appropriate cutoff frequency.**

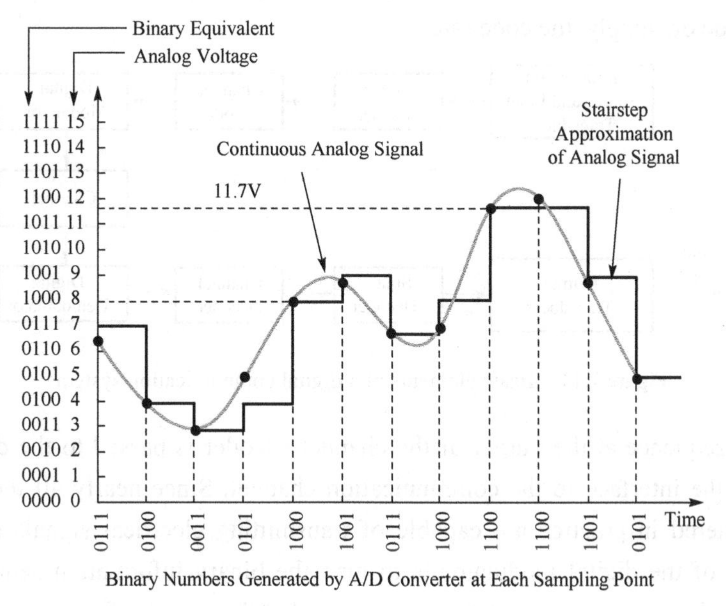

Figure 7.10 The D/A converter produces a stepped approximation of the original signal

7.3.2 Elements of a Digital Communication System

Figure 7.11 illustrates the functional diagram and the basic elements of a digital communication system. **The source output may be either an analog signal, such as an audio or video signal, or a digital signal, such as the output of a computer, that is discrete in time and has a finite number of output characters.** In a digital communication system, the messages produced by the source are converted into a sequence of binary digits. Ideally, we should like to represent the source output (message) by as few binary digits as possible. In other words, we seek an efficient representation of the source output that results in little or no redundancy. The process of efficiently converting the output of either an analog or digital source into a sequence of binary digits is called source encoding or data compression.

The sequence of binary digits from the source encoder, which we call the information sequence, is passed to the channel encoder. **The purpose of the channel encoder is to introduce, in a controlled manner, some redundancy in the binary information sequence that can be used at the receiver to overcome the effects of noise and interference encountered in the transmission of the signal through the channel.** Thus, the added redundancy serves to increase the reliability of the received data and improves the fidelity of the received signal. In effect, redundancy in the information sequence aids the receiver in decoding the desired information sequence. For example, a (trivial) form of encoding of the binary information sequence is simply to repeat each binary digit m times, where m is some positive integer. More sophisticated (nontrivial) encoding involves taking k information bits at a time and mapping each k-bit sequence into a unique n-bit sequence, called a code word. The amount of redundancy introduced by encoding the data in this manner is measured by the ratio n/k. The reciprocal of this ratio, namely k/n, is called the rate of the code or, simply, the code rate.

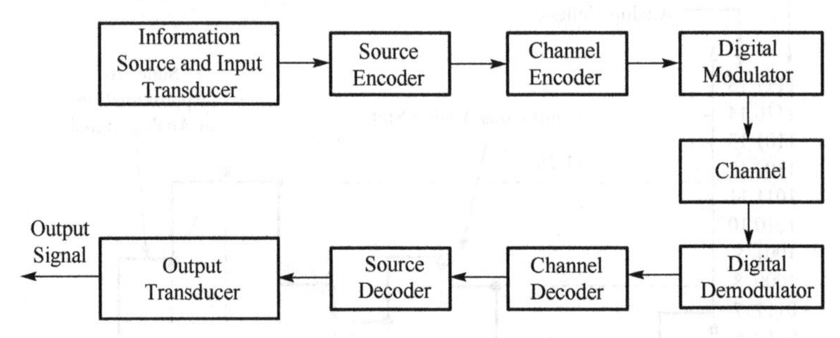

Figure 7.11 Basic elements of a digital communication system

The binary sequence at the output of the channel encoder is passed to the digital modulator, which serves as the interface to the communication channel. Since nearly all the communication channels encountered in practice are capable of transmitting electrical signals (waveforms), the primary purpose of the digital modulator is to map the binary information sequence into signal waveforms. To elaborate on this point, let us suppose that the coded information sequence is to be transmitted one bit at a time at some uniform rate R bits per second (bits/s). The digital modulator may simply map the binary digit 0 into a waveform $s_0(t)$ and the binary digit 1 into a waveform $s_1(t)$. In this manner, each bit from the channel encoder is transmitted separately. We call this binary modulation. Alternatively, the modulator may transmit b coded information bits at a time by using $M = 2^b$ distinct waveforms $s_i(t)$, $i = 0, 1,\cdots, M-1$, one waveform for each of the 2^b possible b-bit sequences. We call this M-ary modulation ($M > 2$). Note that a new b-bit sequence enters the modulator every b/R seconds. Hence, when the channel bit rate R is fixed, the amount of time available to transmit one of the M waveforms corresponding to a b-bit sequence is b times the time period in a system that uses binary modulation.

The communication channel is the physical medium that is used to send the signal from the transmitter to the receiver. In wireless transmission, the channel may be the atmosphere (free space).On the other hand, telephone channels usually employ a variety of physical media, including wire lines, optical fiber cables, and wireless (microwave radio). Whatever the physical medium used for transmission of the information, the essential feature is that the transmitted signal is corrupted in a random manner by a variety of possible mechanisms, such as additive thermal noise generated by electronic devices; man-made noise, e.g., automobile ignition noise; and atmospheric noise, e.g., electrical lightning discharges during thunderstorms.

At the receiving end of a digital communication system, the digital demodulator processes the channel-corrupted transmitted waveform and reduces the waveforms to a sequence of numbers that represent estimates of the transmitted data symbols (binary or M-ary). This sequence of numbers is passed to the channel decoder, which attempts to reconstruct the original information sequence from knowledge of the code used by the channel encoder and the redundancy contained in the received data.

A measure of how well the demodulator and decoder perform is the frequency with which errors occur in the decoded sequence. More precisely, the average probability of a bit-error at the output of the decoder is a measure of the performance of the demodulator-decoder combination. In general, the probability of error is a function of the code characteristics, the types of waveforms used to transmit the information over the channel, the transmitter power, the characteristics of the channel (i.e., the amount of noise, the nature of the interference), and the method of demodulation and decoding.

As a final step, when an analog output is desired, the source decoder accepts the output sequence from the channel decoder and from knowledge of the source encoding method used, attempts to reconstruct the original signal from the source. Because of channel decoding errors and possible distortion introduced by the source encoder, and perhaps, the source decoder, the signal at the output of the source decoder is an approximation to the original source output. The difference or some function of the difference between the original signal and the reconstructed signal is a measure of the distortion introduced by the digital communication system.

Words and Expressions

digitize ['dɪdʒɪtaɪz]	*vt.*（使数据）数字化
encode [ɪn'koʊd]	*vt.* 编码；把……译成电码（或密码）
decoder [ˌdiː'koʊdər]	*n.*（电子信号）解码器，译码器
reciprocal [rɪ'sɪprəkl]	*n.* 倒数
quantization error	量化误差
pulse-code modulation (PCM)	脉冲编码调制
redundancy [rɪ'dʌndənsi]	*n.* 冗余；多余
data compression	数据压缩
source encoder	信源编码器
channel encoder	信道编码器
fidelity [fɪ'deləti]	*n.* 保真度；精确性
elaborate [ɪ'læbəreɪt]	*v.* 详尽阐述；详细描述
optical fiber	光纤

Notes

1. Although theoretically the highest frequency component can be adequately represented by a sampling rate of twice the highest frequency, in practice the sampling rate is much higher than the Nyquist minimum, typically 2.5 to 3 times more.

虽然理论上采用两倍于最高频率分量的采样率，得到的样本信号足以代表原始信号，但在实践中，采样率远高于奈奎斯特最小值，通常是其 2.5 到 3 倍。

2. The physical nature of an A/D converter is such that it divides a voltage range into discrete

increments, each of which is then represented by a binary number. The analog voltage measured during the sampling process is assigned to the increment of voltage closest to it.

A/D 转换器的物理特性是，它将电压范围划分为离散增量，然后每个增量用二进制数表示。采样过程中测量的模拟电压值用最接近它的离散电压增量来表示。

3. However, the stairsteps can be filtered out by passing the D/A converter output through a low-pass filter with an appropriate cutoff frequency.

然而，D/A 转换器输出信号的阶梯形状，可以通过具有适当截止频率的低通滤波器来滤除。

4. The source output may be either an analog signal, such as an audio or video signal, or a digital signal, such as the output of a computer, that is discrete in time and has a finite number of output characters.

信源输出可以是模拟信号，如音频或视频信号，也可以是数字信号，如计算机的输出，其在时间上是离散的，并且具有有限数量的输出字符。

5. The purpose of the channel encoder is to introduce, in a controlled manner, some redundancy in the binary information sequence that can be used at the receiver to overcome the effects of noise and interference encountered in the transmission of the signal through the channel.

信道编码器的目的是以可控的方式在二进制信息序列中引入一些冗余，这些冗余可用于在接收机处消除信号在信道传输中受到的噪声和干扰的影响。

Exercises

1. Match the terms (1)–(6) with the definitions A–F.

(1) channel	A. the original frequency range of a transmission signal before it is modulated
(2) modulation	B. the difference between the analog signal and the closest available digital value at each sampling instant from the A/D converter
(3) noise	C. the act of adding information to an electronic or optical waveform by altering the frequency, amplitude, phase of the waveform or more
(4) transducer	D. the medium by which a signal is transmitted from the sender to the receiver
(5) baseband	E. a device that converts one form of energy into another form of energy
(6) quantization error	F. any electrical signal that interferes with the information signal

2. Translate into Chinese.

(1) If a system uniformly samples an analog signal at a rate that exceeds the signal's highest frequency by at least a factor of two, the original analog signal can be perfectly recovered from the discrete values produced by sampling.

(2) Digital communication systems can be made highly reliable by exploiting powerful error-control coding techniques in such a way that the estimate of a message signal delivered to a user is almost indistinguishable from the message signal delivered by a source of information at the other end of the system.

(3) TDM provides greater flexibility and efficiency, by dynamically allocating more time periods to the signals that need more of the bandwidth, while reducing the time periods to those

signals that do not need it. FDM lacks this type of flexibility, as it cannot dynamically change the width of the allocated frequency.

3. Translate into English.

(1)对于 W 赫兹的信号带宽，每秒 $2W$ 次采样的采样速率称为奈奎斯特速率；其倒数 $1/2W$ 称为奈奎斯特间隔。

(2) 在脉冲编码调制（PCM）中，信号由编码脉冲序列表示，这是通过在时间和幅度上以离散形式表示信号来实现的。

(3) 信源编码器的输出比特速率通常与信道编码器的输入比特速率不完全匹配。

4. Read the following article and write a summary.

Analog communication systems, amplitude modulation (AM) radio being a typifying example, can inexpensively communicate a band limited analog signal from one location to another (point-to-point communication) or from one point to many (broadcast). Although it is not shown here, the coherent receiver provides the largest possible signal-to-noise ratio for the demodulated message. An analysis of this receiver thus indicates that some residual error will always be present in an analog system's output.

Although analog systems are less expensive in many cases than digital ones for the same application, digital systems offer much more efficiency, better performance, and much greater flexibility.

- Efficiency. The Source Coding Theorem allows quantification of just how complex a given message source is and allows us to exploit that complexity by source coding (compression). In analog communication, the only parameters of interest are message bandwidth and amplitude. We cannot exploit signal structure to achieve a more efficient communication system.
- Performance. Because of the Noisy Channel Coding Theorem, we have a specific criterion by which to formulate error-correcting codes that can bring us as close to error-free transmission as we might want. Even though we may send information by way of a noisy channel, digital schemes are capable of error-free transmission while analog ones cannot overcome channel disturbances.
- Flexibility. Digital communication systems can transmit real-valued discrete-time signals, which could be analog ones obtained by analog-to-digital conversion, and symbolic-valued ones (computer data, for example). Any signal that can be transmitted by analog means can be sent by digital means, with the only issue being the number of bits used in A/D conversion (how accurately do we need to represent signal amplitude). Images can be sent by analog means (commercial television), but better communication performance occurs when we use digital systems (HDTV). In addition to digital communication's ability to transmit a wider variety of signals than analog systems, point-to-point digital systems can be organized into global (and beyond as well) systems that provide efficient and flexible information transmission. Computer networks are what we call such systems today. Even analog-based networks, such as the telephone system, employ modern computer networking ideas rather than the purely analog systems of the past.

Consequently, with the increased speed of digital computers, the development of increasingly efficient algorithms, and the ability to interconnect computers to form a communication infrastructure, digital communication is now the best choice for many situations.

5. Language study: Prepositions and time words

Look at the prepositions we normally use with these time expressions:

<table>
<tr><td rowspan="3">at</td><td>7:00</td><td rowspan="3">on</td><td>Monday</td></tr>
<tr><td>the end of the month</td><td>New Year's Day</td></tr>
<tr><td>the weekend</td><td>August 11, 1992</td></tr>
<tr><td rowspan="4">in</td><td>August</td><td rowspan="4">no preposition</td><td>Yesterday</td></tr>
<tr><td>2007</td><td>last week</td></tr>
<tr><td rowspan="2">the 1990s</td><td>next week</td></tr>
<tr><td>Tomorrow</td></tr>
<tr><td colspan="2">by: means not later than.
Example: I need that report by Wednesday.</td><td colspan="2">until: means from time A (often now) to time B.
Example: we will be working with you until the project is completed.</td></tr>
</table>

Complete the following sentences with by, until, in, on, at, or- (no preposition).

(1) Engineers developed optical fiber cables _____ the 1980s.

(2) Bell Labs became the predominant center for communication research in the world, and held that position _____ quite recently.

(3) The full switchover to TCP/IP was performed ____ January 1, 1983, without too many problems.

(4) The project should be completed _____ next March.

Unit 8 Mobile Communication

8.1 From 1G to 5G

Mobile wireless communication system has gone through several evolution stages in the past few decades after the introduction of the first generation mobile network in early 1980s. Due to huge demand for more connections worldwide, mobile communication standards advanced rapidly to support more users. Let's take a look on the evolution stages of wireless technologies for mobile communication.

8.1.1 A Brief History of Wireless Technologies

Marconi, an Italian inventor, transmitted Morse code signals using radio waves wirelessly to a distance of 3.2 km in 1895. It was the first wireless transmission in the history of science. Since then, engineers and scientists were working on an efficient way to communicate using RF waves.

Telephone became popular during the mid of 19th century. Due to wired connection and restricted mobility, engineers started developing a device which doesn't require wired connection and transmits voice using radio waves. Martin Cooper, an engineer at Motorola during 1970s working on a handheld device capable of two way communication wirelessly, invented the first generation mobile phone. It was initially developed to use in a car, the first prototype was tested in 1974. **This invention is considered as a turning point in wireless communication which led to an evolution of many technologies and standards in future.** Figure 8.1 shows the evolution of mobile phone.

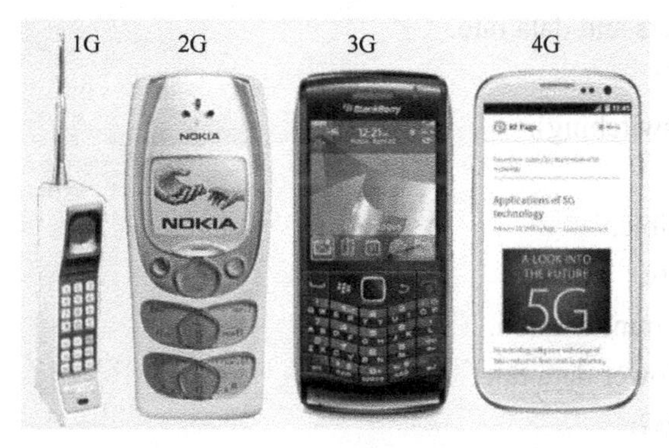

Figure 8.1 The evolution of mobile phone

8.1.2 1G — First Generation Mobile Communication System

The first generation of mobile network was deployed in Japan by Nippon Telephone and

Telegraph Company (NTT) in Tokyo during 1979. In the beginning of 1980s, it gained popularity in the US, Finland, UK and other European countries. This system used analog signals and it had many disadvantages due to technology limitations.

Key features of 1G:

- Frequency 800 MHz and 900 MHz.
- Bandwidth: 10 MHz (666 duplex channels with bandwidth of 30 kHz).
- Technology: analog switching.
- Modulation: frequency modulation (FM).
- Mode of service: voice only.
- Access technique: frequency division multiple access (FDMA).

Disadvantages of 1G:

- Poor voice quality due to interference.
- Poor battery life.
- Large sized mobile phones (not convenient to carry).
- Less security (calls could be decoded using an FM demodulator).
- Limited number of users and cell coverage.
- Roaming was not possible between similar systems.

8.1.3 2G — Second Generation Mobile Communication System

Second generation mobile communication system introduced a new digital technology for wireless transmission also known as global system for mobile communications (GSM). GSM technology became the base standard for further development in wireless standards later. This standard was capable of supporting up to 14.4 to 64 kbps (maximum) data rate which is sufficient for SMS (short message service) and e-mail services.

Code division multiple access (CDMA) system developed by Qualcomm was also introduced and implemented in the mid-1990s. CDMA has more features than GSM in terms of spectral efficiency, number of users and data rate.

Key features of 2G:

- Digital system (switching).
- SMS is possible.
- Roaming is possible.
- Enhanced security.
- Encrypted voice transmission.
- First internet at lower data rate.

Disadvantages of 2G:

- Low data rate.
- Limited mobility.
- Less features on mobile devices.
- Limited number of users and hardware capability.

2.5G and 2.75G:

In order to support higher data rate, general packet radio service (GPRS) was introduced and successfully deployed. GPRS was capable of data rate up to 171 kbps (maximum). EDGE (enhanced data GSM evolution) also developed to improve data rate for GSM network. EDGE was capable to support up to 473.6 kbps (maximum). Another popular technology CDMA2000 was also introduced to support higher data rate for CDMA network. This technology has the ability to provide up to 384 kbps data rate (maximum).

8.1.4 3G — Third Generation Mobile Communication System

Third generation mobile communication started with the introduction of UMTS (universal mobile terrestrial/telecommunications system). UMTS has the data rate of 384 kbps and it supported video calling for the first time on mobile devices.

After the introduction of 3G, smartphones became popular across the globe. Specific applications were developed for smartphones which handle multimedia chat, e-mails, video calling, games, social media and healthcare.

Key features of 3G:

- Higher data rate.
- Video calling.
- Enhanced security, more number of users and coverage.
- Mobile app support.
- Multimedia message support.
- Location tracking and maps.
- Better web browsing.
- TV streaming.
- High quality 3D games.

Disadvantages of 3G:

- Expensive spectrum licenses.
- Costly infrastructure, equipment and implementation.
- Higher bandwidth requirements to support higher data rate.
- Costly mobile devices.
- Compatibility with older generation 2G and frequency bands.

3. 5G and 3.75G:

In order to enhance data rate in existing 3G network, another two technology improvements were introduced to network. HSDPA (high speed downlink packet access) and HSUPA (high speed uplink packet access) were developed and deployed to the 3G network. 3.5G network can support up to 2 Mbps data rate. 3.75G is an improved version of 3G with HSPA+ (high speed packet access plus). Later this system would evolve into more powerful 3.9G known as LTE (long term evolution).

8.1.5 4G — Fourth Generation Mobile Communication System

4G is an enhanced version of 3G developed by IEEE, offering higher data rate and being capable to handle more advanced multimedia services. LTE and LTE advanced wireless technology are used in 4G. Furthermore, it has compatibility with previous versions, and thus easier deployment and upgrade of LTE and LTE advanced networks are possible.

Simultaneous transmission of voice and data is possible with LTE system which significantly improves data rate. All services including voice services can be transmitted over IP packets. Complex modulation schemes and carrier aggregation are used to multiply uplink/downlink capacity.

Wireless transmission technologies like WiMax are introduced in 4G to enhance data rate and network performance.

Key features of 4G:

- Much higher data rate up to 1 Gbps.
- Enhanced security and mobility.
- Reduced latency for mission critical applications.
- High definition video streaming and gaming.
- Voice over LTE network VoLTE (use IP packets for voice).

Disadvantages of 4G:

- Expensive hardware and infrastructure.
- Costly spectrum (most countries, frequency bands are too expensive).
- High end mobile devices compatible with 4G technology required, which are costly.
- Wide deployment and upgrade are time consuming.

8.1.6 5G — Fifth Generation Mobile Communication System

5G network is using advanced technologies to deliver ultra-fast internet and multimedia experience for customers. Existing LTE advanced network will transform into supercharged 5G network in future.

In earlier deployments, 5G network functions in non-standalone mode and standalone mode. In non-standalone mode, both LTE spectrum and 5G-NR (New Radio) spectrum are used together. Control signaling is connected to LTE core network in non-standalone mode.

There is a dedicated 5G core network higher bandwidth 5G-NR spectrum for standalone mode. Sub-6 GHz FR1 spectrum is used in the initial deployments of 5G network.

In order to achieve higher data rate, 5G technology uses millimeter waves and unlicensed spectrum for data transmission. Complex modulation technique has been developed to support massive data rate for internet of things (IoT). **Cloud based network architecture will extend the functionalities and analytical capabilities for industries, autonomous driving, healthcare and security applications.**

Key features of 5G:

- Ultra-fast mobile internet up to 10 Gbps.

- Low latency in milliseconds (significant for mission critical applications).
- Total cost reduction for data.
- Higher security and reliable network.
- Using technologies like small cells, beam forming to improve efficiency.
- Forward compatibility network offers further enhancements in future.
- Cloud based infrastructure offers power efficiency, easy maintenance and upgrade of hardware.

8.1.7 Summary

Wireless technologies have been continuously evolving to meet increasing demands and higher specification requirements. **Since the deployment of first generation mobile network, telecommunication industry faces a lot of new challenges in terms of technology, efficient utilization of spectrum and most importantly security to end users.** Future wireless technologies will provide ultra-fast, feature-rich and highly secure mobile networks.

Words and Expressions

prototype ['proʊtətaɪp] n. 原型；雏形；最初形态
bandwidth ['bændwɪdθ] n. 带宽；频宽；带宽值
switching ['swɪtʃɪŋ] v. 交换；交换技术；转变
access ['ækses] n. 接入；通道；通路；入径
vt. 接入；访问，存取（计算机文件）
cell coverage 小区覆盖；手机覆盖率
roaming ['roʊmɪŋ] n.（移动电话的）漫游
global system for mobile communications (GSM) 全球移动通信系统
short message service (SMS) 短消息业务；（手机的）短信服务；（手机）短信，短消息
spectral efficiency 频谱利用率
encrypt [ɪn'krɪpt] vt. 把……加密（或编码）
capability [ˌkeɪpə'bɪləti] n. 性能；容量；能力；才能
general packet radio service (GPRS) 通用分组无线服务
enhanced data GSM evolution (EDGE) 数据增强型 GSM 演进
universal mobile terrestrial/telecommunications system (UMTS) 通用移动地面/通信系统
video calling 视频电话；视频呼叫功能
mobile app 移动应用程序；手机应用程序
location tracking 定位追踪；位置跟踪
infrastructure ['ɪnfrəstrʌktʃər] n. 基础架构；基础设施；基础建设
compatibility [kəmˌpætə'bɪləti] n. 兼容性；相容性
high speed downlink packet access (HSDPA) 高速下行链路分组接入；高速下行分组接入

high speed uplink packet access (HSUPA)	高速上行链路分组接入；高速上行分组接入
high speed packet access plus (HSPA+)	增强型高速分组接入
long term evolution (LTE)	长期演进技术
IP: Internet protocol	互联网协议；因特网协议；网际协议
carrier aggregation	载波聚合
latency ['leɪtənsɪ]	*n.* 延迟
high definition video streaming	高清视频流
deployment [dɪ'plɔɪmənt]	*n.* 部署；展开；使用；配置
standalone mode	单机模式；独立模式
signaling ['sɪgnəlɪŋ]	*n.* 信令
internet of things (IoT)	物联网
autonomous [ɔː'tɑːnəməs]	*adj.* 自主的；自治的
maintenance ['meɪntənəns]	*n.* 维护；维修；保养；维持

Notes

1. This invention is considered as a turning point in wireless communication which led to an evolution of many technologies and standards in future.

这项发明被认为是无线通信领域的一个转折点，它引领了此后很多技术和标准的演变。

2. Cloud based network architecture will extend the functionalities and analytical capabilities for industries, autonomous driving, healthcare and security applications.

基于云的网络架构将扩展工业、自动驾驶、医疗保健和安全应用程序的功能和分析能力。

3. Since the deployment of first generation mobile network, telecommunication industry faces a lot of new challenges in terms of technology, efficient utilization of spectrum and most importantly security to end users.

自第一代移动网络部署以来，电信业在技术、频谱的有效利用及最重要的终端用户安全方面都面临着许多新的挑战。

8.2 Mobile Networks

Mobile networks, which have a 40-year history that parallels the Internet's, have undergone significant change. The first two generations supported voice and then text, with 3G defining the transition to broadband access, supporting data rates measured in hundreds of kilobits-per-second. Today, the industry is at 4G (supporting data rates typically measured in the few megabits-per-second) and transitioning to 5G, with the promise of a tenfold increase in data rates.

But 5G is about much more than increased bandwidth. **5G represents a fundamental rearchitecting of the access network in a way that leverages several key technology trends and sets it on a path to enable much greater innovation.** In the same way that 3G defined the transition from voice to broadband, 5G's promise is primarily about the transition from a single access service (broadband connectivity) to a richer collection of edge services and devices. 5G is expected to provide support for immersive user interfaces (e.g., AR/VR), mission-critical applications (e.g., public safety, autonomous vehicles), and the internet of things (IoT). **Because these use cases will include everything from home appliances to industrial robots to self-driving cars, 5G won't just support humans accessing the Internet from their smartphones, but also swarms of autonomous devices working together on their behalf.** There is more to supporting these services than just improving bandwidth or latency to individual users. As we will see, a fundamentally different edge network architecture is required. Note that the 5G mobile network is on an evolutionary path and not a point solution, includes standardized specifications, a range of implementation choices, and a long list of aspirational goals.

8.2.1 Main Components

The cellular network provides wireless connectivity to devices that are on the move. These devices, which are known as user equipment (UE), have traditionally corresponded to smartphones and tablets, but will increasingly include cars, drones, industrial and agricultural machines, robots, home appliances, medical devices, and so on.

As shown in Figure 8.2, the cellular network consists of two main subsystems: the radio access network (RAN) and the mobile core. The RAN manages the radio spectrum, making sure it is used efficiently and meets the quality of service (QoS) requirements of every user. It corresponds to a distributed collection of base stations. In 4G, these are (somewhat cryptically) named eNodeB (or eNB), which is short for evolved Node B. In 5G, they are known as gNB. (The g stands for "next Generation".)

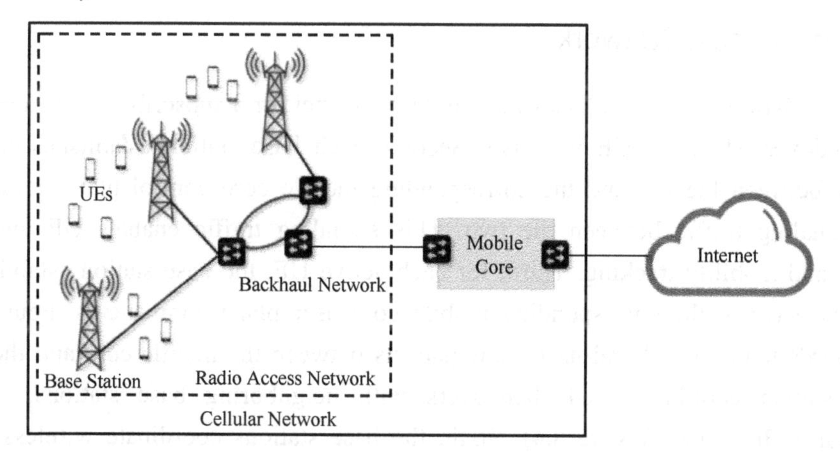

Figure 8.2 The cellular network consists of the radio access network (RAN) and the mobile core

The mobile core is a bundle of functionality (as opposed to a device) that serves several purposes.

- Provide Internet (IP) connectivity for both data and voice services.
- Ensure this connectivity fulfills the promised QoS requirements.
- Track user mobility to ensure uninterrupted service.

- Track subscriber usage for billing and charging.

As shown in Figure 8.3, the mobile core is divided into a control plane and a user plane. Note that mobile core is another example of a generic term. In 4G this is called the evolved packet core (EPC) and in 5G it is called the next generation core (NG-Core). **Even though the word "core" is in its name, from an Internet perspective, the mobile core is still part of the access network, effectively providing a bridge between the RAN in some geographic area and the greater IP-based Internet.**

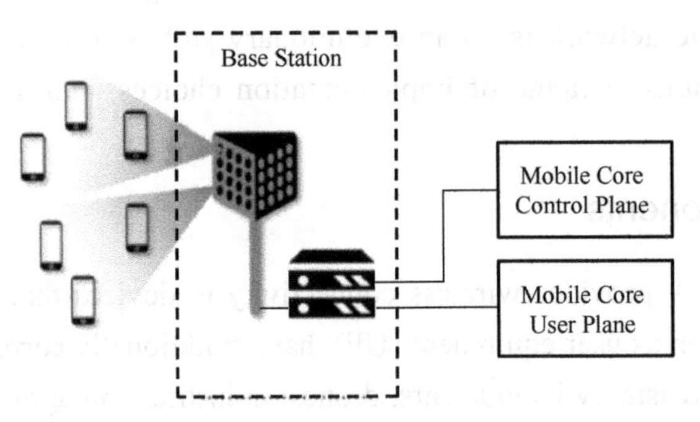

Figure 8.3 The mobile core is divided into a control plane and a user plane

A backhaul network interconnects the base stations that implement the RAN with the mobile core. This network is typically wired, may or may not have the ring topology, and is often constructed from commodity components found elsewhere in the Internet. For example, the passive optical network (PON) that implements fiber to the home (FTTH) is a prime candidate for implementing the RAN backhaul.

8.2.2 Radio Access Network

First, each base station establishes the wireless channel for a subscriber's UE upon power-up or upon handover when the UE is active. Second, each base station establishes control plane connectivity between the UE and the corresponding mobile core control plane component, and forwards signaling traffic between the two. This signaling traffic enables UE authentication, registration, and mobility tracking. Third, for each active UE, the base station establishes one or more tunnels between the corresponding mobile core user plane components. Fourth, the base station forwards both control and user plane packets between the mobile core and the UE. Fifth, each base station coordinates UE handovers with neighboring base stations, using direct station-to-station links (see Figure 8.4). Sixth, the base stations coordinate wireless multi-point transmission to a UE from multiple base stations, which may or may not be part of a UE handover from one base station to another.

Scheduling is complex and multi-faceted, so the RAN as a whole (i.e., not just a single base station) not only supports handovers (an obvious requirement for mobility), but also link aggregation and load balancing.

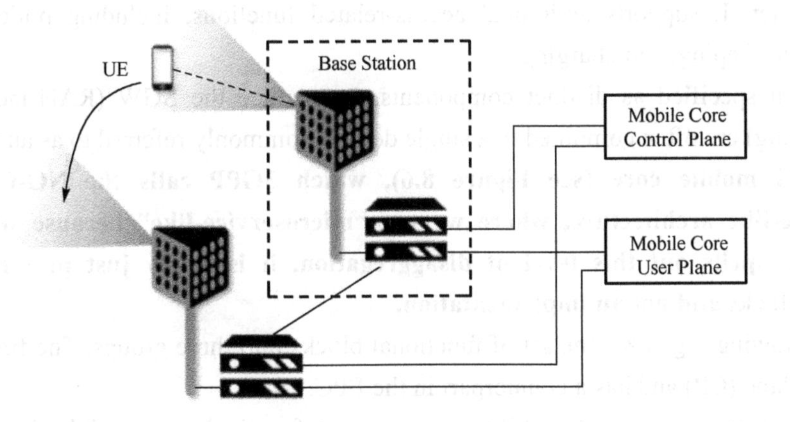

Figure 8.4 Base stations cooperate to implement UE handover

8.2.3 Mobile Core

The main function of the mobile core is to provide external packet data network (i.e., Internet) connectivity to mobile subscribers.

The 4G mobile core (see Figure 8.5), which 3GPP officially refers to as the evolved packet core (EPC), consists of five main components, the first three of which run in the control plane (CP) and the second two of which run in the user plane (UP).

- MME (mobility management entity): tracks and manages the movement of UEs throughout the RAN. This includes recording when the UE is not active.
- HSS (home subscriber server): a database that contains all subscriber-related information.
- PCRF (policy & charging rules function): tracks and manages policy rules and records billing data on subscriber traffic.
- SGW (serving gateway): forwards IP packets to and from the RAN; anchors the mobile core end of the bearer service to a (potentially mobile) UE, and so is involved in handovers from one base station to another.

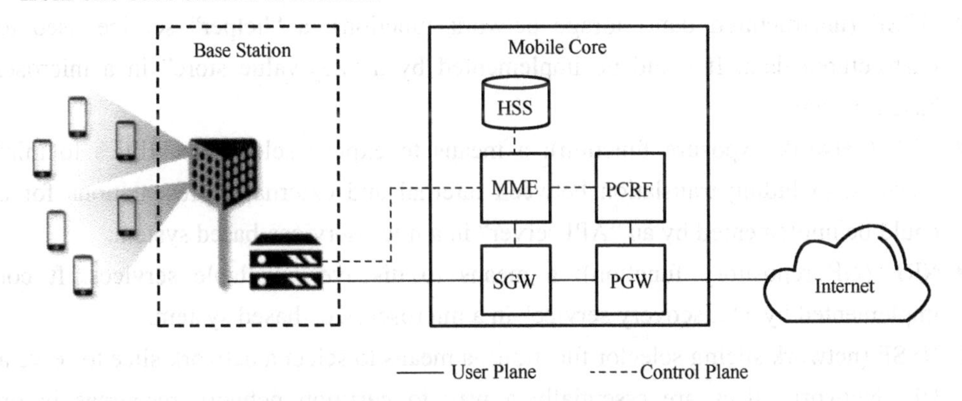

Figure 8.5 4G mobile core (evolved packet core)

- PGW (packet gateway): essentially an IP router, connecting the mobile core to the external

Internet. It supports additional access-related functions, including policy enforcement, traffic shaping, and charging.

Although specified as distinct components, in practice the SGW (RAN-facing) and PGW (Internet-facing) are often combined in a single device, commonly referred to as an S/PGW.

The 5G mobile core (see Figure 8.6), which 3GPP calls the NG-Core, adopts a microservice-like architecture, where we say "microservice-like" because while the 3GPP specification spells out this level of disaggregation, it is really just prescribing a set of functional blocks and not an implementation.

The following organizes the set of functional blocks into three groups. The first group runs in the control plane (CP) and has a counterpart in the EPC.

- AMF (core access and mobility management function): responsible for connection and reachability management, mobility management, access authentication and authorization, and location services. It manages the mobility-related aspects of the EPC's MME.
- SMF (session management function): manages each UE session, including IP address allocation, selection of associated UP function, control aspects of QoS, and control aspects of UP routing. It roughly corresponds to part of the EPC's MME and the control-related aspects of the EPC's PGW.
- PCF (policy control function): manages the policy rules that other CP functions then enforce. It roughly corresponds to the EPC's PCRF.
- UDM (unified data management): manages user identity, including the generation of authentication credentials. It includes part of the functionality in the EPC's HSS.
- AUSF (authentication server function): essentially an authentication server. It includes part of the functionality in the EPC's HSS.

The second group also runs in the control plane (CP) but does not have a direct counterpart in the EPC.

- SDSF (structured data storage network function): a "helper" service used to store structured data. It could be implemented by an "SQL database" in a microservices-based system.
- UDSF (unstructured data storage network function): a "helper" service used to store unstructured data. It could be implemented by a "key/value store" in a microservices-based system.
- NEF (network exposure function): a means to expose select capabilities to third-party services, including translation between internal and external representations for data. It could be implemented by an "API server" in a microservices-based system.
- NRF (NF repository function): a means to discover available services. It could be implemented by a "discovery service" in a microservices-based system.
- NSSF (network slicing selector function): a means to select a network slice to serve a given UE. Network slices are essentially a way to partition network resources in order to differentiate service given to different users. It is a key feature of 5G.

The third group includes the one component that runs in the user plane (UP).

- UPF (user plane function): Forwards traffic between RAN and the Internet, corresponding to the S/PGW combination in EPC. In addition to packet forwarding, it is responsible for policy enforcement, lawful intercept, traffic usage reporting, and QoS policing.

The mobile core can be conceptualized as a graph of services, so it is also called a service graph or service chain.

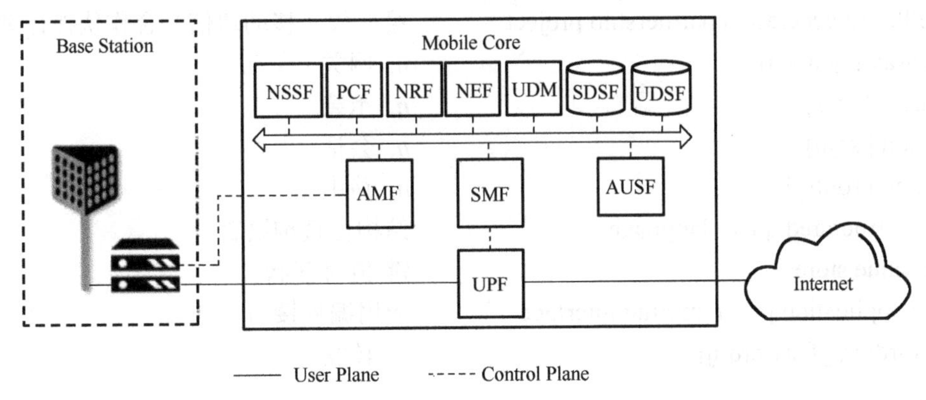

Figure 8.6 5G mobile core (next generation core)

Words and Expressions

tenfold ['tenfoʊld] *adv*. 十倍地
adj. 十倍的
rearchitect [rɪ'ɑːrkɪtekt] *v*. 重新架构；重新构建
leverage ['levərɪdʒ] *v*. 利用；为……融资
n. 影响力；手段；杠杆作用；优势
immersive [ɪ'mɜːrsɪv] *adj*. 沉浸式的；虚拟现实的；身临其境的
AR: augmented reality 增强现实
VR: virtual reality 虚拟现实
autonomous vehicle 无人驾驶汽车；自动驾驶车辆
use case 用例；用况；用例图；使用案例
home appliance 家电；民用；家用
swarm [swɔːrm] *n*. 一大群（昆虫）；一大批（向同方向移动的人）
vi. 成群地来回移动
specification [ˌspesɪfɪ'keɪʃn] *n*. 规格；规范；明细单；说明书
aspirational [ˌæspə'reɪʃənl] *adj*. 有抱负的；渴望成功的
radio access network (RAN) 无线电接入网
mobile core 移动核心网
quality of service (QoS) 服务质量
base station 基站；基地台
subscriber [səb'skraɪbər] *n*. 消费者；用户；订阅人；订购者
backhaul network 回程网络；回程网

passive optical network (PON)	无源光网络
fiber to the home (FTTH)	光纤到户；光纤入户
handover ['hændoʊvər]	*n.* 移交；切换
authentication [ɔːˌθentɪ'keɪʃn]	*n.* 身份验证；认证；鉴定
3GPP: 3rd generation partnership project	第三代（移动通信）合作伙伴计划
gateway ['ɡeɪtweɪ]	*n.* 网关
bearer ['berər]	*n.* 承载
session ['seʃn]	*n.* 会话
routing ['raʊtɪŋ]	*n.* 路由
SQL: structured query language	结构化查询语言
key/value store	键/值对存储
API: application programming interface	应用编程接口
forwarding ['fɔːrwərdɪŋ]	*n.* 转发

Notes

1. 5G represents a fundamental rearchitecting of the access network in a way that leverages several key technology trends and sets it on a path to enable much greater innovation.

5G 代表着接入网的根本性重组，利用了几个关键技术趋势，并实现了更大的创新。

2. Because these use cases will include everything from home appliances to industrial robots to self-driving cars, 5G won't just support humans accessing the Internet from their smartphones, but also swarms of autonomous devices working together on their behalf.

因为这些用例将包括从家用电器到工业机器人再到自动驾驶汽车的所有领域，5G 将不仅支持用户通过智能手机访问因特网，还将支持成群的自主设备实现自动联网协作。

3. Even though the word "core" is in its name, from an Internet perspective, the mobile core is still part of the access network, effectively providing a bridge between the RAN in some geographic area and the greater IP-based Internet.

尽管移动核心网名字当中包含了"核心"一词，但从因特网的角度来看，移动核心网仍然是接入网的一部分，它有效地把指定地理区域的无线接入网（RAN）连接到基于 IP 的因特网。

4. Scheduling is complex and multi-faceted, so the RAN as a whole (i.e., not just a single base station) not only supports handovers (an obvious requirement for mobility), but also link aggregation and load balancing.

调度是复杂和多方面的，因此 RAN 作为一个整体（不仅仅是一个基站）不仅支持基站间的用户移交（对移动性的明显要求），而且还支持链路聚合和负载平衡。

5. The 5G mobile core, which 3GPP calls the NG-Core, adopts a microservice-like architecture, where we say "microservice-like" because while the 3GPP specification spells out this level of disaggregation, it is really just prescribing a set of functional blocks and not an implementation.

5G 移动核心网（3GPP 称为 NG-Core）采用了类似微服务的架构，我们称之为"类似微服务"，因为尽管 3GPP 规范指明了服务分解的级别，但它实际上只是规定了一组功能模块，而不是具体实现。

Exercises

1. Match the terms (1)–(6) with the definitions A–F.

(1) location tracking	A. a major component of a wireless telecommunications system that connects individual devices to the mobile core
(2) RAN	B. a standard of the protocols for second generation digital cellular network
(3) base station	C. a transceiver with a radio antenna tower that is responsible for connecting mobile devices in a geographical area
(4) GSM	D. a process in which a connected cellular call is transferred from one cell site to another without disconnecting the session
(5) handover	E. the network of physical objects that are embedded with sensors, software, and other technologies for the purpose of connecting and exchanging data with other devices and systems over the internet
(6) IoT	F. following the progress of a moving vehicle or person

2. Translate into Chinese.

(1) The specific frequency bands that are licensed for mobile communication networks vary around the world, and are complicated by the fact that network operators often simultaneously support both old technologies and new technologies, each of which occupies a different frequency band.

(2) 2G used time division multiple access (TDMA) and 3G used code division multiple access (CDMA), while 4G and 5G are based on orthogonal frequency division multiplexing (OFDM), which is a method of multiplexing data on multiple orthogonal subcarrier frequencies, and each orthogonal subcarrier frequency is modulated independently.

(3) As standards bodies like 3GPP and infrastructure vendors like Nokia and Ericsson architected the 5G New Radio (5G-NR) core, they broke apart the monolithic EPC (evolved packet core) and implemented each function, so that it can run independently from each other on common, off-the-shelf server hardware. This allows the 5G core to become decentralized 5G nodes and very flexible.

3. Translate into English.

（1）低延迟和高可靠性对于工业自动化、自动驾驶及其他物联网应用至关重要。

（2）5G 可以在一种新的高频频谱（毫米波）上工作，其频率在 30 GHz 到 300 GHz 之间，而 4G LTE 的工作频率在 6 GHz 以下。

（3）移动应用程序，通常称为 app，是一种设计用于在移动设备（如智能手机或平板电脑）上运行的应用软件。

4. Read the following article and write a summary.

The most important software in any smartphone is its operating system (OS). An operating system manages the hardware and software resources of smartphones. Some platforms cover the entire range of the software stack. Others may only include the lower levels (typically the kernel and middleware layers) and rely on additional software platforms to provide a user interface framework.

Designed primarily for touch-screen mobile devices, Android, or Droid, technology is the operating system that most mobile telephones used as of Comscore's February 2014 numbers. Developed by Google, most people consider the Droid technology revolutionary because its open source technology allows people to write program codes and applications for the operating system, which means Android is evolving constantly. Smartphone users can decide whether to download the applications. Moreover, Android operating systems can run multiple applications, allowing users to be multitasking mavens. And get this: any hardware manufacturer is free to produce its own Android phone by using the operating system. In fact, many smartphone companies do just that. Android app's store has hundreds of thousands of apps.

Apple is always innovating, and iOS allows iPhone screens to be used simply and logically. Touted by Apple as the "world's most advance mobile operating system", iOS supports more access from sports scores to restaurant recommendations. As of publication, its version iOS7 allows for automatic updates and a control center that gives users access to their most used features. It also makes surfing the net easier with an overhaul to the Safari browser.

Reviewers say that Windows Phone 8 (WP8) is as simple to use as iOS and as easy to customize as Android. Its crowning achievement is LiveTiles, which are programmed squares that users can rearrange on their screen to easily access the information they want. WP8 works well with other Microsoft products, including Office and Exchange.

At first glance, experts say, Ubuntu 13.10 Touch might seem like an ordinary operating system, but it's not. Experts say Ubuntu Touch one of the easiest systems to use, allowing seamless navigation with multiple scopes. There are no hardware buttons on the bottom, for example. Instead, Ubuntu works from the edges. Developed by Canonical, the Ubuntu Touch allows users to unlock the phone from the right edge. You can swipe down from the top edge to access the phone's indicators, including date, time, messages and wireless networks. The phone also makes it easy for people to organize and share photos. Every shot is automatically uploaded to a personal cloud account, which makes it available on all devices, including iOS, Android and Windows.

5. Language study: Simple past versus present perfect

We use the simple past for events which took place in the past and are complete. Sometimes a day, date or time is given, e.g., *in the mid*-1990*s*, *during* 1895.

We use the present perfect for past events which have present results. This tense links the past with the present. Sometimes we use expressions such as *in the last twenty years*, *Since the mid*-1970*s*, *in the past few decades* to show the link. Using the present perfect shows that we think the past events are of current relevance.

Now look through the text for examples of the simple past and present perfect, which tense is used most often? Why?

Unit 9 Optical Communication

9.1 Optical Communication Systems

Optical communication systems use light to transmit information from one place to another. Light is a type of electromagnetic radiation like radio waves. Today, infrared light is being used increasingly as the carrier for information in a communication system. The transmission medium is either free space as with radio waves or a special light-carrying cable or waveguide known as fiber-optic cable. Both media are used, although the fiber-optic cable is far more practical and more widely used. Because the frequency of light is extremely high, it can accommodate very high rates of data transmission with excellent reliability.

9.1.1 Light Wave Communication in Free Space

Figure 9.1 shows the elements of an optical communication system using free space. **It consists of a light source modulated by the signal to be transmitted, a modulator, a photodetector to pick up the light and convert it back into an electric signal, an amplifier, and a demodulator to recover the original information signal.**

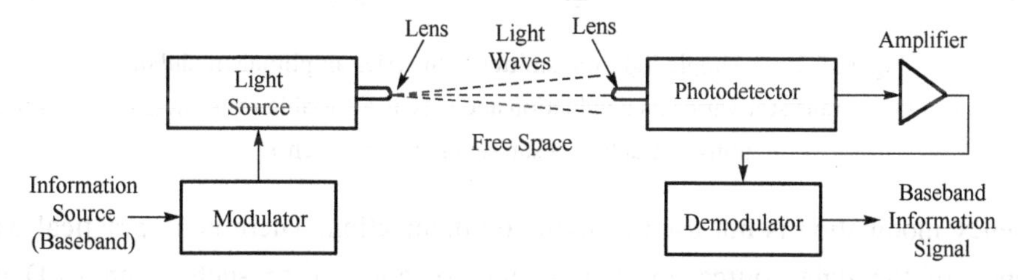

Figure 9.1 Light wave communication in free space

1. Light Source

A transmitter is a light source. Other common light sources are light emitting diodes (LEDs) and lasers. These sources can follow electric signal changes as fast as 100 GHz or more. **Lasers generate monochromatic, or single-frequency, light that is fully coherent; i.e., all the light waves are lined up in sync with one another and as a result produce a very narrow and intense light beam.**

2. Modulator

A modulator is used to vary the intensity of the light beam in accordance with the modulating baseband signal. Amplitude modulation, also referred to as intensity modulation, is used where the information or intelligence signal controls the brightness of the light. Analog signals vary the brightness continuously over a specified range. This technique is used in some cable TV systems. Digital signals simply turn the light beam off and on at the data rate. Digital modulation is usually NRZ-formatted binary data that turns a laser on or off to produce on-off keying (OOK) or amplitude-shift keying (ASK).

A modulator for analog signals can be a power transistor in series with the light source and its DC power supply (see Figure 9.2). The voice, video, or other information signal is applied to an amplifier that drives the class A modulator transistor. **As the analog signal goes positive, the base drive on the transistor increases, turning the transistor on harder and decreasing its collector-to-emitter voltage. This applies more of the supply voltage to the light source, making it brighter.** A negative-going or decreasing signal amplitude drives the transistor toward cutoff, thereby reducing its collector current and increasing the voltage drop across the transistor. This decreases the voltage to the light source.

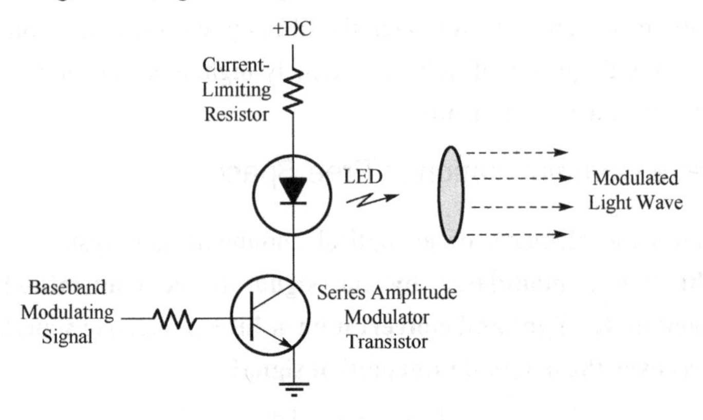

Figure 9.2 A simple light transmitter with series amplitude modulator

(Analog signals: transistor varies its conduction and acts as a variable resistance. Pulse signals: transistor acts as a saturated on/off switch.)

Frequency modulation is not used in light communication. There is no practical way to vary the frequency of the light source, even a monochromatic source such as an LED or a laser. Amplitude modulation is used with analog signals, but otherwise most light wave communication is accomplished by pulse modulation.

Pulse modulation refers to turning the light source off and on in accordance with some serial binary signal. The most common type of pulse modulation is pulse-code modulation (PCM), which is serial binary data organized into bytes or longer words. NRZ, RZ, and Manchester formats are common.

3. Receiver

The modulated light wave is picked up by a photodetector. This is usually a photodiode or transistor whose conduction is varied by the light. The small signal is amplified and then demodulated to recover the originally transmitted signal. Digital processing may be necessary. For example, if the original signal is voice that was digitized by an A/D converter before being transmitted as a PCM signal, then a D/A converter will be needed at the receiver to recover the voice signal.

Communication by light beam in free space is impractical over very long distances because of the great attenuation of the light due to atmospheric effects. **Fog, haze, smog, rain, snow, and other conditions absorb, reflect, refract, and disperse the light, greatly attenuating it and thereby limiting the transmission distance.** Artificial light beams used to carry information are obliterated during daylight hours by the sun. And they can be interfered with by any other light source that points in the direction of the receiver. Distances are normally limited to several hundred feet with low-power LEDs and lasers. If high-power lasers are used, a distance of several miles may be possible. Light beam communication has become far more practical with the invention of the laser, a special high-intensity, single-frequency light source. It produces a very narrow beam of brilliant light of a specific wavelength (color). Because of its great intensity, a laser beam can penetrate atmospheric obstacles better than other types of light can, thereby making light beam communication more reliable over longer distances. When the laser is used, the light beam is so narrow that the transmitter and receiver must be perfectly aligned with each other for communication to occur. Although this causes initial installation alignment problems, it also helps to eliminate external interfering light sources.

9.1.2 Fiber-Optic Communication System

Instead of free space, some type of light-carrying cable can be used. Today, fiber-optic cables have been highly refined. Cables many miles long can be constructed and then interconnected for the purpose of transmitting information. Thanks to these fiber-optic cables, a new transmission medium is now available. Its great advantage is its immense information-carrying capacity (wide bandwidth). Whereas hundreds of telephone conversations may be transmitted simultaneously at microwave frequencies, many thousands of signals can be carried on a light beam through a fiber-optic cable. When multiplexing techniques similar to those in telephone and radio systems are used, fiber-optic communication systems have an almost limitless capacity for information transfer.

The components of a typical fiber-optic communication system are shown in Figure 9.3. The information signal to be transmitted may be voice, video, or computer data. The first step is to convert the information to a form compatible with the communication medium, usually by converting continuous analog signals such as voice and video (TV) signals to a series of digital pulses. An A/D converter is used for this purpose.

These digital pulses are then used to flash a powerful light source off and on very rapidly. In simple low-cost systems that transmit over short distances, the light source is usually a light-emitting diode that emits a low-intensity infrared light beam. Infrared beams such as those used in TV remote controls are also used in transmission.

The light beam pulses are then fed into a fiber-optic cable, which can transmit them over long distances. At the receiving end, a light-sensitive device known as a photocell, or light detector, is used to detect the light pulses. It converts the light pulses to an electric signal. The electrical pulses are amplified and reshaped back into digital form. They are fed to a decoder, such as a D/A converter, where the original voice or video is recovered.

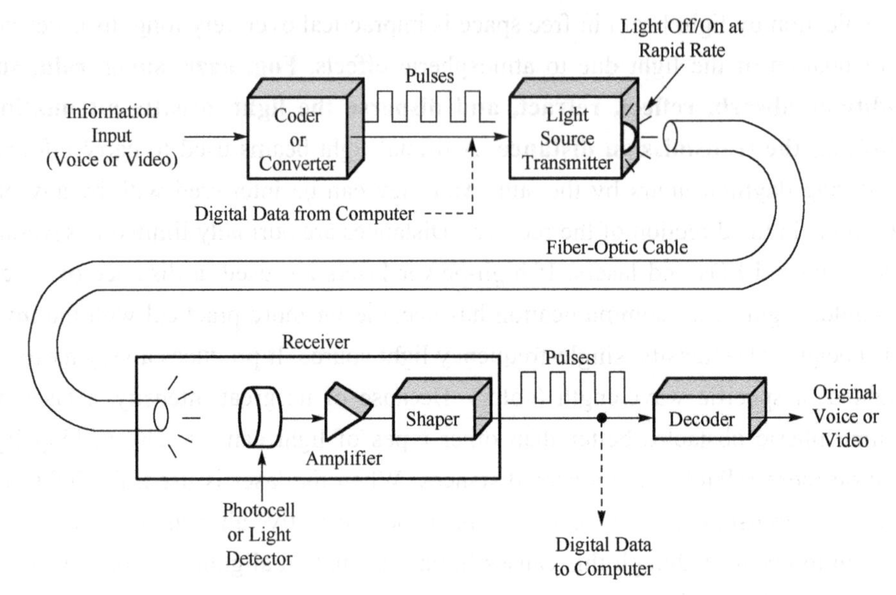

Figure 9.3 The components of a typical fiber-optic communication system

In very long transmission systems, repeater units must be used along the way. Since the light is greatly attenuated when it travels over long runs of cable, at some point it may be too weak to be received reliably. To overcome this problem, special relay stations are used to pick up the light beam, convert it to electrical pulses that are amplified, and then retransmit on another light beam.

9.1.3 Applications of Fiber Optics

Fiber-optic communication systems are being used more and more each day. Their primary use is in long-distance telephone systems and cable TV systems. Fiber-optic networks also form the core or backbone of the Internet. **Fiber-optic cables are no more expensive or complex to install than standard electrical cables, yet their information-carrying capacity is many times greater.**

Fiber-optic communication systems are being used in other applications. For example, they are being used to interconnect computers in networks within a large building, to carry control signals in airplanes and in ships, and in TV systems because of the wide bandwidth. In all cases, the fiber-optic cables replace conventional coaxial or twisted-pair cables. Table 9.1 lists some of the applications in which fiber-optic cables are being used.

Table 9.1 Applications of fiber-optic cables

1. TV studio to transmitter interconnection eliminating a microwave radio link.	9. Nuclear plant instrumentation.
2. Closed-circuit TV systems used in buildings for security.	10. College campus communication.
3. Secure communication systems at military bases.	11. Utilities (electric, gas, and so on) station communication.

Continued

4. Computer networks, wide area, metro and local area.	12. Cable TV systems, replacing coaxial cable.
	13. The Internet
5. Shipboard communication.	14. Backhaul for cellular base stations.
6. Aircraft communication/controls.	15. Distributed antenna systems (DAS).
7. Interconnection of measuring and monitoring instruments in plants and laboratories.	16. Connections between servers, routers, and switches in data centers.
8. Data acquisition and control signal communication in industrial process control systems.	17. Connection between the cellular base station and the remote radio head mounted at the antenna.

Words and Expressions

waveguide ['weɪvgaɪd] *n.* 波导
accommodate [ə'kɑːmədeɪt] *v.* 容纳；为……提供空间；顺应，适应（新情况）
photodetector [ˌfoʊtoʊdɪ'tektər] *n.* 光电探测器；光检测器；光度感应器
laser ['leɪzər] *n.* 激光；激光器
monochromatic [ˌmɑːnəkroʊ'mætɪk] *adj.* 单色的；单色光的；单频的；单能的
coherent [koʊ'hɪrənt] *adj.* 相干的；连贯的；一致的；合乎逻辑的
in sync (synchronization) 同步；一致；协调
intensity modulation 强度调制
NRZ: non-return to zero 不归零；不归零制
on-off keying (OOK) 开关键控；通断键控
attenuation [əˌtenjʊ'eɪʃn] *n.* 衰减；变薄；稀释
reflect [rɪ'flekt] *v.* 反射（声、光、热等）；反映；认真思考
refract [rɪ'frækt] *v.* 使（光波、声波、能量波等）折射；使产生折射
disperse [dɪ'spɜːrs] *v.* 散射；（使）分散，散开；驱散；疏散；传播
obliterate [ə'blɪtəreɪt] *v.* 毁掉；覆盖；清除
refine [rɪ'faɪn] *v.* 改进；改善；精炼；提纯；去除杂质
immense [ɪ'mens] *adj.* 极大的；巨大的
photocell ['foʊtoʊsel] *n.* 光电池；光电管；光电元件
repeater [rɪ'piːtər] *n.* 网络中继器；转发器；无线中继
relay station 中继站
backbone ['bækboʊn] *n.* 骨干；网络骨干；主干网
shipboard communication 船上通信

Notes

1. It consists of a light source modulated by the signal to be transmitted, a modulator, a photodetector to pick up the light and convert it back into an electric signal, an amplifier, and a

demodulator to recover the original information signal.

它（光通信系统）由被待传输信号调制的光源、调制器、用于拾取光信号并将其转换回电信号的光电探测器、放大器和用于恢复原始信息信号的解调器组成。

2. Lasers generate monochromatic, or single-frequency, light that is fully coherent; i.e., all the light waves are lined up in sync with one another and as a result produce a very narrow and intense light beam.

激光产生完全相干的单色光或单频光，即所有光波彼此同步排列，产生高强度的窄光束。

3. As the analog signal goes positive, the base drive on the transistor increases, turning the transistor on harder and decreasing its collector-to-emitter voltage. This applies more of the supply voltage to the light source, making it brighter.

当模拟信号变为正时，晶体管的基极电压增大，使晶体管导通，降低了集电极到发射极间的电压。这会导致电源施加在光源上的电压变大，从而使其更亮。

4. Fog, haze, smog, rain, snow, and other conditions absorb, reflect, refract, and disperse the light, greatly attenuating it and thereby limiting the transmission distance.

雾、霾、烟雾、雨、雪和其他条件会吸收、反射、折射和散射光线，大大衰减光强，从而限制传输距离。

5. Fiber-optic cables are no more expensive or complex to install than standard electrical cables, yet their information-carrying capacity is many times greater.

光缆的安装并不比标准电缆更昂贵或更复杂，但其信息承载能力要强很多倍。

9.2 Passive Optical Networks

The primary applications for fiber-optic networks are in wide-area networks such as long-distance telephone service and the Internet backbone. As speeds have increased and prices have declined, fiber-optic technology has been adopted into metropolitan area networks (MANs), storage area networks (SANs), and local area networks.

Passive optical network (PON) is a type of metropolitan area network technology. This technology is also referred to as fiber to the home (FTTH). Similar terms are fiber to the premises or fiber to the curb, designated as FTTP or FTTC. The term FTTx is used, where x represents the destination. PONs are already widely used in many countries.

9.2.1 The PON Concept

Most optical networking uses active components to perform optical-to-electrical and electrical-to-optical (OEO) conversions during transmission and reception. These conversions are expensive because each requires a pair of transceivers and the related power supply. Over long distances, usually 10 to 40 km or more, repeaters or optical amplifiers are necessary to overcome the attenuation, restore signal strength, and reshape the signal. These OEO repeaters and amplifiers are a nuisance as well as expensive and power-hungry. One solution to this problem is to use a passive optical network. The term passive implies no OEO repeaters, amplifiers, or any other device that uses power. Instead, the transmitter sends the signal out over the network cable, and a receiver at the destination picks it up. There are no intervening repeaters or amplifiers. Only passive optical devices such as splitters and combiners are used. **By using low-attenuation fiber-optic cables, powerful lasers, and sensitive receivers, it is possible to achieve distances of up to about 20 km without intervening active equipment. This makes PONs ideal for metropolitan area networks.**

The PON method has been adopted by telecommunications carriers as the medium of choice for their very high-speed broadband Internet connections to consumers and businesses. These metropolitan area networks cover a portion of a city or a similar-size area. PONs will be competitive with cable TV and DSL connections but faster than both. Using optical techniques, the consumer can have an Internet connection speed of 100 Mbps to 1 Gbps or higher. This is much faster than the typical 1- to 20-Mbps cable TV and DSL connections. And PONs make digital TV distribution more practical. Furthermore, they provide the extra bandwidth to carry Internet phone calls (VoIP).

9.2.2 PON Technologies

There are several different types of PONs and standards. The earliest standard, called APON, was based upon ATM packets and featured speeds of 155.52 and 622.08 Mbps. A more advanced version, called BPON, features a data rate up to 1.25 Gbps. The most recent version is a superset of BPON, called GPON, for Gigabit PON. It provides download speeds up to 2.5 Gbps and upload speeds up to 1.25 Gbps. One of the key features of GPON is that it uses encapsulation, a technique that makes it protocol-agnostic. Any type of data protocol including TDM (such as T1 or SONET) or Ethernet can be transmitted.

Figure 9.4 shows a basic block diagram of a BPON/GPON network. The carrier central office (CO) serves as the Internet service provider (ISP), TV supplier or telephone carrier as the case may be. The equipment at the carrier central office is referred to as the optical line terminal (OLT). It develops a signal on 1,490 nm for transmission (download) to the remote terminals. This signal carries all Internet data and any voice signals as in voice over IP (VoIP). If TV is transmitted, it is modulated on to a 1,550 nm laser. The 1,550 and 1,490 nm outputs are mixed or added in a passive combiner to create a coarse wavelength division multiplexed (CWDM) signal. This master signal is then sent to a passive splitter that divides the signal into four equal power levels for transmission over the first part of the network. In BPON the data rate is 622 Mbps or 1.25 Gbps but with GPON it is 2.5 Gbps.

Additional splitters are used along the way to further split the signals for distribution to multiple homes. Splitters are available to divide the power by 2:1, 4:1, 8:1, 16:1, 32:1, 64:1, and 128:1. One OLT can transmit to up to 64 destinations up to about 20 km. The upper limit may be 16 devices depending upon the ranges involved. Just remember that each time the signal is split, its power is decreased by the split ratio. The power out of each port on a 4:1 splitter is only one-fourth of the input power. Splitters are passive demultiplexers (DEMUX).

Note also, because the splitters are optical devices made of glass or silicon, they are bidirectional. They also serve as combiners in the opposite direction. In this capacity, they serve as multiplexers (MUX).

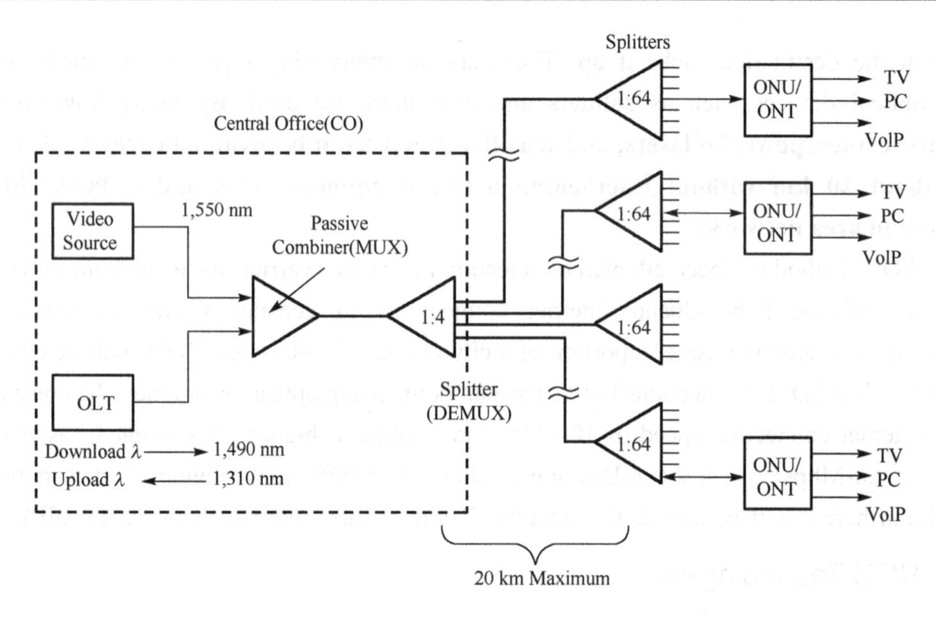

Figure 9.4 A passive optical network (PON) used as a high-speed Internet connection and for TV distribution in fiber to the home systems

At the receiving end, each subscriber has an optical networking unit (ONU) or optical networking terminal (ONT). These boxes connect to your PC, TV set, and/or VoIP telephone. The ONU/ONT is a two-way device, meaning that it can transmit as well as receive. In VoIP or Internet applications, the subscriber needs to transmit voice and dialing data back to the OLT. This is done over a separate 1,310 nm laser using the same fiber-optic path. The splitters are bidirectional and also work as combiners or multiplexers. The upload speed is 155 Mbps in BPON and 1.25 Gbps in GPON.

The latest version of GPON is 10-Gigabit GPON, called XGPON or 10G-PON. **It can achieve 10 Gbps downstream on 1,577 nm and 2.5 Gbps upstream on 1,270 nm. These different wavelengths allow it to coexist with existing GPON services.** The split ratio is 128:1. Most new installations will use XGPON. Data formatting is the same as for standard GPON.

Another widely used standard is EPON, or Ethernet PON. While BPON and GPON have been adopted as the North American PON standard, EPON is the de facto PON standard in Japan, South Korea, and some European nations. EPON is one part of the popular IEEE Ethernet standard and is designated 802.3ah. You will also hear it referred to as Ethernet in the first mile (EFM). The term first mile refers to that distance between the subscriber and any central office. Sometimes it is also called the last mile.

The topology of EPON is similar to that of GPON, but the downstream and upstream data rates are symmetric at 1.25 Gbps. The downstream is on 1,490 nm while the upstream is on 1,310 nm. Standard Ethernet packets are transmitted with a data payload to 1,518 bytes. A real plus for EPON is that because it is Ethernet, it is fully compatible with any other Ethernet LAN.

The latest version of EPON is designated 802.3av. It can achieve 10 Gbps upstream and downstream on wavelengths of 1,575–1,580 nm and 1,260–1,280 nm respectively. A 1 Gbps upstream version is available.

While PONs provide the ultimate in bandwidth and data rate for home broadband connections, they are expensive. Carriers must invest in a huge infrastructure that requires rewiring the area served, again. **Fiber-optic cables are mostly laid underground, but that process is expensive because right-of-way must be bought and trenches dug. Cables can be carried overhead on existing poles at less expense, but the effect is less pleasing.**

Words and Expressions

metropolitan area network (MAN)	城域网
storage area network (SAN)	存储区域网
transceiver [træn'siːvər]	*n.* 无线电收发机；收发器
nuisance ['nuːsns]	*n.* 麻烦事；讨厌的人（或东西）；损害
intervening [ˌɪntər'viːnɪŋ]	*adj.* 介于中间的；发生于其间的
splitter ['splɪtə]	*n.* 光分路器；分光器
combiner [kəm'bɑɪnə]	*n.* 合路器；合波器；合成器；复合器；组合器
DSL: digital subscriber loop	数字用户环路
distribution [ˌdɪstrɪ'bjuːʃn]	*n.* 分布；分配；分发
ATM: asynchronous transfer mode	异步传输模式
BPON: broadband passive optical network	宽带无源光网络
GPON: gigabit PON	吉比特无源光网络；千兆无源光网络
encapsulation [inˌkæpsju'leiʃn]	*n.* 封装；封闭；胶囊化作用；包封
SONET: synchronous optical network	同步光纤网络；同步光网络
central office (CO)	电话总局；网络中心局；端局
internet service provider (ISP)	因特网服务提供者
telephone carrier	电话运营商；电话载波
optical line terminal (OLT)	光线路终端
voice over IP (VoIP)	IP 电话；互联网电话
optical networking terminal (ONT)	光网络终端
Ethernet passive optical network (EPON)	以太网无源光网络
de facto [ˌdeɪ 'fæktoʊ]	*adj.* 事实上的；实际上的

Notes

1. By using low-attenuation fiber-optic cables, powerful lasers, and sensitive receivers, it is possible to achieve distances of up to about 20 km without intervening active equipment. This makes PONs ideal for metropolitan area networks.

通过使用低损耗光缆、高能激光器和灵敏的接收器，可以在不使用有源设备的情况下达到约 20 公里的传播距离。这使得无源光网络（PON）成为城域网的理想选择。

2. It can achieve 10 Gbps downstream on 1,577 nm and 2.5 Gbps upstream on 1,270 nm. These different wavelengths allow it to coexist with existing GPON services.

它（10G-PON）的下载带宽达到 10 Gbps，波长采用 1 577 nm，上传带宽达到 2.5 Gbps，

波长采用 1 270 nm。使用不同的波长允许它与现有的 GPON 服务共存。

3. Fiber-optic cables are mostly laid underground, but that process is expensive because right-of-way must be bought and trenches dug. Cables can be carried overhead on existing poles at less expense, but the effect is less pleasing.

光缆大多铺设在地下，但这一过程成本高昂，因为必须购买路权并挖沟。而电缆可以驾设在现有电线杆上，费用较低，但效果不太令人满意。

Exercises

1. Match the terms (1)-(6) with the definitions A-F.

(1) DEMUX	A. a device that emits light through a process of optical amplification based on the stimulated emission of electromagnetic radiation
(2) photodetector	B. the delivery of a communication signal over optical fiber from the operator's switching equipment all the way to a home or business
(3) laser	C. a device that separates the two or more signals that have been combined into a common signal
(4) attenuation	D. an electro-optic device that transforms light energy into electrical energy
(5) FTTH	E. a device inserted at intervals along a circuit that detects a weak signal, amplifies it, cleans it up, and retransmits it in optical form
(6) repeater	F. the decrease in signal strength along a fiber-optic waveguide caused by absorption and scattering, normally measured in decibels

2. Translate into Chinese.

(1) The International Telecommunications Union (ITU) has designated six spectral bands for use in intermediate-range and long-distance optical fiber communications within the 1,260- to 1,675-nm region.

(2) An especially important feature of optical fibers relates to the fact that they consist of dielectric materials, which means they do not conduct electricity and will be immune to the electromagnetic interference effects.

(3) Because the optical signal is usually weakened and distorted when it is sent from the end of the optical fiber, the photodetector must meet strict performance requirements, such as high sensitivity, low noise and fast response.

3. Translate into English.

（1）无源光网络（PON）使用光缆和无源组件（如分路器和合路器），不需要有源元件（如放大器或中继器）。

（2）在无线媒体中，信号强度的衰减与传输距离和大气成分的函数关系更复杂。

（3）采用部分相干光作为激光光源，可以减小系统误码率，提高系统性能。

4. Read the following article and write a summary.

The end of 2020 saw many satellite launches, paving the way for a decade in which the satellite race will intensify. Many companies are joining forces with space agencies to launch satellites into orbit to meet various challenges, such as worldwide internet access or Earth observation. Data transmission needs to be reliable, so inter-satellite and satellite-ground station links must be ensured at all times. This satellite race therefore generates a technological race with major challenges such as robustness of communication links and very high-throughput data transmission.

Today, inter-satellite communication is facing a new technological challenge. Many satellites already in orbit communicate mainly by radio frequency electromagnetic waves. However, several companies planning to launch major satellite constellations — such as Space X or Telesat — are now turning to free space optical communication to transmit their data since it provides many advantages. Optical communications provide higher data throughput rates and are essential for handling the continuous increase in data exchange. Another advantage of optical links is the narrowness of the light beam, which makes it more complicated to intercept and helps ensure information security. Finally, the incoherence of the sources prevents the various communication systems in the area from interfering with each other, which is essential for large constellations.

However, this technology seems complicated to implement, especially for large constellations. Space X had planned to launch all of the satellites in its Starlink constellation equipped with laser terminals, but it turns out that the first satellites sent into space communicate by radio frequency. The primary inherent difficulty of optical links is the relative positioning of two satellites. Optical links are point to point links, so the satellites must locate each other in space and keep track of each other so as not to interrupt the data exchange. Also, to prevent data loss the inter-satellite link must be precise and the pointing from one satellite to the other as fine as possible.

Two types of point-to-point alignment are used successively in the optical terminal. The first is coarse. It allows the beam to be directed to an accuracy of a few degrees and with an actuation speed of the order of several tens of hertz. A second, more precise and faster pointing system is required to ensure injection into an optical fiber or towards a photodiode.

5. Language study: Reduced relative clauses

One way of adding extra information to an explanation, or any other text, is to use relative clauses. For example:

(1) A transmitter is a light source.

(2) The light source follows electric signal changes.

(1)+(2) A transmitter is a light source **which follows electric signal changes.**

We can make this sentence shorter by omitting *which* and using an *-ing* clause:

A transmitter is a light source ***following electric signal changes.***

Study this example:

(1) The microprocessor converts the information into signals.

(2) The signals are called analog tones.

(3) The signals are suitable for telephone transmission.

(1)+(2)+(3) The microprocessor converts the information into signals, which are called analog tones, which are suitable for telephone transmission.

We can make this sentence shorter by omitting *which* + *to be*:

The microprocessor converts the information into signals, ***called analog tones, suitable for telephone transmission.***

The following paragraph is about transmission lines, please shorten it by reducing the relative clauses where possible.

The lines which connect telephones within a building are the simplest type of transmission line, which consists of parallel wires. Those which link telephones to a local exchange may be twisted pairs, although these are being replaced. Coaxial cable, which is formed from a copper core which is surrounded by a copper braid, is used to carry a large number of signals over long distances. The cables which provide connections between telephone exchanges are often coaxial. Waveguides, which are made of copper, are used to carry microwave signals between dish aerials and receivers. They are suitable for frequencies which are between 1 GHz and 300 GHz. Optical fibers, which are made from very pure silica fiber, are the form of transmission line which is most often used these days.

Unit 10 Telecommunication Networks

10.1 A Telephone Network

Telecommunications are today widely understood to mean the electrical means of communicating over a distance. Alexander Graham Bell invented the telephone in 1876. Remarkably soon afterwards, the world's first telephone exchange was opened in 1878 in New Haven, Connecticut, United States. Since then, telephony has become the ubiquitous means of communicating for mankind.

Of course, telecommunications are now global and include not only telephony over fixed and mobile networks, but also data — the World Wide Web, e-mail, broadcast TV, video, social networking, etc. **In setting out how all of this works, it is helpful to understand how the basic telephone networks are structured since they form the basis of all today's telecommunication networks.**

10.1.1 Basic Telephony

Figure 10.1 illustrates a basic simple one-way telephone circuit between two people. The set-up comprises of a microphone associated with the speaker, which is connected via an electrical circuit with a receiver at the remote end associated with the listener. A battery provides power for the operation of the microphone and receiver. During talking, variations in air pressure are generated by the vocal tract of the speaker. These variations in air pressure, known as sound waves, travel from the speaker to the microphone, which converts them into an electrical signal varying in sympathy with the pattern of the sound waves.

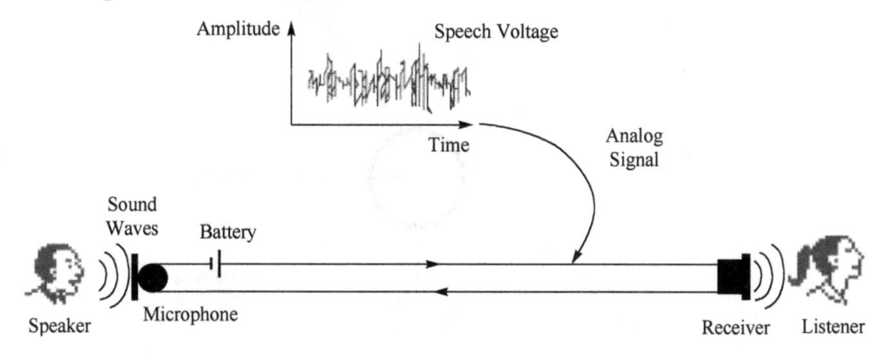

Figure 10.1 Simple one-way telephone circuit between two people

At the receiving end of the circuit, the analog electrical signal energizes the receiver (i.e.,

an earpiece), generating a set of sound waves, which are an approximate reproduction of the sound of the speaker.

Obviously for conversation to be possible, it is necessary to have transmission in both directions, and a telephone would need to include a mechanism for the caller to dial and indicate to the recipient that he wished to speak: a touch-tone keypad and a bell.

We can now extend this basic two-person scenario to the more general case of several people with telephones wishing to be able to talk to each other. For example, the logical extension to a four-telephone scenario is shown in Figure 10.2.

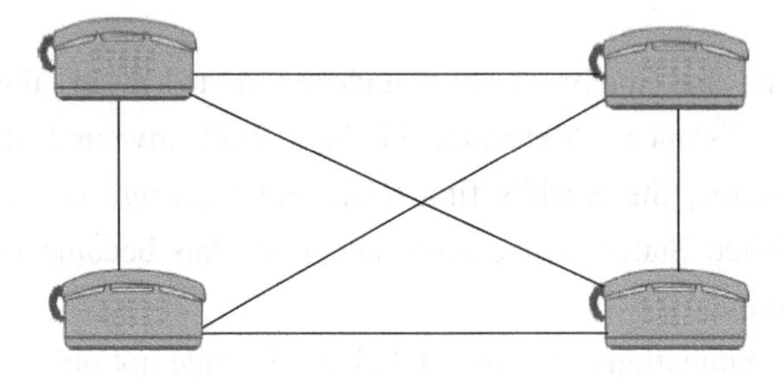

Figure 10.2 Direct interconnection of several telephones. For four telephones, number of links required = 6 Generally: for *n* telephones, number of links required = $n(n-1)/2$

Whilst this arrangement is quite practical for networks of just a few telephones, it does not scale up well. In general, the number of links to fully interconnect *n* telephones is given by $n(n-1)/2$. As the number of telephones becomes large, the number of directly connected links approaches $n^2/2$. Clearly, providing the necessary 5,000 direct links in a system serving just 100 telephones would not be an economical or practical design! (In addition, the complexity of the selection mechanism in each telephone would increase for it to be capable of switching 1-out-of-99 lines.) The solution to the scaling problem is to introduce a central hub (commonly called an exchange or central office) onto which each telephone is linked directly, and which can provide connectivity between any two telephone lines, on demand (see Figure 10.3).

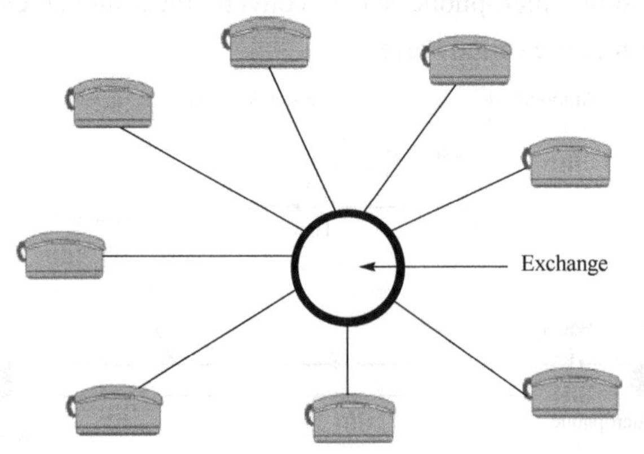

Figure 10.3 Interconnection: use of an exchange. For *n* telephones, number of links to each exchange = *n*

When telephone exchanges were first introduced, the method of connecting two telephone lines together was through a human operator. Today, telephone exchanges are fully automatic, the exchange control is provided by computers with the procedures captured in call-control software. However, there are still occasions when a human intervention is required, for example, providing various forms of assistance and emergency calls, and special automanual exchanges with operators provide such services.

10.1.2 Telephone System

When we refer to the telephone system, we are talking about the organizations and facilities involved in connecting your telephone to the called telephone regardless of where it might be in the world. The telephone system is called the public switched telephone network (PSTN). You will sometimes hear the telephone system referred to as the plain old telephone service (POTS).

1. Subscriber Interface

Most telephones are connected to a local central office by way of the two-line, twisted-pair local loop cable. The central office contains all the equipment that operates the telephone and connects it to the telephone system that makes the connection to any other telephone. Each telephone connected to the central office is provided with a group of basic circuits that power the telephone and provide all the basic functions, such as ringing, dial tone, and dialing supervision. These circuits are collectively referred to as the subscriber interface or the subscriber line interface circuits (SLIC). In older central office systems, the subscriber interface circuits used discrete components. Today, most functions of the subscriber line interface are implemented by one or perhaps two integrated circuits plus supporting equipment. The subscriber line interface is also referred to as the line side interface.

The SLIC provides seven basic functions generally referred to as BORSCHT (representing the first letters of the functions battery feed, overvoltage protection, ringing, supervision, codec, hybrid, and test). A general block diagram of the subscriber line interface and BORSCHT functions is given in Figure 10.4.

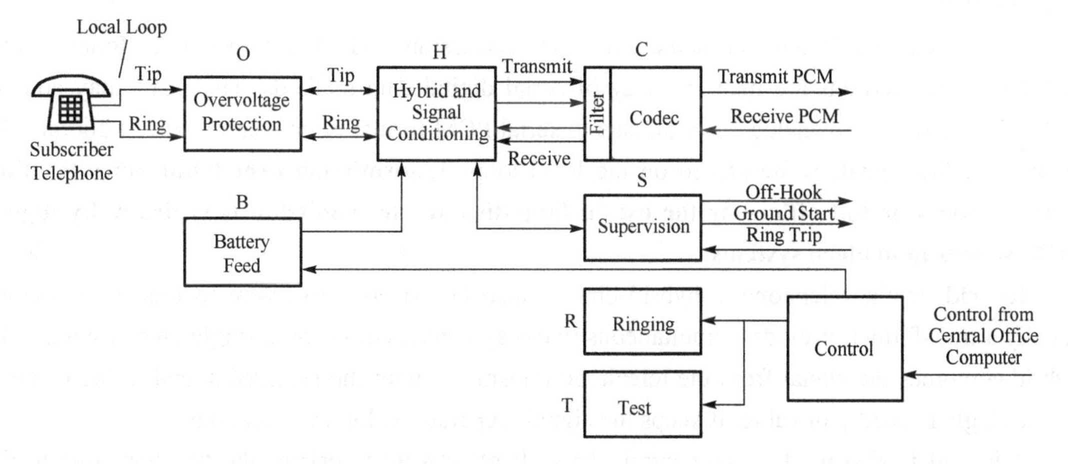

Figure 10.4 BORSCHT functions in the subscriber line interface at the central office

Battery feed. The subscriber line interface at the central office must provide a DC voltage to the subscriber to operate the telephone. In the United States, this is typically –48 V DC with respect to ground. The actual voltage can be anything between approximately –20 and –80 V when the telephone is on the hook, i.e., disconnected. The voltage at the telephone drops to approximately 6 V when the telephone is taken off the hook. The large difference between the on-hook and off-hook voltages has to do with the large voltage drop that occurs across the components in the telephone and the long local loop cable.

Overvoltage protection. The circuits and components that protect the subscriber line interface circuits from electrical damage are referred to collectively as overvoltage protection. The telephone lines are vulnerable to many types of electrical problems. Lightning is by far the worst threat, although other hazards exist, including accidental connection to an electric power line or some type of misconnection that would occur during installation. Induced disturbances from other sources of noise can also cause problems. Overvoltage protection ensures reliable telephone operation even under such conditions.

Ringing. When a specific telephone is receiving a call, the telephone local office must provide a ringing signal. This is commonly a 90-Vrms AC signal at approximately 20 Hz. The SLIC must connect the ringing signal to the local loop when a call is received. This is usually done by closing relay contacts that connect the ringing signal to the line. The SLIC must also detect when the telephone is picked up (off-hook) so that the ringing signal can be disconnected.

Supervision. Supervision refers to a group of functions within the subscriber line interface that monitor local loop conditions and provide various services. For example, the supervision circuits in the SLIC detect when a telephone is picked up to initiate a new call. A sensing circuit recognizes the off-hook condition and signals circuits within the SLIC to connect a dial tone. The caller then dials the desired number, which causes interconnection through the telephone system.

The supervision circuits continuously monitor the line during the telephone call. The circuits sense when the call is terminated and provide the connection of a busy signal if the called number is not available.

Codec. Coding is another name for A/D conversion and D/A conversion. Today, many telephone transmissions are made by way of serial digital data methods. The SLIC may contain codec that converts the analog voice signals to serial PCM format or converts received digital calls back to analog signals to be placed on the local loop. **Transmission over trunk lines to other central offices or toll offices or for use in long-distance transmission is typically by digital PCM signals in modern systems.**

Hybrid. In the telephone, a hybrid circuit (also known as a two-wire to four-wire circuit), usually a transformer, provides simultaneous two-way conversations on a single pair of wires. The hybrid combines the signal from the telephone transmitter with the received signal to the receiver on the single twisted-pair cable. It keeps the signals separate within the telephone.

A hybrid is also used at the central office. It effectively translates the two-wire line to the subscriber back into four lines, two each for the transmitted and received signals. The hybrid provides separate transmit and receive signals. Although a single pair of lines is used in the local loop to the subscriber, all other connections to the telephone system treat the transmitted and received signals separately and have independent circuits for dealing with them along the way.

Test. To check the status and quality of subscriber lines, the phone company often puts special test tones on the local loop and receives resulting tones in return. These can give information about the overall performance of the local loop. The SLIC provides a way to connect the test signals to the local loop and to receive the resulting signals for measurement.

2. Telephone Hierarchy

Whenever you make a telephone call, your voice is connected through your local exchange to the telephone system. From there it passes through at least one other local exchange, which is connected to the telephone you are calling. Several other facilities may provide switching, multiplexing, and other services required to transmit your voice. The organization of the telephone hierarchy in the United States is shown in Figure 10.5.

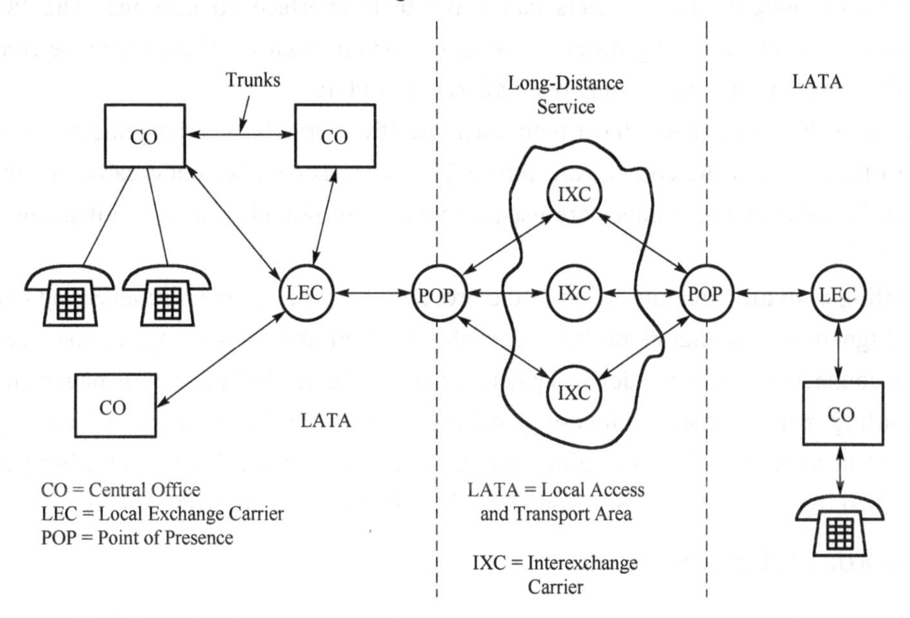

Figure 10.5 The organization of the telephone hierarchy in the United States

Central office. The central office or local exchange is the facility to which your telephone is directly connected by a twisted-pair cable. **Also known as an end office (EO), the local exchange can serve up to 10,000 subscribers, each of whom is identified by a four-digit number from 0000 through 9999 (the last four digits of the telephone number).**

The local exchange also has an exchange number. These are the three additional digits that make up a telephone number. Obviously, there can be as many as 1,000 exchanges with numbers from 000 through 999. These exchanges become part of an area code region, which is defined by an additional three-digit number. Each area code is fully contained within one of the geographic areas assigned to one of the regional operating companies.

These companies are called local exchange carriers, or local exchange companies (LEC).

Operational relationships. The LECs provide telephone services to designated geographic areas referred to as local access and transport areas (LATA). The United States is divided into approximately 200 LATAs. The LATAs are defined within the individual states making up the seven operating regions. The LECs provide telephone services for the LATAs within their regions but do not provide long-distance services for the LATAs.

Long-distance services are provided by long-distance carriers known as interexchange carriers (IXC). Long-distance carriers must be used for the interconnection for any inter-LATA connections. The LECs can provide telephone services within the LATAs that are part of their operating regions, but links between LATAs within a region, even though they may be directly adjacent to one another, must be made through an IXC.

Each LATA contains a serving, or point of presence (POP), office that is used to provide the interconnections to the IXCs. The local exchanges communicate with one another via individual trunks. And all local exchanges connect to an LEC central office, which provides trunks to the POP. At the POP, the long-distance carriers can make their interface connections. The POPs must provide equal access for any long-distance carrier desiring to connect. Many POPs are connected to multiple IXCs, but in many areas, only one IXC serves a POP.

Most long-distance carriers have their own specific hierarchical arrangements. A variety of switching offices across the country are linked by trunks using fiber-optic cables or microwave relay links. Multiplexing techniques are used throughout to provide many simultaneous paths for telephone calls.

Signaling system. Signaling refers to the process of setting up and disconnecting calls on the network. Signaling uses digital packets to perform all of the various operations necessary to establish a connection and tear it down. Typical functions include billing, call management (such as call-forwarding, number display, three-way calling, and 800 and 900 calls), and routing of calls from one point to another. The signaling system used in the United States and other parts of the world is called Signaling System No. 7 or SS7, which is standardized by the ITU-T.

Words and Expressions

telephone exchange	电话局；总机；电话交换机；电信交换
ubiquitous [juːˈbɪkwɪtəs]	*adj*. 无处不在的；十分普遍的
vocal tract	声道
touch-tone keypad	按键式键盘
scaling problem	缩放问题；尺度问题
call-control software	呼叫控制软件
automanual exchange	半自动交换台；自动手动切换
public switched telephone network (PSTN)	公用电话交换网
plain old telephone service (POTS)	普通传统电话业务
dial tone	拨号音
subscriber line interface circuit (SLIC)	用户专线接口滤波路；用户线接口电路；用户接口电路
battery feed	馈电

overvoltage protection 过压保护
ringing ['rɪŋɪŋ] *n.* 振铃
supervision[ˌsuːpər'vɪʒn] *n.* 监视
codec ['koʊdek] *n.* 编解码器
hybrid ['hɑɪbrɪd] *n.* 混合（用户线 2/4 线转换）
test [test] *n.* 测试
Vrms: root-mean-square voltage 电压均方根值；电压有效值
trunk line 干线
toll office 长途局；长途电话局；长途端局；收费站
exchange number 交换局号
area code 区号
local exchange carrier/company (LEC) 本地交换运营商/公司
local access and transport area (LATA) 本地接入和传输区域
interexchange carrier (IXC) 局间通信公司；长途电话公司
point of presence (POP) 入网点；出现点；存在点
hierarchical [ˌhɑɪə'rɑːrkɪkl] *adj.* 等级制的；按等级划分的；等级制度的
signaling system 信令系统
call-forwarding 呼叫转移
number display 来电显示
three-way calling 三方通话；三方通信；电话会议

Notes

1. In setting out how all of this works, it is helpful to understand how the basic telephone networks are structured since they form the basis of all today's telecommunications networks.

在阐述所有这些工作原理时，了解基本电话网络的结构是有帮助的，因为它们构成了当今所有电信网络的基础。

2. At the receiving end of the circuit, the analog electrical signal energizes the receiver (i.e., an earpiece), generating a set of sound waves, which are an approximate reproduction of the sound of the speaker.

在电路的接收端，模拟电信号为激励接收器（即听筒），产生一组声波，它们是说话者声音的近似再现。

3. Transmission over trunk lines to other central offices or toll offices or for use in long-distance transmission is typically by digital PCM signals in modern systems.

在现代系统中，(语音信号）通过干线传输到其他电话端局或长途电话局，或用于远距离传输，通常采用数字 PCM 信号。

4. Also known as an end office (EO), the local exchange can serve up to 10,000 subscribers, each of whom is identified by a four-digit number from 0000 through 9999 (the last four digits of the telephone number).

本地交换局，也称为终端局（EO），可为多达 10 000 个用户提供服务，每个用户由 0000 到 9999（电话号码的最后 4 位）之间的一个 4 位数字进行标识。

10.2 Internet Telephony

Internet telephony, also called Internet protocol (IP) telephony or voice over Internet protocol (VoIP), uses the Internet to carry digital voice telephone calls. VoIP, in effect, for the most part, bypasses the existing telephone system, but not completely. It has been in development for over a decade, but only recently has it become practical and popular. VoIP is a highly complex digital voice system that relies on high-speed Internet connections from cable TV companies, phone companies supplying DSL, and other broadband systems including wireless. It uses the Internet's vast fiber-optic cabling network to carry phone calls without phone company charges. This new telephony system is slowly replacing traditional phones, especially in large companies. It offers the benefits of lower long-distance calling charges and reduces the amount of new equipment needed, because phone service is essentially provided over the same local area network (LAN) that interconnects the PCs in an organization. VoIP is rapidly growing in use and in the future is expected to replace standard phones in many companies and homes. **While the legacy PSTN will virtually never go away, over time it will play a smaller and smaller role as VoIP is more widely adopted or as more and more individuals choose a cell phone as their main telephone service.**

10.2.1 VoIP Fundamentals

There are two basic parts to an IP phone call: the dialing process, which establishes an initial connection, and the voice signal flow.

1. Voice Signal Flow

Figure 10.6 shows the signal flow and major operations in a VoIP system. The voice signal is first amplified and digitized by an analog-to-digital converter (ADC) that is part of a coder-decoder (codec) circuit, which also includes a digital-to-analog converter (DAC). The ADC usually samples the voice signal at 8 kHz and produces an 8-bit word for each sample. These samples occur one after another serially and therefore produce a 64-kbps digital signal. A relatively wide bandwidth is needed to transmit this bit stream (64 kHz or more). To reduce the data rate and the need for bandwidth, the bit stream is processed by a voice encoder that compresses the voice signal. This compression is usually done by DSP either in a separate DSP processor chip or as hardwired logic on a larger chip. The output is at a greatly reduced serial digital data rate.

The type of compression used is determined by International Telecommunications Union (ITU) standards. Various mathematical algorithms beyond the scope of this text are used. The 64-kbps digital signal is designated as standard G.711 and is better known as pulse-code modulation (PCM), covered earlier in this book. Standard G.729a is probably the most common compression standard used and results in an 8-kbps digital voice signal. Another popular standard is G.723, which produces an even more highly compressed 5.3-kbps signal at the expense of some voice quality. Numerous other compression standards are used, and they are selected based upon the application.

Most VoIP phones contain all the common compression standard algorithms in the DSP memory for use as called for. The signal is also processed in the DSP to provide echo cancellation, a problem in digital telephony.

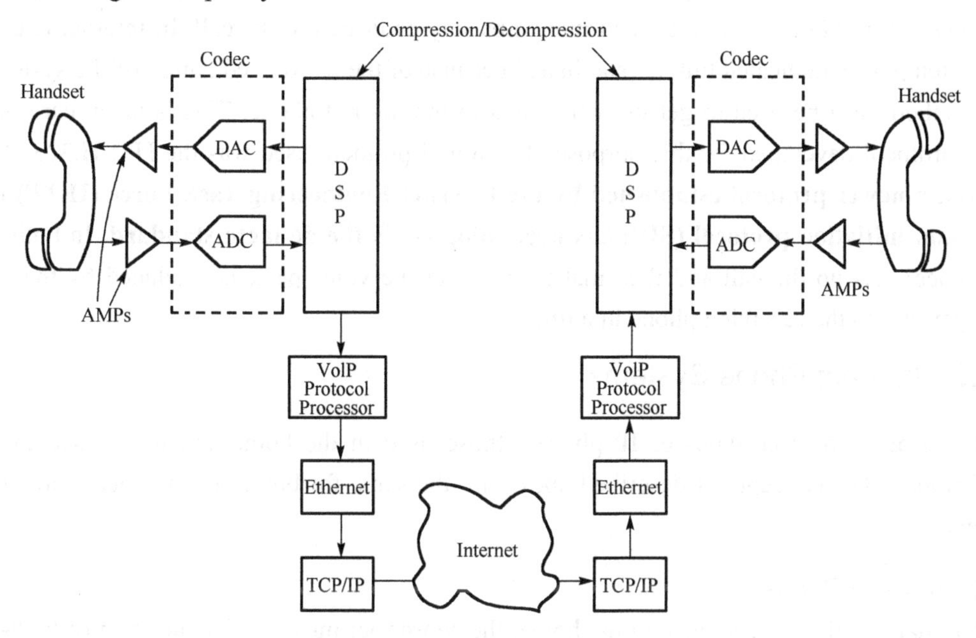

Figure 10.6 The signal flow and major operations in a VoIP system

The resulting serial digital signal is put into a special packet by a microcomputer processor running a VoIP protocol and then transmitted by Ethernet over a LAN or via a high-speed Internet connection such as is available from a cable TV company or on DSL. From there, the signal travels over standard available Internet connections using TCP/IP through multiple servers and routers until it comes to the desired location.

At the receiving phone, the process is reversed. The Internet signal gets converted back to Ethernet, and then the VoIP processor recovers the original packet. From there, the compressed data is extracted, decompressed by a DSP, and sent to the DAC in the codec where the original voice is heard.

One of the main problems with VoIP is that it takes a relatively long time to transmit the voice data over the Internet. The packets may take different routes through the Internet. They all do eventually arrive at their intended destination, but often the packets are out of sequence. The receiving phone must put them back together in the correct sequence. This takes time.

Furthermore, even though the signals traverse the high-speed optical Internet lines at gigabit speeds, the packets pass through numerous routers and servers, each adding transit time or latency. Latency is the delay between the time the signal is transmitted and the time it is received. It has been determined that the maximum acceptable latency is about 150 ms. Any longer time is noticeable by the user. One party may have to wait a short time before responding to avoid talking while the signal is still be received. This annoying wait is unacceptable to most. Keeping the latency below 150 ms minimizes this problem.

2. Link Establishment

In the PSTN, the dialing process initiates multiple levels of switching that literally connects the calling phone to the called phone. That link is maintained for the duration of the call because the switches stay in place and the electronic paths stay dedicated to the call. In Internet telephony, no such temporary dedicated link is established because of the packetized nature of the system. Yet some method must be used to get the voice data to the desired phone. This is taken care of by a special protocol developed for this purpose. The initial protocol used was the ITU H.323. **Today, however, a newer protocol established by the Internet Engineering Task Force (IETF) called the session initiation protocol (SIP) has been adopted as the de facto standard.** In both cases, the protocol sets up the call and then makes sure that the voice packets produced by the calling phone get sent to the receiving phone in a timely manner.

10.2.2 Internet Phone Systems

There are two basic types of IP phones: those used in the home and those used in larger organizations. The concepts as described above are the same for both, but the details are slightly different.

1. Home IP Phones

To establish IP phone service in the home, the subscriber must have some form of high-speed Internet service. Cable TV provides this service in most homes, but it can also be provided over the standard POTS local loop with DSL. In addition, the subscriber must have a VoIP interface. This is called different things by the different service providers. A common example is the analog terminal adapter (ATA). This device connects the standard home telephone to the existing broadband Internet modem. Another configuration is a VoIP gateway that contains the ATA circuitry as well as the broadband modem.

A general block diagram of an ATA is shown in Figure 10.7. Notice that the ATA allows standard telephones and cordless phones to attach to the ATA via the usual RJ-11 modular plug. In fact, the input to the ATA is the phone wiring in the home. The home wiring is disconnected from the subscriber interface at the connection provided outside the home by the phone company. In this way, any of the available home phones can be used with the ATA over installed wiring. Note in the figure that because standard phones are used, they must be provided with SLIC BORSCHT functions. The SLIC is usually packaged in a single IC chip, and often the codec is also contained on this chip.

The codec inputs and outputs go to one or more processors where the H.323 or SIP protocol is implemented and where the DSP functions for compression and decompression reside. An Ethernet interface is also provided. The Ethernet signal connects to the broadband modem for cable TV service or DSL. If the cable modem is used, the POTS and last mile local loop are simply not used. However, if DSL service provides the broadband connection, the POTS connection is used for the DSL modem. The home phone wiring must be disconnected from the POTS line.

2. Enterprise IP Phones

IP phones in companies or large organizations are especially designed for VoIP service. The telephone set contains all the ATA circuitry except for the SLIC and connects directly to the available Ethernet connection usually supplied to each desk. No broadband modem is needed. Since most employees will also have a PC connected to the LAN, a two-port Ethernet switch in the phone or PC provides a single Ethernet connection to the LAN that the phone and PC share.

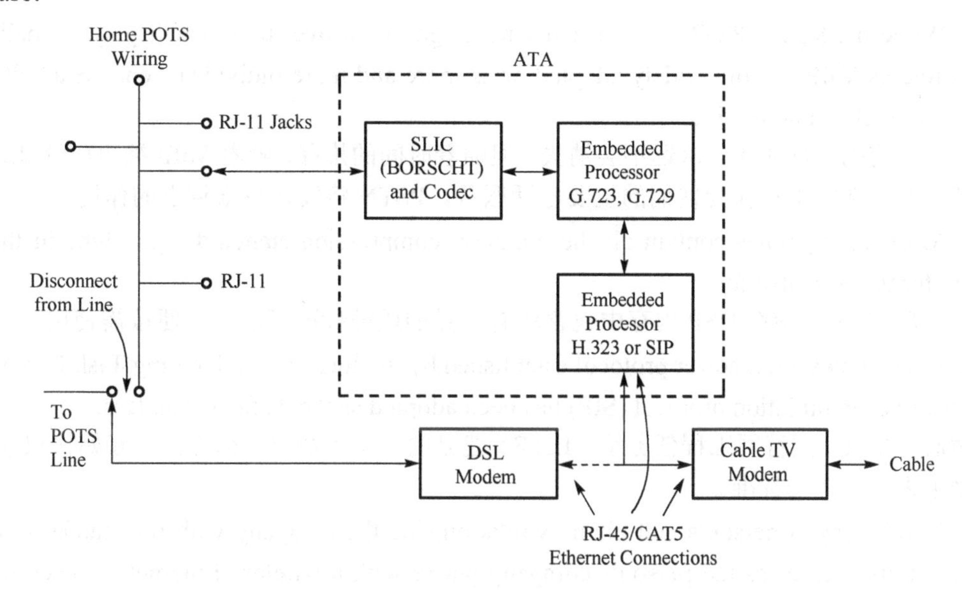

Figure 10.7 Analog terminal adapter (ATA) or VoIP gateway

A major benefit of IP phones is that they may also use wireless Ethernet connections. Wireless Ethernet, generally called WiFi or the IEEE standard designation 802.11, is widely used to extend the LANs in most companies. If the IP phone is equipped with a wireless Ethernet transceiver, then no wired connection is needed. Already some cell phone manufacturers are including WiFi VoIP in some models. **In this way, a person's cell phone works outside the company with the standard cell site service but also serves as the person's company phone with a wireless Ethernet connection when inside the company.**

Words and Expressions

Internet telephony/IP telephony/VoIP 网络电话；互联网电话
bypass ['baɪpæs] *vt.* 绕过；避开
International Telecommunications Union (ITU) 国际电信联盟
algorithm ['ælgərɪðəm] *n.* 算法；计算程序
designate ['dezɪgneɪt] *vt.* 命名；指定；选定；标明；指明
echo cancellation 回音消除；回波抵消
extract ['ekstrækt, ɪk'strækt] *v.* 提取；摘录；提炼；获得
n. 提取物；摘录；节录

decompress [ˌdiːkəm'pres] *v.* 解压缩（将压缩文件等恢复到原大小）
out of sequence 失序；顺序混乱；次序错误
Internet Engineering Task Force (IETF) 因特网工程任务组
session initiation protocol (SIP) 会话起始协议
analog terminal adapter (ATA) 模拟终端适配器

Notes

1. While the legacy PSTN will virtually never go away, over time it will play a smaller and smaller role as VoIP is more widely adopted or as more and more individuals choose a cell phone as their main telephone service.

虽然传统的PSTN几乎永远不会消失，但随着时间的推移，随着VoIP被更广泛地采用或越来越多的人选择手机作为他们的主要电话服务，PSTN将发挥越来越小的作用。

2. Most VoIP phones contain all the common compression standard algorithms in the DSP memory for use as called for.

大多数VoIP电话在DSP内存中包含所有常见的压缩标准算法，以便按需使用。

3. Today, however, a newer protocol established by the Internet Engineering Task Force (IETF) called the session initiation protocol (SIP) has been adopted as the de facto standard.

然而，今天，因特网工程任务组（IETF）建立的一个新协议，称为会话起始协议（SIP），已被采纳为事实上的标准。

4. In this way, a person's cell phone works outside the company with the standard cell site service but also serves as the person's company phone with a wireless Ethernet connection when inside the company.

通过这种方式，一个人的手机可以在公司外使用标准的手机站点服务，但在公司内也可以作为个人的公司电话，通过无线以太网连接。

Exercises

1. Match the terms (1)-(6) with the definitions A-F.

(1) VoIP	A. a telecommunication network which allows subscribers at different sites to communicate by voice
(2) PSTN	B. a local telephone company switching center
(3) SS7	C. a method and group of technologies for the delivery of voice communications and multimedia sessions over Internet protocol networks, such as the Internet
(4) central office	D. an international telecommunication protocol standard that defines how the network elements in PSTN exchange information and control signals
(5) trunk line	E. the signal that a person hears on a landline telephone before he dials a phone number
(6) dial tone	F. a direct link between two telephone exchanges or switchboards that are a considerable distance apart

2. Translate into Chinese.

(1) Originally, the PSTNs around the world were fully analog; they comprised of analog transmission systems in the access and core parts of the networks, using a variety of FDM-

based systems to gain economies, together with analog switching in the local and trunk exchanges.

(2) A major component of any telephone system is signaling, in which electric pulses or audible tones are used for alerting (requesting service), addressing (e.g., dialing the called party's number at the subscriber set), supervision (monitoring idle lines), and information (providing dial tones, busy signals, and recordings).

(3) The human speaking voice typically has frequencies of up to 4,000 Hz, so a sampling rate of 8,000 Hz is sufficient to sample it with an acceptable quality, using Nyquist's theorem, and with a bit depth of 8 bits, resulting in a required bandwidth of 8,000 Hz×8 bits = 64 kbps.

3. Translate into English.

（1）用户电话机和电话端局之间的 2 线双绞线连接被称为本地环路或用户环路。

（2）在 20 世纪 90 年代，电话网发生了重大变化，由于大量新增电话用户和使用电话网络访问互联网，导致电话网通信量迅速增加。

（3）与其他实时应用程序一样，VoIP 对带宽和延迟非常敏感。

4. Read the following article and write a summary.

In the early days of VoIP (voice over IP), the limited bandwidth of Internet telephony led to low quality calls that frequently got dropped. But VoIP quality has since improved with the increase in available bandwidth and enhancements in the technology. Today, VoIP is capable of higher call quality than the public switched telephone network (PSTN), the legacy circuit-switched telephony system that connected callers for years before VoIP came along.

In fact, PSTN has been losing voice traffic to the IP despite being a stable and reliable global telephone network that connects callers all over the world. But how did IP get to be a reliable voice protocol when the Internet wasn't really designed for the real-time communication requirements of a phone call? And how should businesses think about the two protocols in their voice communications with customers?

A quick history lesson will provide a better understanding.

The PSTN works by establishing a dedicated circuit between two parties for the duration of a call. The analog voice data is carried over the dedicated circuits via copper wires. In contrast, VoIP uses packet-switched telephony where the voice data is transmitted in multiple individual network packets across the Internet.

A series of related technology advancements have culminated in a robust IP voice channel. First, the ubiquitous availability of high-speed Internet has provided the bandwidth for stable connections. Next, IP voice calls have become richer with the support of technologies such as high definition encoding through codecs and HD processors for mobile phones, which deliver superior audio fidelity.

And all this easily accessible voice quality is delivered at a lower cost than consumers and businesses were paying carriers and telcos for PSTN calls. Pioneering companies brought VoIP products to market such as a software-based VoIP phone (Netfone, 1991), peer-to-peer Internet calling (Skype, 2003), a residential VoIP service (Vonage, 2004), and mobile VoIP apps for smartphones (Truphone, 2006). Consumers and businesses, lured by the quality and cost benefits, flocked to VoIP. Persistence Market Research anticipates strong growth in the global market for VoIP services during the next several years, projecting the market will surpass US$190B in revenue by 2024.

Callers likely don't give a second thought to which protocol is transferring their voice to the person —or people—on the other end. Whether it's a traditional PSTN call or an IP voice call via WebRTC or a mobile app, they care most about the quality and—perhaps too often forgotten—the utility of the call interface.

For brands, it's imperative to offer that utility by delivering contextual call experiences in the ways customers have grown accustomed to during the PSTN-to-IP transition. Rather than viewing VoIP as a replacement for PSTN, they should see it as an additional voice channel that will provide the broadest reach to customers—not everyone has a smartphone or access to a reliable Internet connection; PSTN still has its place.

The smart approach is simply adding the VoIP channel to the existing channels already supported. Again, users likely aren't making the choice between PSTN and IP for their voice interactions but rather choosing the most convenient interface in the given use case. Brands should support whichever protocol allows customers to make calls within the contexts of their given interactions.

For example, an online marketplace where a buyer needs to resolve an issue with the seller and can't wait for an e-mail response, he or she may opt for the immediacy of a phone call. Which protocol connects the call (suppose it's an in-app voice call via IP for the buyer and an inbound PSTN call for the seller) is not as important a consideration as having a simple click-to-call button within the marketplace app for convenient calling.

That feature saves the user from having to jot down or memorize the seller's number, open the phone app, manually dial the number, and then quickly explain who he is and why they're calling when the seller answers. Having the ability to terminate the call in the protocol that best meets the user's needs in that moment—as well as provide context for the call—should be the priority.

The same principle applies when calling a company's contact center for service or support. A phone call is often the last step in an escalation of communication mode that may have begun with an e-mail or a website or mobile chat. By the time a customer calls, he is expecting to resolve his issue or complete his transaction in a timely manner. But too often the customer has to call multiple times and repeat the reason for his call each time he connect because context doesn't always flow from one interaction to the next. All the audio fidelity advancements in the world won't help alleviate a customer's frustration when his experience is that poor.

5. Language study: Describing a process

In English, the passive is often used to describe processes. Study these examples:

(1) A general block diagram of an ATA **is shown** in Figure 10.7.

(2) Because standard phones **are used**, they must **be provided** with SLIC BORSCHT functions.

The passive is made using the verb *to be* (*be*, *is*, *are*, etc.) and the past participle of the verb. Most technical verbs are regular so the past participle is made simply by adding *-ed*. Watch the spelling of the past participle of verbs like *control* (*controlled*) and *use* (*used*). The passive infinitive is used in the same place as ordinary infinitives, for example after verbs like *must* and *can*.

Complete the call setup procedure in PSTN, by putting each of the verbs in brackets in the correct form.

When the handset _____ (lift), indicating the telephone system that a call is to ____ (make). A signal ____ (call) the dial tone ____ (send) back to notify the subscriber that the phone number can ____ (enter). Then the numbers dialed by the subscriber ____ (collect) by a terminal and _____ (transmit) to the central office. Then, the call will ____ (route) to the appropriate receiver.

Unit 11 Electromagnetics

11.1 Introduction of Electromagnetics

The discipline of electromagnetic field theory and its pertinent technologies is also known as electromagnetics. It has been based on Maxwell's equations, which are the result of the significant work of James Clerk Maxwell completed in 1865, after his presentation to the British Royal Society in 1864. It has been over 150 years ago now, and this is a long time compared to the leaps and bounds progress we have made in technological advancements. But despite, research in electromagnetics has continued unabated despite its age. The reason is that electromagnetics is extremely useful, and has impacted a large sector of modern technologies(see Figure 11.1).

11.1.1 Importance of Electromagnetics

To understand why electromagnetics is so useful, we have to understand a few points about Maxwell's equations.

- First, Maxwell's equations are valid over a vast length scale from subatomic dimensions to galactic dimensions. Hence, these equations are valid over a vast range of wavelengths, going from static to ultraviolet wavelengths.

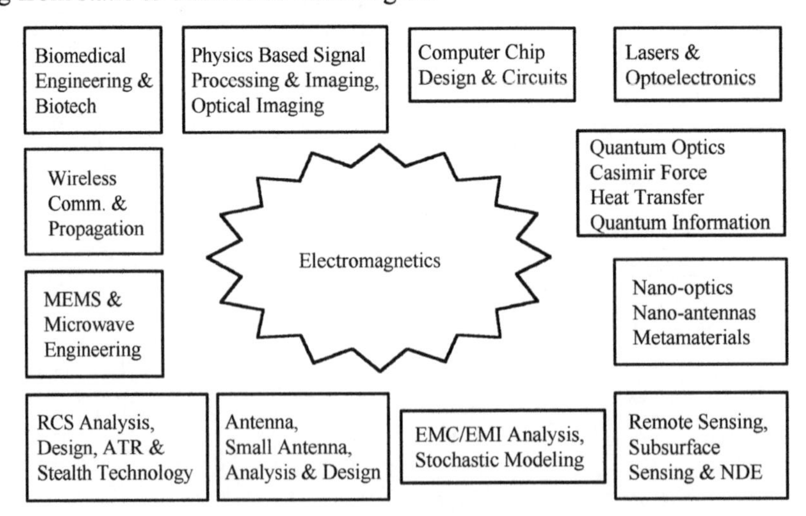

Figure 11.1 The impact of electromagnetics in many technologies

- Maxwell's equations are relativistic invariant in the parlance of special relativity. In fact, Einstein was motivated with the theory of special relativity in 1905 by Maxwell's equations. These equations look the same, irrespective of what inertial reference frame one is in.

- Maxwell's equations are valid in the quantum regime, as it was demonstrated by Paul Dirac in 1927. Hence, many methods of calculating the response of a medium to classical field can be applied in the quantum regime also. When electromagnetic theory is combined with quantum theory, the field of quantum optics came about. Roy Glauber won a Nobel Prize in 2005 because of his work in this area.
- Maxwell's equations and the pertinent gauge theory has inspired Yang-Mills theory (1954), which is also known as a generalized electromagnetic theory. Yang-Mills theory is motivated by differential forms in differential geometry. To quote from Misner, Thorne, and Wheeler, "Differential forms illuminate electromagnetic theory, and electromagnetic theory illuminates differential forms."
- Maxwell's equations are some of the most accurate physical equations that have been validated by experiments. In 1985, Richard Feynman wrote that electromagnetic theory has been validated to one part in a billion. Now, it has been validated to one part in a trillion.

As a consequence, electromagnetics has had a tremendous impact in science and technology. This is manifested in electrical engineering, optics, wireless and optical communications, computers, remote sensing, biomedical engineering, etc.

11.1.2 A Brief History of Electromagnetics

Electricity and magnetism have been known to humans for a long time. Also, the physical properties of light has been known. **But electricity and magnetism, now termed electromagnetics in the modern world, has been thought to be governed by different physical laws as opposed to optics.** This is understandable as the physics of electricity and magnetism is quite different from the physics of optics as they were known to humans.

For example, lodestone was known to the ancient Greek and Chinese around 600 BC to 400 BC. Compass was used in China since 200 BC. Static electricity was reported by the Greek as early as 400 BC. But these curiosities did not make an impact until the age of telegraphy. The coming about of telegraphy was due to the invention of the voltaic cell or the galvanic cell in the late 1700s, by Luigi Galvani and Alesandro Volta. It was soon discovered that two pieces of wire, connected to a voltaic cell, can be used to transmit information.

So by the early 1800s, this possibility had spurred the development of telegraphy. Both André-Marie Ampère (1823) and Michael Faraday (1838) did experiments to better understand the properties of electricity and magnetism. And hence, Ampère's law and Faraday's law are named after them. Kirchhoff's voltage and current laws were also developed in 1845 to help better understand telegraphy. Despite these laws, the technology of telegraphy was poorly understood. It was not known as to why the telegraphy signal was distorted. Ideally, the signal should be a digital signal switching between 1s and 0s, but the digital signal lost its shape rapidly along a telegraphy line.

It was not until 1865 that James Clerk Maxwell put in the missing term in Ampère's law, the term that involves displacement current, only then the mathematical theory for electricity and magnetism was complete. Ampère's law is now known as generalized Ampère's law. The complete set of equations are now named Maxwell's equations in honor of James Clerk Maxwell.

The rousing success of Maxwell's theory was that it predicted wave phenomena, as they have been observed along telegraphy lines. Heinrich Hertz in 1888 did experiment to proof that electromagnetic field can propagate through space across a room. Moreover, from experimental measurement of the permittivity and permeability of matter, it was decided that electromagnetic wave moves at a tremendous speed. But the velocity of light has been known for a long while from astronomical observations (Roemer, 1676). The observation of interference phenomena in light has been known as well. When these pieces of information were pieced together, it was decided that electricity and magnetism, and optics, are actually governed by the same physical law or Maxwell's equations. And optics and electromagnetics are unified into one field.

In the beginning, it was thought that electricity and magnetism, and optics were governed by different physical laws. Low frequency electromagnetics was governed by the understanding of fields and their interaction with media. Optical phenomena were governed by ray optics, reflection and refraction of light. But the advent of Maxwell's equations in 1865 revealed that they could be unified by electromagnetic theory. Then solving Maxwell's equations becomes a mathematical endeavor.

The photoelectric effect, and Planck radiation law point to the fact that electromagnetic energy is manifested in terms of packets of energy. Each unit of this energy is now known as the photon. A photon carries an energy packet equal to $h\omega$, where ω is the angular frequency of the photon and $h \approx 6.626 \times 10^{-34}$ J·s, the Planck constant, which is a very small constant. Hence, the higher the frequency, the easier it is to detect this packet of energy, or feel the graininess of electromagnetic energy. Eventually, in 1927, quantum theory was incorporated into electromagnetics, and the quantum nature of light gives rise to the field of quantum optics. Recently, even microwave photons have been measured. **It is a difficult measurement because of the low frequency of microwave (10^9 Hz) compared to optics (10^{15} Hz): microwave photon has a packet of energy about a million times smaller than that of optical photon.**

The progress in nano-fabrication allows one to make optical components that are subwavelength as the wavelength of blue light is about 450 nm. As a result, interaction of light with nano-scale optical components requires the solution of Maxwell's equations in its full glory.

In 1980s, Bell's theorem (by John Steward Bell) was experimentally verified in favor of the Copenhagen school of quantum interpretation (led by Niel Bohr). This interpretation says that a quantum state is in a linear superposition of states before a measurement. But after a measurement, a quantum state collapses to the state that is measured. This implies that quantum information can be hidden in a quantum state. Hence, a quantum particle, such as a photon, its state can remain incognito until after its measurement. In other words, quantum theory is "spooky". This leads to growing interest in quantum information and quantum communication using photons. Quantum technology with the use of photons, an electromagnetic quantum particle, is a subject of growing interest.

11.1.3 Applications of Electromagnetism

A branch of physics that deals with electric current or fields and magnetic fields and their interaction on substance or matter is called Electromagnetism. Electromagnetism has created a great revolution in the field of engineering applications. In addition, this caused a great impact on various fields such as medical, industrial, space, etc.

We can find many practical applications of electromagnetism in everyday life from household applications to research applications.

1. Household Applications

Electromagnetism serves as a basic principle of working for many of the home appliances in household applications. These applications include lighting systems, kitchen appliances, air conditioning systems, etc.

The most dominant use of power in homes as well as commercial buildings is lighting systems. These lighting systems used numerous fluorescent lighting fixtures(see Figure 11.2). Ballasts used in the fluorescent lamps use electromagnetism principle, so that at the time of turn ON of the light it produces a high voltage.

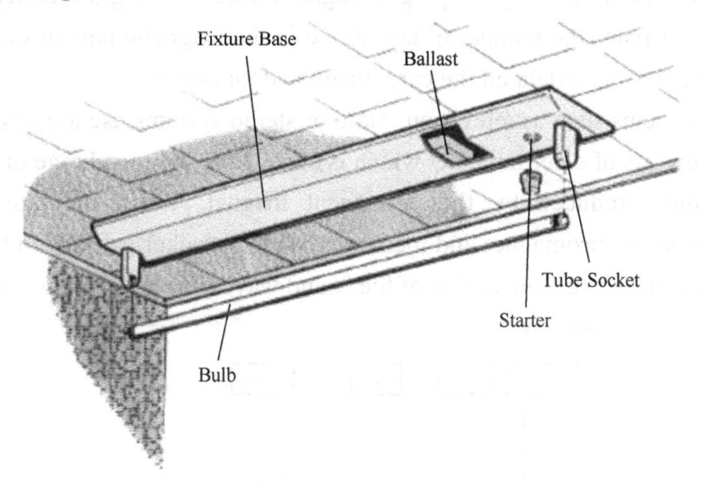

Figure 11.2 Fluorescent lighting fixtures

Electric fans, blowers and other cooling systems use electric motors. These motors work on the principle of electromagnetic induction which is the branch of electromagnetism. In any electrical appliance, electric motor is moved by the magnetic field produced by the electric current according to the Lorenz force principle. These motors are vary in size, rating and cost based on the application.

Kitchen appliances like induction cookers, microwave ovens, electric mixers and grinders, bread toasters, etc. use the electromagnetism for their operations.

Alarming systems use electric bells(see Figure 11.3) which work based on electromagnetic principle. In these bells, the sound is produced by an electromagnetic coil which moves the striker against the bell. As long as the coil is energized, iron striker gets attracted by it, hence it strikes the bell. **The electromagnet will be demagnetized when the striker gets contact with bell and by the spring tension striker comes back to its original position, and again electrical contact will be made once again.** This process repeats until the switch gets opened.

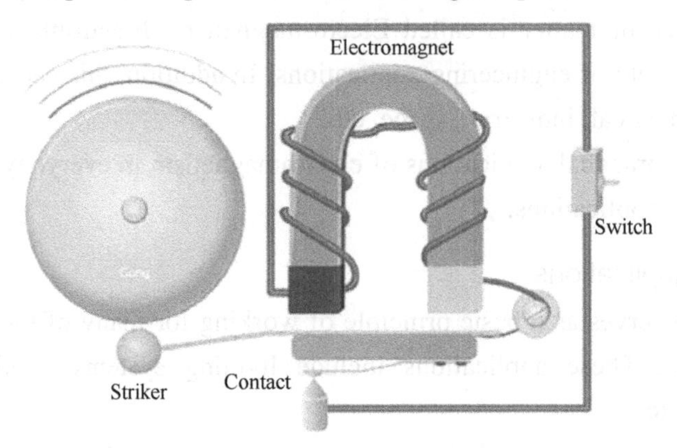

Figure 11.3 Electric bell

Security systems use locking systems for doors which are generally magnetic locking systems. These systems are unlocked either by swiping a magnetic card or using a security code. Magnetic card reader on the door reads the number of key stored in the magnetic tape of the card. When key stored in memory matches with data on the card, then the door opens.

Entertainment systems such as television, radio or stereo systems use loudspeakers(see Figure 11.4). This device consists of electromagnet which is attached to the membrane or cone surrounded by the magnetic flux produced by the permanent magnet. When the current through the electromagnet is varied, electromagnet and membrane of the speaker get moved back and forth. If the current is varied at the same frequencies of the sound waves, it results a vibration of the speaker which will further create sound waves.

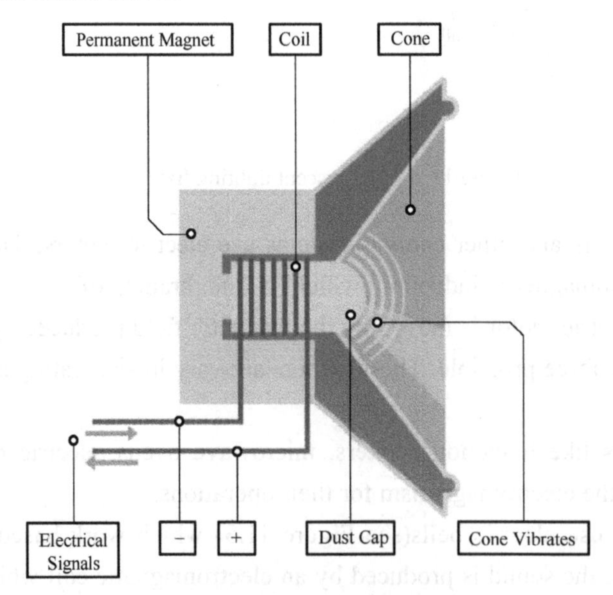

Figure 11.4 Loudspeaker

2. Industrial Applications

Almost all of the instruments or devices used in industries are based on the electromagnetism. Materials used in constructing such devices include iron, cobalt, nickel, etc. which naturally responds to the magnetic fields. Starting from small control instruments to the large power equipment, the electromagnetism is used at least at one stage of their workings.

Generators and motors dominate in most of the industries which are the primary power sources and driving systems respectively. Generators convert mechanical energy to electrical energy, whereas motors convert electrical energy to mechanical energy. Generators supply the electrical energy in the time of mains power interruption and most of the cases, these are driven by the injection combustion(IC) engines. There are different classes of motors which are employed in industries. These are used for cranes, hoists, lifts, conveyor systems, etc.

Various sensors and actuating devices work based on electromagnetism. Electromagnetic sensors include Hall-effect sensors, magnetoresistive sensors, fluxgate sensors, etc. These sensors convert the physical quantities such as flow, pressure, level, proximity, etc. into electrical signals. Actuators are the final control elements which drive the loads at specific conditions. These actuator devices include solenoid valves, relays, motors, etc. and all these work on the principle of electromagnetism.

3. Communication System

Communication is the process of transmitting information from a source to a receiver. The transmission of energy over long distances is carried out through electromagnetic waves at high frequencies. These waves are also called as microwaves or high frequency radio waves.

Suppose in case of mobile phones, sound energy is converted into electromagnetic energy. By using the radio transmitter, this electromagnetic energy is transferred to the receiver. At the receiver, these electromagnetic waves are again transformed back into sound energy.

Electromagnetic fields produced by time-varying sources propagate through waveguides or transmission lines. Electromagnetic wave radiation is formed when the electromagnetic field propagates away from the source without any connection or conducting medium to the source.

4. Medical Applications

Nowadays electromagnetic fields play a key role in advanced medical equipment such as hyperthermia treatments for cancer, x-ray scanner and magnetic resonance imaging (MRI).

RF range frequencies are mostly used in medical applications. In MRI scans, sophisticated equipment working based on the electromagnetism can scan minute details of the human body.

The electromagnetic therapy is an alternate form of medicine which claims to treat disease by applying pulsed electromagnetic fields or electromagnetic radiation to the body. This type of treatment is used for wide range of ailments such as nervous disorders, diabetes, spinal cord injuries, ulcers, asthma, etc.

Words and Expressions

electromagnetics [ɪˌlektroʊmæg'netɪks] *n.* 电磁（学）

electromagnetic field	电磁场
pertinent ['pɜːrtnənt]	*adj*. 有关的；恰当的；相宜的
leaps and bounds	突飞猛进；跨越式发展；跳跃
unabated [ˌʌnə'beɪtɪd]	*adj*. 未减弱的；有增无减的；未变弱的
MEMS: micro-electromechanical system	微电子机械系统；微型制造机电系统
RCS: radar cross section	雷达截面积
ATR: automatic target recognition	自动目标识别
stealth technology	隐身技术；隐形技术；低可侦测性技术
EMC: electromagnetic compatibility	电磁兼容性
EMI: electromagnetic interference	电磁干扰
stochastic modeling	随机建模
NDE: nondestructive evaluation	无损评估；无损评价；无损检测；非破坏检测原理与应用
metamaterial [ˌmetəmə'tɪriəl]	*n*. 超材料；特异材料；超常材料；人工材料
subatomic [ˌsʌbə'tɑːmɪk]	*adj*. 亚原子的；比原子小的；原子内的
galactic [ɡə'læktɪk]	*adj*. 银河的；星系的
ultraviolet['ʌltrə'vaɪələt]	*adj*. 紫外线的；利用紫外线的
relativistic [ˌrelətɪ'vɪstɪk]	*adj*. 相对主义；相对论的；相对性的；相对的
parlance ['pɑːrləns]	*n*. 用语；说法；术语
special relativity	狭义相对论
inertial reference frame	惯性参考坐标系；惯性基准坐标系
quantum regime	量子体系；量子领域
gauge theory	规范理论；规范场理论
Yang-Mills theory	杨-米尔斯理论
differential geometry	微分几何
biomedical engineering	生物医学工程
electricity [ɪˌlek'trɪsəti]	*n*. 电；电能
magnetism ['mæɡnətɪzəm]	*n*. 磁性；磁力
lodestone ['loʊdstoʊn]	*n*. 磁铁矿
displacement current	位移电流
permittivity [ˌpɜːrmɪ'tɪvəti]	*n*.介电常数；电容率
permeability [ˌpɜːrmiə'bɪləti]	*n*. 磁导率
electromagnetic wave	电磁波
ray optics	几何光学；射线光学；光线光学
photoelectric effect	光电效应
photon ['foʊtɑːn]	*n*. 光子；光量子
graininess ['ɡreɪninis]	*n*. 颗粒度；（多）粒状；颗粒性
quantum optics	量子光学
superposition [ˌsuːpərpə'zɪʃn]	*n*. 叠加态；叠加性；叠加法
incognito [ˌɪnkɑːɡ'niːtoʊ]	*adj*. 伪装的；匿名的；隐姓埋名的

home appliance	家用电器；家用器具；家电
fluorescent lighting fixture	荧光灯设备；荧光照明灯具
ballast ['bæləst]	*n.* 镇流器；电子整流器；稳压器
electric motor	电动机
electromagnetic induction	电磁感应
induction cooker	电磁炉；电磁灶
electromagnetic coil	电磁线圈
demagnetize [ˌdiː'mægnətaɪz]	*v.* 消磁，使退磁
magnetic flux	磁通量；磁力线
cobalt ['koʊbɔːlt]	*n.* 钴
nickel ['nɪkl]	*n.* 镍
injection combustion(IC) engine	直喷式内燃机
Hall-effect sensor	霍尔效应传感器
magnetoresistive sensor	磁阻传感器
fluxgate sensor	磁通门传感器
solenoid valve	电磁阀；螺线管操纵阀
hyperthermia[ˌhaɪpər'θɜːrmɪə]	*n.* 温热疗法；体温过高；发热；高烧
magnetic resonance imaging(MRI)	磁共振成像

Notes

1. But electricity and magnetism, now termed electromagnetics in the modern world, has been thought to be governed by different physical laws as opposed to optics.

但是电和磁，现在在现代世界中被称为电磁学，被认为是受不同于光学的物理定律支配的。

2. It was not until 1865 that James Clerk Maxwell put in the missing term in Ampère's law, the term that involves displacement current, only then the mathematical theory for electricity and magnetism was complete.

直到 1865 年，詹姆斯·克拉克·麦克斯韦 (James Clerk Maxwell) 才在安培定律中加入了与位移电流有关的缺失项，直到那时，电学和磁学的数学理论才完整。

3. It is a difficult measurement because of the low frequency of microwave (10^9 Hz) compared to optics (10^{15} Hz): microwave photon has a packet of energy about a million times smaller than that of optical photon.

由于微波（10^9 Hz）的频率比光学（10^{15} Hz）的频率低，因此很难测量：微波光子的能量比光学光子的能量小约一百万倍。

4. The progress in nano-fabrication allows one to make optical components that are subwavelength as the wavelength of blue light is about 450 nm.

纳米制造技术的进步使人们能够制造亚波长的光学元件，因为蓝光的波长约为 450 纳米。

5. The electromagnet will be demagnetized when the striker gets contact with bell and by the spring tension striker comes back to its original position, and again electrical contact will be made once again.

当撞锤与铃接触时，电磁铁将退磁，通过弹簧张力，撞锤回到其原始位置，并再次进行电气接触。

11.2 Engineering Electromagnetics and Waves

The subject of electromagnetics encompasses electricity, magnetism, and electrodynamics, including all electric and magnetic phenomena and their practical applications. **A branch of electromagnetics, dealing with electric charges at rest (static electricity) named electrostatics, provides a framework within which we can understand the simple fact that a piece of amber, when rubbed, attracts itself to other small objects.** Another branch dealing with static magnetism, namely magnetostatics, is based on the facts that some mineral ores (e.g., lodestone) attract iron and that current-carrying wires produce magnetic fields. The branch of electromagnetics known as electrodynamics deals with the time variations of electricity and magnetism and is based on the fact that magnetic fields that change with time produce electric fields.

Electromagnetic phenomena are governed by a compact set of principles known as Maxwell's equations, the most fundamental consequence of which is that electromagnetic energy can propagate, or travel from one point to another, as waves. The propagation of electromagnetic waves results in the phenomenon of delayed action at a distance; in other words, electromagnetic fields can exert forces, and hence can do work, at distances far away from the places where they are generated and at later times. Electromagnetic radiation is thus a means of transporting energy and momentum from one set of electric charges and currents (at the source end) to another (those at the receiving end). Since whatever can carry energy can also convey information, electromagnetic waves thus provide the means of transmitting energy and information at a distance.

11.2.1 Lumped Versus Distributed Electrical Circuits

A typical electrical engineering student is familiar with circuits, which are described as lumped, linear, and time-invariant systems and which can be modeled by ordinary, linear, and constant-coefficient differential equations. The concepts of linearity and time invariance refer to the relationships between the inputs and outputs of the system. The concept of a lumped circuit refers to the assumption that the entire circuit (or system) is at a single point (or in one "lump"), so that the dimensions of the system components (e.g., individual resistors or capacitors) are negligible. In other words, current and voltage do not vary with space across or between circuit elements, so that when a voltage or current is applied at one point in the circuit, currents and voltages of all other points in the circuit react instantaneously. Lumped circuits consist of interconnections of lumped elements. A circuit element is said to be lumped if the instantaneous current entering one of its terminals is equal to the instantaneous current throughout the element and leaving the other terminal. Typical lumped circuit elements are resistors, capacitors, and inductors. **In a lumped circuit, the individual lumped circuit elements are connected to each other and to sources and loads within or outside the circuit by conducting paths of negligible electrical length.** Figure 11.5(a) illustrates a lumped electrical circuit to which an input voltage of $V_{\text{in}} = V_0$ is applied at $t = 0$. **Since the entire circuit is considered in one lump, the effect of the**

input excitation is instantaneously felt at all points in the circuit, and all currents and voltages (such as I_1, I_2, V_1, and the load current I_L) either attain new values or respond by starting to change at $t = 0$, in accordance with the natural response of the circuit to a step excitation as determined by the solution of its corresponding differential equation. Many powerful techniques of analysis, design, and computer-aided optimization of lumped circuits are available and widely used.

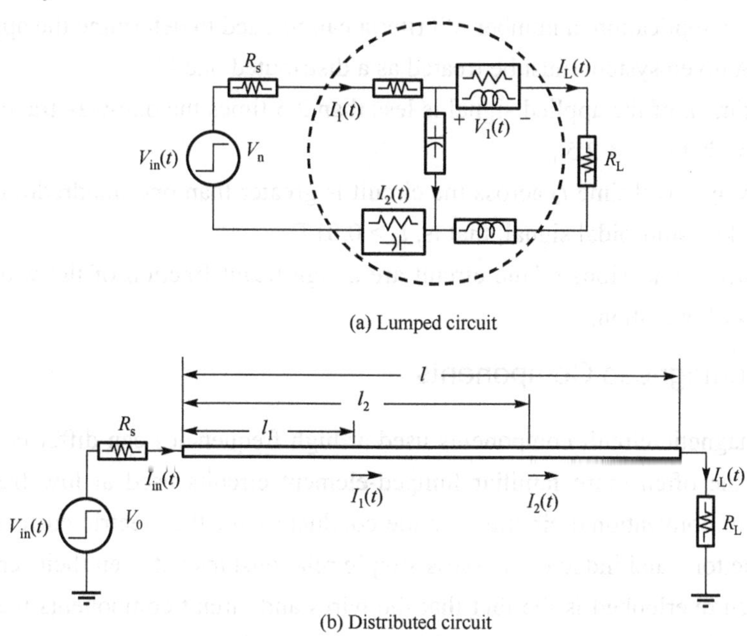

Figure 11.5 Lumped versus distributed electrical circuits

With the "lumped" assumption, one does not have to consider the travel time of the signal from one point to another. In reality, however, disturbances or signals caused by any applied energy travel from one point to another in a nonzero time. For electromagnetic signals, this travel time is determined by the speed of light, $c \approx 3 \times 10^8$ m/s = 30 cm/ns. In practical transmission systems, the speed of signal propagation is determined by the electrical and magnetic properties of the surrounding media and the geometrical configuration of the conductors and may in general be different from c, but it is nevertheless of the same order of magnitude as c. Circuits for which this nonzero travel time cannot be neglected are known as distributed circuits. An example of a distributed circuit is a long wire, as shown in Figure 11.5(b). When an input voltage V_0 is applied at the input terminals of such a distributed circuit (i.e., between the input end of the wire and the electrical ground) at $t = 0$, the voltages and currents at all points of the wire cannot respond simultaneously to the applied excitation because the energy corresponding to the applied voltage propagates down the wire with a finite velocity v. Thus, while the input current $I_{in}(t)$ may change from zero to I_0 at $t = 0$, the current $I_1(t)$ does not flow until after $t = l_1/v$, $I_2(t)$ does not flow until after $t = l_2/v$, and no load current $I_L(t)$ can flow until after $t = l/v$. Similarly, when a harmonically (i.e., sinusoidally) time-varying voltage is connected to such a line, the successive rises and falls of the source voltage propagate along the line with a finite velocity, so that the currents and voltages at other points on the line do not reach their maxima and minima at the same time as the input voltage.

In view of the fundamentally different behavior of lumped and distributed circuits as illustrated in Figure 11.5, it is important, in practice, to determine correctly whether a lumped treatment is sufficiently accurate or whether the circuit in hand has to be treated as a distributed circuit. In any given application, a number of criteria can be used to determine the applicability of a lumped analysis. A given system must be treated as a distributed one if:

- The rise time t_r of the applied signal is less than 2.5 times the one-way travel time t_d across the circuit, that is, $t_r < 2.5t_d$.
- The one-way travel time t_d across the circuit is greater than one-hundredth of the period T of the applied sinusoidal signal, that is, $t_d > 0.01T$.
- The physical dimensions of the circuit are a significant fraction of the wavelength at the frequency of operation.

11.2.2 Electromagnetic Components

The electromagnetic circuit components used at high frequencies can differ conspicuously in appearance from the often more familiar lumped-element circuits used at low frequencies. The connecting wires of conventional circuits provide conductors for the electric currents to flow, and the resistors, capacitors, and inductors possess simple relationships between their terminal currents and voltages. Often overlooked is the fact that the wires and circuit components merely provide a framework over which charges move and disperse. These charges set up electric and magnetic fields that permeate the circuit, often having almost indescribably complicated configurations. It would, in principle, be possible to treat the behavior of circuits entirely in terms of these electromagnetic fields instead of the usual practice of working in terms of circuit voltages and currents. It can be argued, however, that much of the progress in modern electrical and electronic applications would not have come about if it were not for the simple but powerful circuit theory.

As the operating frequency of circuits increases, however, the effects of stray capacitances and inductances alter the effective circuit behavior radically compared with its low-frequency characteristics. Radiation from the circuit also increases rapidly with frequency, which can cause significant power loss. This radiative power loss may be prevented by confining the fields to the interior of metallic enclosures. For example, in hollow metallic tubes (referred to as waveguides) the charges move exclusively on the interior surfaces of conductors, and because of the simple geometry of the enclosures, the electromagnetic fields can have simple analytical forms. Since it is often not possible to define voltages and currents uniquely within waveguides, analysis of waveguides is usually carried out on the basis of the full electromagnetic theory. At even higher frequencies, metallic enclosures become too lossy and impractical. Efficient guiding of electromagnetic energy at optical frequencies occurs in optical fibers, consisting of hair-thin glass strands. The light wave in an optical fiber is guided along the fiber by means of multiple reflections from its walls.

In summary, component sizes in electromagnetic applications can be categorized as shown in Figure 11.6:

- Component sizes much smaller than λ (lumped elements). Examples: Resistors, capacitors, inductors, ICs, transistors, and interconnects used for <~30 MHz.
- Component sizes comparable to λ. Examples: Hollow waveguides, long coaxial cables, some optical fibers, and some antennas.
- Component sizes much greater than λ. Examples: Graded index optical fibers and some antennas.

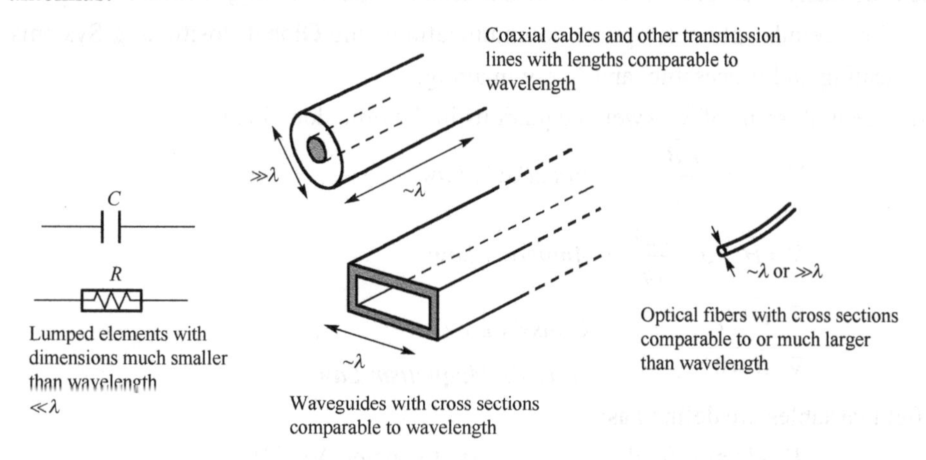

Figure 11.6 Different categories of electromagnetic circuit components

11.2.3 Maxwell's Equations and Electromagnetic Waves

The principles of guiding and propagation of electromagnetic energy in these widely different regimes differ in detail, but are all governed by one set of equations, known as Maxwell's equations, which are based on experimental observations and which provide the foundations of all electromagnetic phenomena and their applications. The sequence of events in the late 19th century that led to the development of these fundamental physical laws is quite interesting in its own right. Many of the underlying concepts were developed by earlier scientists, especially Michael Faraday, who was a visual and physical thinker but not enough of a mathematician to express his ideas in a complete and self-consistent form to provide a theoretical framework. James Clerk Maxwell translated Faraday's ideas into strict mathematical form and thus established a theory that predicted the existence of electromagnetic waves. The experimental proof that electromagnetic waves do actually propagate through empty space was supplied by the experiments of Heinrich Hertz, carried out many years after Maxwell's brilliant theoretical work.

Maxwell's equations are based on experimentally established facts, namely Coulomb's law, Ampère's law, Faraday's law, and the principle of conservation of electric charge. When most of classical physics was fundamentally revised as a result of Einstein's introduction of the special theory of relativity, Maxwell's equations remained intact. To this day, they stand as the most general mathematical statements of the fundamental natural laws that govern all of classical electrodynamics. **The basic justification and validity of Maxwell's equations lies in their consistency with physical experiments over the entire range of the experimentally observed electromagnetic spectrum, extending from cosmic rays at frequencies greater than 10^{22} Hz to the so-called micropulsations at frequencies of 10^{-3} Hz.** The associated practical applications cover an equally wide range, from the use of gamma rays (10^{18}–10^{22} Hz) for cancer therapy to the use of waves at frequencies of a few hertz and below for geophysical prospecting. Electromagnetic wave theory as embodied in Maxwell's equations has provided the underpinning for the development of many vital practical tools of our technological society, including broadcast radio, radar, television, cellular phones, optical communications, the Global Positioning Systems (GPS), microwave heating and processing, and X-ray imaging.

The differential forms of Maxwell's equations in the time domain are:

$$\nabla \times \overline{E} = -\frac{\partial \overline{B}}{\partial t} \qquad \text{Faraday's Law}$$

$$\nabla \times \overline{H} = \overline{J} + \frac{\partial \overline{D}}{\partial t} \qquad \text{Ampère's Law}$$

$$\nabla \cdot \overline{D} = \rho \qquad \text{Gauss's Law}$$

$$\nabla \cdot \overline{B} = 0 \qquad \text{Gauss's Magnetism Law}$$

The field variables are defined as:

$\overline{E}$	electric field	[volts/meter; $V{\cdot}m^{-1}$]
$\overline{H}$	magnetic field	[amperes/meter; $A{\cdot}m^{-1}$]
$\overline{B}$	magnetic flux density	[Tesla; T]
$\overline{D}$	electric displacement	[coulombs/m^2; $C{\cdot}m^{-2}$]
$\overline{J}$	electric current density	[amperes/m^2; $A{\cdot}m^{-2}$]
ρ	electric charge density	[coulombs/ m^3; $C{\cdot}m^{-3}$]

- Faraday's Law shows that a changing magnetic field within a loop gives rise to an induced current, which is due to an electromotive force or voltage within that circuit.
- Ampère's law with Maxwell's addition states that magnetic fields can be generated in two ways: by electric current (this was the original "Ampère's law") and by changing electric fields (this was "Maxwell's addition", which he called displacement current).
- Gauss's law describes the relationship between a static electric field and electric charges: a static electric field points away from positive charges and towards negative charges, and the net outflow of the electric field through a closed surface is proportional to the enclosed charge.
- Gauss's law for magnetism states that electric charges have no magnetic analogs, called magnetic monopoles. Instead, the magnetic field of a material is attributed to a dipole, and the net outflow of the magnetic field through a closed surface is zero.

Maxwell's equations embody all of the essential features of electromagnetics, including the ideas that light is electromagnetic in nature, that electric fields that change in time create magnetic fields in the same way as time-varying voltages induce electric currents in wires,

and that the source of electric and magnetic energy resides not only on the body that is electrified or magnetized but also, and to a far greater extent, in the surrounding medium. However, arguably the most important and far-reaching implication of Maxwell's equations is the idea that electric and magnetic effects can be transmitted from one point to another through the intervening space whether that be empty or filled with matter.

To appreciate the concept of propagation of electromagnetic waves in empty space, it is useful to think of other wave phenomena that we may observe in nature. When a pebble is dropped into a body of water, the water particles in the vicinity of the pebble are immediately displaced from their equilibrium positions. The motion of these particles disturbs adjacent particles, causing them to move, and the process continues, creating a wave. Because of the finite velocity of the wave, a finite time elapses before the disturbance causes the water particles at distant points to move. Thus the initial disturbance produces, at distant points, effects that are retarded in time. The water wave consists of ripples that move along the surface away from the initial disturbance. Although the motion of any particular water particle is essentially a small up-and-down movement, the cumulative effects of all the particles produce a wave that moves radially outward from the point at which the pebble is dropped. Another excellent example of wave propagation is the motion of sound through a medium. In air, this motion occurs through the to-and-fro movement of the air molecules, but these molecules do not actually move along with the wave.

Electromagnetic waves consist of time-varying electric and magnetic fields. Suppose an electrical disturbance, such as a change in the current through a conductor, occurs at a point in a region. The time-varying electric field resulting from the disturbance generates a time-varying magnetic field. The time-varying magnetic field, in turn, produces an electric field. These time-varying fields continue to generate one another in an ever-expanding region, and the resulting wave propagates away from the location of the initial disturbance. When electromagnetic waves propagate in vacuum, there is no vibration of physical particles as in the case of water and sound waves. Nevertheless, the velocity of wave propagation is limited by the speed of light, so that the fields produced at distant points are retarded in time with respect to those near the source.

Words and Expressions

encompass [ɪn'kʌmpəs]	*vt.* 包含，包括，涉及（大量事物）；包围；围绕
electrodynamics [ɪˌlektroʊdaɪ'næmɪks]	*n.* 电动力学
electrostatics [ɪˌlektroʊ'stætɪks]	*n.*静电学
amber ['æmbər]	*n.* 琥珀；琥珀色
magnetostatics [mægˌniːtoʊ'stætɪks]	*n.* 静磁学
mineral ore	矿石；矿砂
compact ['kɑːmpækt]	*adj.* 紧密的；紧凑的；简洁的
	vt. 使简洁；使装满；压实
exert [ɪg'zɜːrt]	*vt.* 运用；施加
momentum [moʊ'mentəm]	*n.* 动量；推进力；动力；势头
lumped circuit	集总参数电路；集总电路

input excitation 输入激励
step excitation 阶跃激励
computer-aided optimization 计算机辅助优化
distributed circuit 分布参数电路
conspicuously [kən'spɪkjuəsli] *adv.* 明显地；惹人注目地
permeate ['pɜːrmieɪt] *v.* 弥漫；渗透；普及
stray capacitance 杂散电容
stray inductance 杂散电感；漏电感
radically ['rædɪkli] *adv.* 根本地；彻底地
power loss 功率损耗；功率损失；功率消耗；动力损失
graded index optical fiber 渐变折射率光纤
principle of conservation of electric charge 电荷守恒原理
micropulsation [ˌmaɪkroʊpʌl'seɪʃn] *n.* 微脉动
geophysical prospecting 地球物理勘探
underpin [ˌʌndər'pɪn] *v.* 加强，巩固，构成（……的基础等）；支撑
embody [ɪm'bɑːdi] *vt.* 使具体化；包含；代表；体现
equilibrium [ˌiːkwɪ'lɪbriəm, ekwɪ'lɪbriəm] *n.* 平衡；均衡；均势
molecule ['mɑːlɪkjuːl] *n.* 分子

Notes

1. A branch of electromagnetics, dealing with electric charges at rest (static electricity) named electrostatics, provides a framework within which we can understand the simple fact that a piece of amber, when rubbed, attracts itself to other small objects.

处理静止电荷（静电）的电磁学分支称为静电学，它提供了一个框架，我们可以在其中理解一个简单的事实，即一块琥珀在被摩擦时会吸引其他小物体。

2. In a lumped circuit, the individual lumped circuit elements are connected to each other and to sources and loads within or outside the circuit by conducting paths of negligible electrical length.

在集总电路中，各个集总电路元件通过电气长度可忽略的导电路径相互连接，并连接到电路内部或外部的电源和负载。

3. Since the entire circuit is considered in one lump, the effect of the input excitation is instantaneously felt at all points in the circuit, and all currents and voltages (such as I_1, I_2, V_1, and the load current I_L) either attain new values or respond by starting to change at $t = 0$, in accordance with the natural response of the circuit to a step excitation as determined by the solution of its corresponding differential equation.

由于将整个电路视为一个整体，因此输入激励的影响同时到达电路中的所有节点，求解相应的微分方程可得电路对阶跃激励的自然响应，从而决定所有电流和电压（如 I_1、I_2、V_1 和负载电流 I_L）是达到新值，还是在 $t = 0$ 时开始变化。

4. The basic justification and validity of Maxwell's equations lies in their consistency with physical experiments over the entire range of the experimentally observed electromagnetic spectrum, extending from cosmic rays at frequencies greater than 10^{22} Hz to the so-called

micropulsations at frequencies of 10^{-3} Hz.

麦克斯韦方程组的基本合理性和有效性在于它们在整个实验观测的电磁频谱范围内与物理实验结果保持一致，从频率大于 10^{22} Hz 的宇宙射线延伸到频率为 10^{-3} Hz 的所谓微脉动。

5. Maxwell's equations embody all of the essential features of electromagnetics, including the ideas that light is electromagnetic in nature, that electric fields that change in time create magnetic fields in the same way as time-varying voltages induce electric currents in wires, and that the source of electric and magnetic energy resides not only on the body that is electrified or magnetized but also, and to a far greater extent, in the surrounding medium.

麦克斯韦方程组体现了电磁学的所有基本特征，包括光在本质上是电磁的，随时间变化的电场产生磁场的方式与时变电压在电线中感应电流的方式相同，以及电能和磁能的来源不仅存在于带电或磁化的物体上，而且在更大程度上存在于周围的介质中。

Exercises

1. Match the terms (1)–(6) with the definitions A–F.

(1) stray/parasitic capacitance	A. the process of using magnetic fields to produce voltage, and in a closed circuit, a current
(2) electromagnetic induction	B. the phenomenon in which the absorption of electromagnetic radiation, as light, of sufficiently high frequency by a surface, usually metallic, induces the emission of electrons from the surface
(3) photoelectric effect	C. an unavoidable and usually unwanted capacitance that exists between the parts of an electronic component or circuit simply because of their proximity to each other
(4) permittivity	D. a material's property that affects the Coulomb force between two point charges in the material
(5) photon	E. one of the waves that are propagated by simultaneous periodic variations of electric and magnetic field intensity and that include radio waves, infrared, visible light, ultraviolet, X-rays, and gamma rays
(6) electromagnetic wave	F. an elementary particle, the quantum of the electromagnetic field, the basic "unit" of light and all other forms of electromagnetic radiation

2. Translate into Chinese.

(1) Electromagnetic engineering problems generally involve the design and use of materials that can generate, transmit, guide, store, distribute, scatter, absorb, and detect electromagnetic waves and energy.

(2) Applications of electrostatics in industry encompass diverse areas—cathode ray tubes (CRT) and flat panel displays, which are widely used as display devices for computers and oscilloscopes; ink jet printers, which can produce good-quality printing at very fast speeds; and photocopy machines are all based on electrostatic fields.

(3) Hertz not only showed the existence of electromagnetic waves, but also demonstrated that the waves, which had wavelength ten million times that of the light waves, could be diffracted, refracted and polarised.

3. Translate into English.

（1）电磁波可以通过真空、空气或其他传输介质传播，取决于使用电磁频谱的哪一部分。

（2）电磁元件基本可分为3类，即其尺寸远小于、可比于或远大于波长。

（3）基本上，高斯定理描述了封闭表面内的总电荷与其在整个表面上产生的场之间的关系。

4. Read the following article and write a summary.

Electromagnetic interference (EMI) is a phenomenon that occurs when the operation of an electronic device is disturbed by an electromagnetic (EM) field and typically occurs when the device is close to an EM field, which disrupts the radio frequency spectrum. EMI is a common issue for electronic components used in various industries, such as military, defence, communication systems, appliances, and aerospace.

EMI can arise from various sources, both natural and man-made. Naturally occurring interference can arise from various natural sources and phenomena such as atmospheric types of noise like lightning or electrical storms. Man-made interference generally occurs due to the activities of other electronic devices in the vicinity of the device (also known as the receiver) experiencing the interference. In a perfect world, electromagnetic interference would not be present; it is an unwanted signal at the signal receiver.

There are three different methods to help reduce or eliminate EMI: filtering, shielding, and grounding.

(1) Filtering

A direct way to get rid of unwanted signals is through filtering them out, and in this instance, passive filters work well, and they're used in most new equipment to minimise EMI. Filtering usually starts with an AC line filter that prevents bad signals from entering the power supply or powered circuits. It keeps internal signals from being added to the AC line.

Filtering is commonly used with cables and connectors on lines into and out of a circuit, and some special connectors can have built-in low-pass filters whose main job is to soften digital waveforms to increase the rise and fall times and reduce harmonic generation, according to Electronic Design.

Low-voltage analog signals will typically need to be amplified and subsequently filtered to reduce background noise before digitization. Signal conditioning often requires the input signal to be filtered and isolated to remove unwanted background noise and remove voltage signals far beyond the in-line digitiser's range. Filtering is commonly used to reject noise outside of a pre-defined frequency range.

(2) Shielding

On the other hand, shielding is the preferred way to contain radiation or coupling in source or victim devices, and it usually consists of encasing the circuit inside a completely sealed enclosure, such as a metallic box. Shielding is crucial because it reflects electromagnetic waves into the enclosure and absorbs waves that aren't reflected.

In most cases, a small amount of radiation ends up penetrating the shield if it's not thick enough. Practically any common metal can be used for shielding (e.g., scopper, steel, aluminium).

(3) Grounding

Grounding is the establishment of an electrically conductive path between an electrical or electronic element of a system and a reference point or plane referenced to ground, according to

DAU, and it can refer to an electrical connection made to Earth as well.

Some best practices to keep in mind to achieve the best possible ground include:

- Keep leads away from internal circuits or other components to ground as short as possible to reduce inductance.
- Use multiple grounding points on a large ground plane for best results.
- Try to isolate circuits from ground if ground loop voltages can't be controlled any other way.
- Maintain separate grounds for analog and digital circuits—you can combine them later at a single point.

Utilising any one of these three methods above can help you not only reduce EMI but can help ensure your equipment is less vulnerable to future interference and can assist with reducing emissions.

5. Language study: Grammar links

Sentences in a text are held together by grammar links. Note the links in this paragraph:

> Metal detectors are used to locate hidden metal objects such as water pipes. They contain a search coil and a control box. The coil is mounted in the search head. When an AC voltage from the box is applied to the coil, a magnetic field is created around it. In turn this induces a current in any metal object the head passed over.

This text illustrates some common grammar links:

- Nouns become pronouns:
  metal detectors becomes *they*.
- Repeated nouns change from *a* to *the* and sometimes words are dropped:
  a search coil becomes *the coil.*
- Clauses and even sentences become *this* or *that*:
  a magnetic field is created around it becomes *this*.

Now mark the grammar links in this paragraph by joining the words in italics with the words they refer to.

> When an AC voltage is applied to the search coil, a magnetic field is produced around *it*. If there is a metal object under the ground. *The field* induces an electric current *in the object*. The induced current in turn creates a magnetic field around the object. *This* induces a voltage in the search coil. *The induced voltage* is converted into an audible note by the circuitry in the control box. *This sound* guides the treasure hunter to *the buried object*.

Unit 12 Microwave Communication

12.1 Microwave Concepts

Microwaves are the ultrahigh, superhigh, and extremely high frequencies directly above the lower frequency ranges where most radio communication now takes place and below the optical frequencies that cover infrared, visible, and ultraviolet light. **The outstanding benefits for radio communication of these extremely high frequencies and accompanying short wavelengths more than offset any problems connected with their use.** Today, most new communication services and equipment use microwaves or the millimeter-wave bands.

12.1.1 Microwave Frequencies and Bands

The practical microwave region is generally considered to extend from 1 to 30 GHz, although some definitions include frequencies up to 300 GHz. Microwave signals of 1 to 30 GHz have wavelengths of 30 cm (about 1 ft) to 1 cm (or about 0.4 in).

The microwave frequency spectrum is divided up into groups of frequencies, or bands, as shown in Table 12.1. Frequencies above 30 GHz are referred to as millimeter waves because their wavelength is only millimeters (mm). Note that parts of the L and S bands overlap part of the UHF band, which is 300 to 3,000 MHz. Recent developments in semiconductor technology, such as smaller-geometry silicon and GaN, have made millimeter waves practical and useful. Frequencies above 300 GHz are in the submillimeter band. Currently the only communication in the submillimeter ranges is for research and experimental activities.

Table 12.1 Microwave and millimeter-wave frequency bands

Band Designation	Frequency Range
L band	1–2 GHz
S band	2–4 GHz
C band	4–8 GHz
X band	8–12 GHz
Ku band	12–18 GHz
K band	18–26.5 GHz
Ka band	26.5–40 GHz

Continued

Band Designation	Frequency Range	
Q band	30–50 GHz	Millimeter Waves
U band	40–60 GHz	
V band	50–75 GHz	
E band	60–90 GHz	
W band	75–110 GHz	
F band	90–140 GHz	
D band	110–170 GHz	
Submillimeter	>300 GHz	

12.1.2 Benefits of Microwaves

Every electronic signal used in communication has a finite bandwidth. When a carrier is modulated by an information signal, sidebands are produced. The resulting signal occupies a certain amount of bandwidth, called a channel, in the radio-frequency (RF) spectrum. **Channel center frequencies are assigned in such a way that the signals using each channel do not overlap and interfere with signals in adjacent channels.** As the number of communication signals and channels increases, more and more of the spectrum space is used up. Over the years as the need for electronic communication has increased, the number of radio communication stations has increased dramatically. As a result, the radio spectrum has become extremely crowded.

Use of the radio-frequency spectrum is regulated by the government. The various classes of radio communication are assigned specific areas in the spectrum within which they can operate. Over the years, the available spectrum space, especially below 300 MHz, has essentially been used up. In many cases, communication services must share frequency assignments. In some areas, new licenses are no longer being granted because the spectrum space for that service is completely full. In spite of this, the demand for new electronic communication channels continues.

Technological advances have helped solve some problems connected with overcrowding. For example, the selectivity of receivers has been improved, so that adjacent channel interference is not as great. This permits stations to operate on more closely spaced frequencies.

On the transmitting side, new techniques have helped squeeze more signals into the same frequency spectrum. A classic example is the use of SSB, where only one sideband is used rather than two, thereby cutting the spectrum usage in half. Limiting the deviation of FM signals also helps to reduce bandwidth. In data communication, new modulation techniques such as PSK and QAM have been used to narrow the required bandwidth of transmitted information or to transmit at higher speeds in narrower bandwidths. Digital compression methods also transmit more information through a narrow channel. Multiplexing techniques help put more signals or information into a given bandwidth. Broadband schemes such as spread spectrum and orthogonal frequency division multiplexing (OFDM) allow many radios to share a single bandwidth.

The other major approach to solving the problem of spectrum crowding has been to move into the higher frequency ranges. Initially, the VHF and UHF bands were tapped. Today, most new communication services are assigned to the microwave and millimeter-wave regions.

To give you some idea why more bandwidth is available at the higher frequencies, let's take an example. Consider a standard AM broadcast station operating on 1,000 kHz. The station is permitted to use modulating frequencies up to 5 kHz, thus producing upper and lower sidebands 5 kHz above and below the carrier frequency, or 995 and 1,005 kHz. This gives a maximum channel bandwidth of 1,005 kHz–995 kHz = 10 kHz. This bandwidth represents 10/1,000 = 0.01 or 1 percent of the spectrum space at that frequency.

Now consider a microwave carrier frequency of 4 GHz. One percent of 4 GHz is 0.01×4,000,000,000 Hz = 40,000,000 Hz or 40 MHz. A bandwidth of 40 MHz is incredibly wide. In fact, it represents all the low-frequency, medium-frequency, and high-frequency portions of the spectrum plus 10 MHz. This is the space that might be occupied by a 4-GHz carrier modulated by a 20-MHz information signal. Obviously, most information signals do not require that kind of bandwidth. A voice signal, e.g., would take up only a tiny fraction of that. A 10-kHz AM signal represents only 10,000/4,000,000,000 = 0.000,25 percent of 4 GHz. **Up to 4,000 AM broadcast stations with 10-kHz bandwidths could be accommodated within the 40-MHz (1 percent) bandwidth.**

Obviously, then, the higher the frequency, the greater the bandwidth available for the transmission of information. This not only gives more space for individual stations, but also allows wide-bandwidth information signals such as video and high-speed digital data to be accommodated. The average TV signal has a bandwidth of approximately 6 MHz. It is impractical to transmit video signals on low frequencies because they use up entirely too much spectrum space. That is why most TV transmission is in the VHF and UHF ranges. There is even more space for video in the microwave region.

Wide bandwidth also makes it possible to use various multiplexing techniques to transmit more information. Multiplexed signals generally have wide bandwidths, but these can be easily handled in the microwave region. Finally, transmission of high-speed binary information usually requires wide bandwidths, and these are also easily transmitted on microwave frequencies.

12.1.3 Disadvantages of Microwaves

The higher the frequency, the more difficult it becomes to analyze electronic circuits. The analysis of electronic circuits at lower frequencies, say, those below 30 MHz, is based upon current-voltage relationships (circuit analysis). Such relationships are simply not usable at microwave frequencies. Instead, most components and circuits are analyzed in terms of electric and magnetic fields (wave analysis). Thus techniques commonly used for analyzing antennas and transmission lines can also be used in designing microwave circuits. Measuring techniques are, of course, also different. In low-frequency electronics, currents and voltages are calculated. In microwave circuits, measurements are of electric and magnetic fields. Power measurements are more common than voltage and current measurements.

Another problem is that at microwave frequencies, conventional components become difficult to implement. For example, a common resistor that looks like pure resistance at low frequencies does not exhibit the same characteristics at microwave frequencies. The short leads of a resistor, although they may be less than an inch, represent a significant amount of inductive reactance at very high frequencies. A small capacitance also exists between the leads. These small stray and distributed reactances are sometimes called residuals. Because of these effects, at microwave frequencies a simple resistor looks like a complex *RLC* circuit. This is also true of inductors and capacitors. Figure 12.1 shows equivalent circuits of components at microwave frequencies.

To physically realize resonant circuits at microwave frequencies, the values of inductance and capacitance must be smaller and smaller. Physical limits become a problem. Even a 0.5-in piece of wire represents a significant amount of inductance at microwave frequencies. Tiny surface-mounted chip resistors, capacitors, and inductors have partially solved this problem. Furthermore, as integrated-circuit dimensions have continued to decrease, smaller and smaller on-chip inductors and capacitors have been made successfully.

Another solution is to use distributed circuit elements, such as transmission lines, rather than lumped components, at microwave frequencies. When transmission lines are cut to the appropriate length, they act as inductors, capacitors, and resonant circuits. Special versions of transmission lines known as striplines, microstrips, waveguides, and cavity resonators are widely used to implement tuned circuits and reactances.

In addition, because of inherent capacitances and inductances, conventional semiconductor devices such as diodes and transistors simply will not function as amplifiers, oscillators, or switches at microwave frequencies.

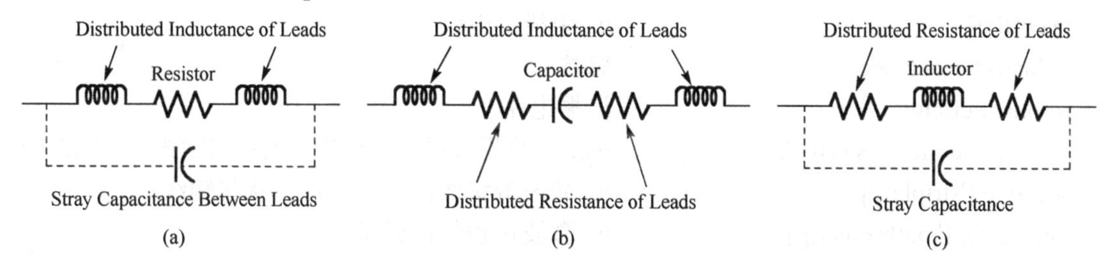

Figure 12.1 Equivalent circuits of components at microwave frequencies: (a) resistor; (b) capacitor; (c) inductor

Another serious problem is transistor transit time—the amount of time it takes for the current carriers (holes or electrons) to move through a device. At low frequencies, transit times can be neglected; but at microwave frequencies, they are a high percentage of the actual signal period.

This problem has been solved by designing smaller and smaller microwave diodes, transistors, and ICs and using special materials such as gallium arsenide (GaAs), indium phosphide (InP), and silicon germanium (SiGe) in which transit time is significantly less than in silicon. In addition, specialized components have been designed for microwave applications. This is particularly true for power amplification, where special vacuum tubes known as klystrons, magnetrons, and traveling-wave tubes are the primary components used for power amplification. However, newer GaN semiconductor transistors are now useful well into the millimeter-wave range.

Another problem is that microwave signals, as do light waves, travel in perfectly straight lines.

This means that the communication distance is usually limited to line-of-sight range. Antennas must be very high for long-distance transmission. Microwave signals penetrate the ionosphere, so multiple-hop communication is not possible. **The physics of electromagnetic waves indicates that the shorter the wavelength and the higher the frequency, the shorter the transmission range for a given power or antenna gain.**

Words and Expressions

visible ['vɪzəbl] *adj*. 看得见的；可见的；显而易见的
offset ['ɔːfset, 'ɑːfset] *v*. 抵消；弥补；补偿；形成分支
n. 偏移量；抵消；补偿
UHF: ultrahigh frequency 特高频
sideband ['saɪdˌbænd] *n*. 边带（频带；能带）
regulate ['regjuleɪt] *v*. 调整；校准；管理；控制
grant [grænt] *vt*. 授予；同意；承认；认为
SSB: single sideband 单边带
deviation [ˌdiːvi'eɪʃn] *n*. 偏差
QAM: quadrature amplitude modulation 正交振幅调制
spread spectrum 扩频
orthogonal frequency division multiplexing(OFDM) 正交频分复用
VHF: very high frequency 甚高频
tap [tæp] *v*. 开发；利用；窃听；轻拍；轻击
lead [liːd] *n*. 引线；引脚
inductive reactance 感抗
resonant circuit 谐振电路
surface-mounted ['sɜːrfɪs 'maʊntɪd] *adj*. 表面贴装的；表面安装的；表贴式的；贴面的
stripline ['strɪplaɪn] *n*. 微波带状线；微带线；带状传输线
microstrip ['maɪkroʊstrɪp] *n*. 微波传输带；微带
cavity resonator 谐振腔；空腔谐振器
tuned circuit 调谐电路；谐振腔
inherent [ɪn'hɪrənt] *adj*. 固有的；内在的
transit time 渡越时间
indium phosphide 磷化铟
klystron ['klaɪstrɑːn] *n*. 极超短波用电子管；速调管；克莱斯管
magnetron ['mægnəˌtrɑn] *n*. 磁控管
traveling-wave tube 行波管；行波电子管
line-of-sight [ˌlaɪn əv'saɪt] *n*. 视线

Notes

1. The outstanding benefits for radio communication of these extremely high frequencies and accompanying short wavelengths more than offset any problems connected with their use.

这些极高频率和伴随而来的短波对无线电通信的显著好处，足以抵消其使用带来的任何问题。

2. Channel center frequencies are assigned in such a way that the signals using each channel do not overlap and interfere with signals in adjacent channels.

信道中心频率的分配方式使得每个信道中的信号不会重叠和干扰相邻信道中的信号。

3. Up to 4,000 AM broadcast stations with 10-kHz bandwidths could be accommodated within the 40-MHz (1 percent) bandwidth.

40 MHz (1%) 带宽内可容纳多达 4 000 个带宽为 10 kHz 的 AM 广播电台。

4. Another solution is to use distributed circuit elements, such as transmission lines, rather than lumped components, at microwave frequencies.

另一种解决方案是在微波频率下使用分布参数电路元件，如传输线，而不是集总元件。

5. The physics of electromagnetic waves indicates that the shorter the wavelength and the higher the frequency, the shorter the transmission range for a given power or antenna gain.

电磁波的物理性质表明，波长越短，频率越高，对于给定的功率或天线增益，其传输范围就越短。

12.2 Microwave Communication Systems

Like any other communication system, a microwave communication system uses transmitters, receivers, and antennas. The same modulation and multiplexing techniques used at lower frequencies are also used in the microwave range. But the RF part of the equipment is physically different because of the special circuits and components that are used to implement it.

12.2.1 Transmitters

Like any other transmitter, a microwave transmitter starts with a carrier generator and a series of amplifiers. It also includes a modulator followed by more stages of power amplification. The final power amplifier applies the signal to the transmission line and antenna. The carrier generation and modulation stages of a microwave application are similar to those of lower-frequency transmitters. Only in the later power amplification stages are special components used.

Figure 12.2 shows several ways that microwave transmitters are implemented.

In the transmitter circuit shown in Figure 12.2(a), a microwave frequency is first generated in the last multiplier stage. The operating frequency is 1,680 MHz, where special microwave components and techniques must be used. **Instead of tuned circuits made of loops of wire for inductors and discrete capacitors, microstrip transmission lines are used as tuned circuits and as impedance-matching circuits.** SAW filters are the most commonly used filters in low-power circuits. One or more additional power amplifiers are then used to boost the signal to the desired power level. Both bipolar and MOSFET microwave power transistors are available that give power levels up to several hundred watts. When FM is used, the remaining power amplifiers can also be class C, which provides maximum efficiency. For phase modulation and QAM, linear amplifiers are needed. If more power is desired, several transistor power amplifiers can be paralleled, as in Figure 12.2(a).

If AM is used in a circuit like that in Figure 12.2(a), an amplitude modulator can be used to modulate one of the lower-power amplifier stages after the multiplier chain. When this is done, the remaining power amplifier stages must be linear amplifiers to preserve the signal modulation.

For very high-power output levels—beyond several hundred watts—a special amplifier must be used, e.g., the klystron.

Figure 12.2(b) shows another possible transmitter arrangement, in which a mixer is used to up-convert an initial carrier signal with or without modulation to the final microwave frequency.

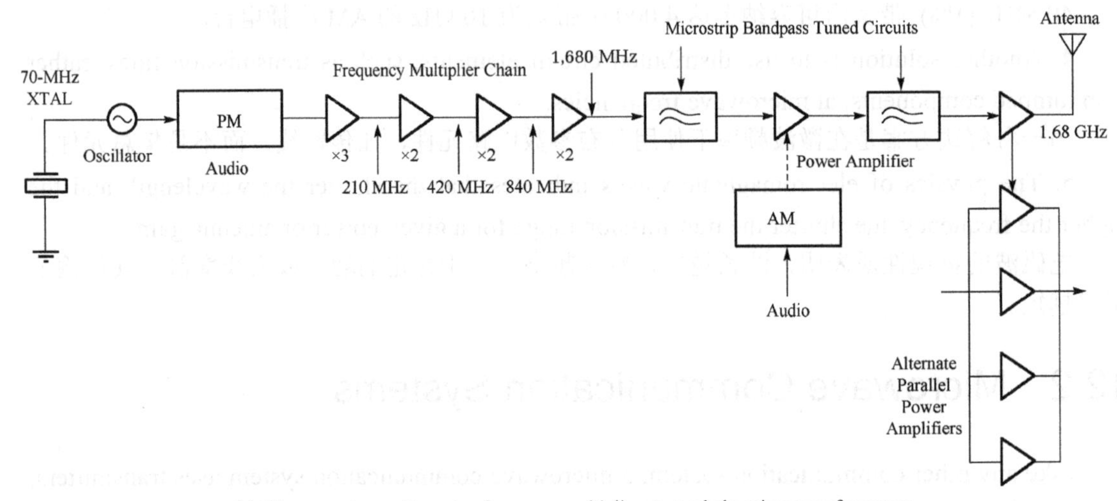

(a) Microwave transmitter using frequency multipliers to reach the microwave frequency

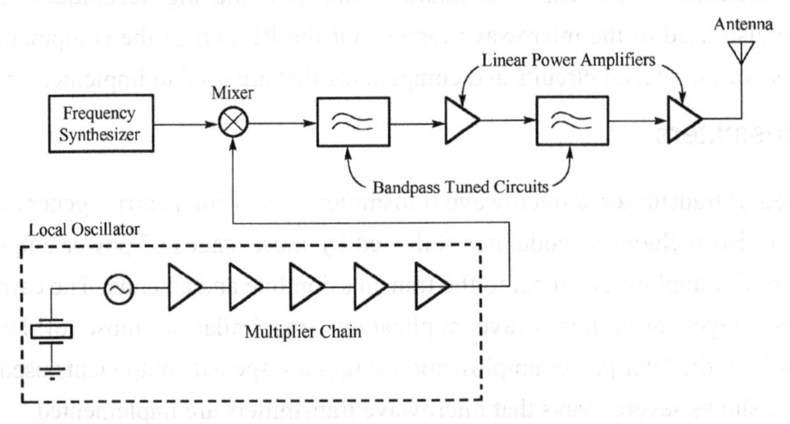

(b) Microwave transmitter using up-conversion with a mixer to achieve an output in the microwave range

Figure 12.2 Microwave transmitters

The synthesizer output and a microwave local oscillator signal are applied to the mixer. The mixer then translates the signal up to the desired final microwave frequency. **A conventional crystal oscillator using fifth-overtone VHF crystals followed by a chain of frequency multipliers can be used to develop the local oscillator frequency.** Alternatively, one of several special microwave oscillators could be used, e.g., a Gunn diode, a microwave semiconductor in a cavity resonator, or a dielectric resonator oscillator(DRO).

The output of the mixer is the desired final frequency at a relatively low power level, usually tens or hundreds of milliwatts at most. Linear power amplifiers are used to boost the signal to its final power level. At frequencies less than about 10 GHz, a microwave transistor can be used. At the higher frequencies, special microwave power tubes are used.

The bandpass tuned circuits shown in Figure 12.2(b) can be microstrip transmission lines when transistor circuits are used, or cavity resonators when the special microwave tubes are used.

Modulation could occur at several places in the circuit in Figure 12.2(b). An indirect FM modulator might be used at the output of the frequency synthesizer; for some applications, a PSK modulator would be appropriate.

12.2.2 Receivers

Microwave receivers, like low-frequency receivers, are the superheterodyne type. Their front ends are made up of microwave components. Most receivers use double conversion. A first down-conversion gets the signal into the UHF or VHF range, where it can be more easily processed by standard methods. A second conversion reduces the frequency to an IF appropriate for the desired selectivity.

Figure 12.3 is a general block diagram of a double-conversion microwave receiver. The antenna is connected to a tuned circuit, which could be a cavity resonator or a microstrip or stripline tuned circuit. The signal is then applied to a special RF amplifier known as a low-noise amplifier (LNA). Special low-noise transistors, usually gallium arsenide FET amplifiers, must be used to provide some initial amplification. Another tuned circuit connects the amplified input signal to the mixer. Most mixers are of the doubly balanced diode type, although some simple single-diode mixers are also used.

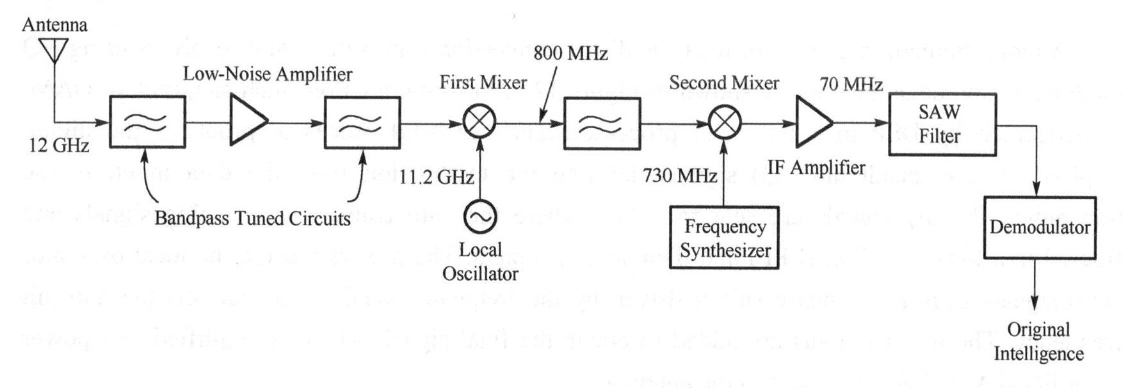

Figure 12.3 A microwave receiver

The local oscillator signal is applied to the mixer. The mixer output is usually within the UHF or VHF range. The 700- to 800-MHz range is typical. A SAW filter selects out the difference signal, which in Figure 12.3 is 12 GHz−11.2 GHz = 0.8 GHz, or 800 MHz.

The remainder of the receiver is typical of other superheterodynes. Note that the desired selectivity is obtained with a SAW filter, which is sometimes used to provide a specially shaped IF response. Many of the newer microwave cell phones and LAN receivers are of the direct conversion type, and selectivity is obtained with *RC* and DSP filters.

In more recent microwave equipment such as cell phones and wireless networking interfaces, the microwave frequencies are generated by a phase-locked loop (PLL) operating as a multiplier. See Figure 12.4. The VCO produces the desired local oscillator (LO) or final transmitting frequency directly. The VCO frequency is controlled by the phase detector and its low-pass loop filter. The frequency divider and input crystal determine the output frequency. Recent advances in quartz crystal design permit input oscillator frequencies up to 200 MHz. In Figure 12.4, the 155-MHz crystal combined with a frequency divider of 20 produces a VCO output at 155 MHz×20 = 3,100 MHz, or 3.1 GHz. The ÷20 divider reduces the 3.1-GHz output to 155 MHz to match the input crystal signal at the phase detector, as required for closed-loop control. Of course, the divider can also be part of a microcontroller-based frequency synthesizer that is designed to permit setting the output to multiple channel frequencies as required by the application.

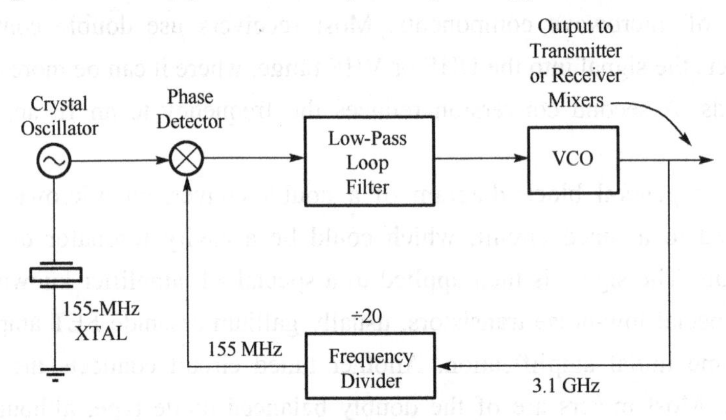

Figure 12.4 A phase-locked loop (PLL) multiplier is the primary signal source in modern microwave transceivers

A more common architecture today is direct conversion transmitters and receivers using I/Q modulators and demodulators, as shown in Figure 12.5. Most modulation, such as QPSK or QAM, is performed by DSP in a baseband processor. The baseband processor produces the digital in-phase (I) and quadrature (Q) signals defining the modulation from the data input. In the transmitter, the I/Q signals are sent to DACs where they are converted to analog signals and filtered in a low-pass filter (LPF) and then sent to mixers. The mixers receive the local oscillator (LO) signals from a 90° phase shifter driven by the frequency synthesizer that sets the transmit frequency. The mixer outputs are added to create the final signal, which is amplified in a power amplifier (PA) before being sent to the antenna.

The I/Q modulator is in IC form and can usually handle signals in the 200-MHz to 6-GHz range. If a higher final frequency is needed, the modulator output is sent to an up-converting mixer with its own local oscillator. The resulting higher frequency is then sent to the PA.

At the receiver, the signal from the antenna is fed to a low-noise amplifier (LNA) and a bandpass filter (BPF) to define the bandwidth. The signal is then sent to the mixers in the I/Q demodulator. If higher frequencies are involved, the LNA signal may go to a down-converting mixer first and then to the I/Q demodulator mixers. The signal is then mixed with the LO signal from the synthesizer at the signal frequency. The mixer outputs are then filtered and sent to ADCs, where the digital I/Q signals are generated. These I/Q signals are then processed in the baseband circuits to recover the original data.

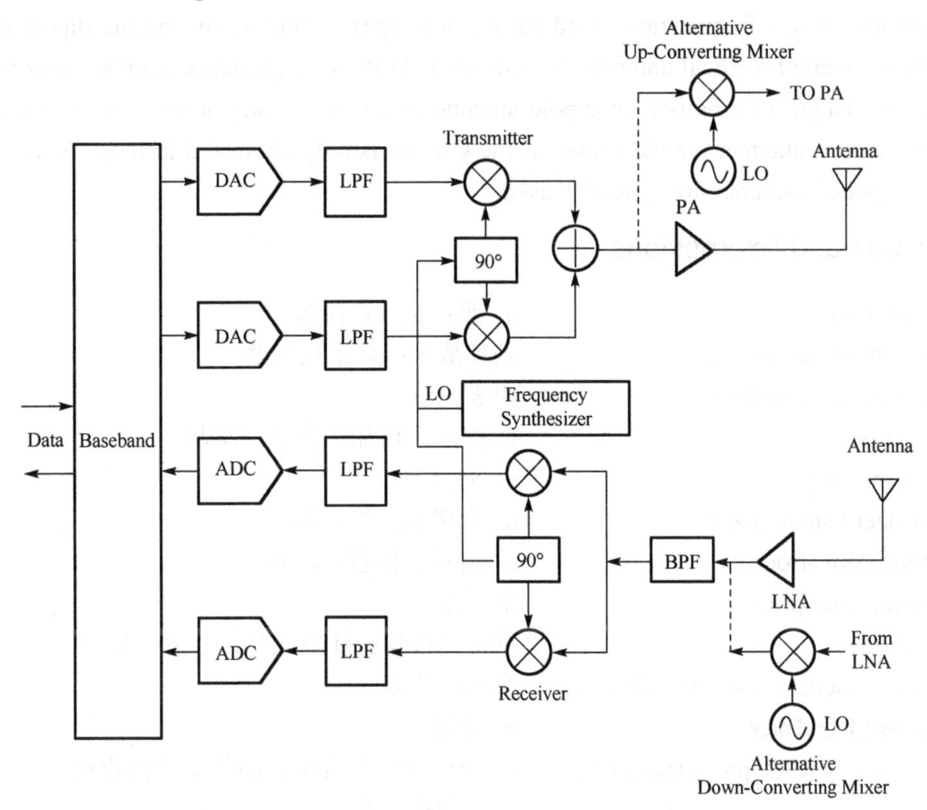

Figure 12.5 A direct conversion microwave transceiver

12.2.3 Transmission Lines

The transmission line most commonly used in lower-frequency radio communication is coaxial cable. **However, coaxial cable has very high attenuation at microwave frequencies, and conventional cable is unsuitable for carrying microwave signals except for very short runs, usually several feet or less.** Newer types of coaxial cables permit lengths of up to 100 ft at frequencies to 10 GHz.

Special microwave coaxial cable that can be used on the lower microwave bands—L, S, and C—is made of hard tubing rather than wire with an insulating cover and a flexible braid shield. The stiff inner conductor is separated from the outer tubing with spacers or washers, forming low-loss coaxial cable known as hard line cable. **The insulation between the inner conductor and the outer tubing can be air; in some cases, a gas such as nitrogen is pumped into the cable to minimize moisture buildup, which causes excessive power loss.** This type of cable is used for long runs of transmission lines to an antenna on a tower. At higher microwave frequencies, C band and upward, a special hollow rectangular or circular pipe called waveguide is used for the transmission line.

12.2.4 Antennas

At low microwave frequencies, standard antenna types, including the simple dipole and the one-quarter wavelength vertical antenna, are still used. At these frequencies, antenna sizes are very small; e.g., the length of a half-wave dipole antenna at 2 GHz is only about 3 in. A one-quarter wavelength vertical antenna for the center of the C band is only about 0.6 in long. At the higher frequencies, special antennas are generally used.

 Words and Expressions

stage [steɪdʒ]	*n.* 级；级数；阶段
power amplification	功率放大；功率放大器
SAW: surface acoustic wave	声表面波
boost [buːst]	*v.* 提高；增加；鼓励；举起
mixer ['mɪksər]	*n.* 混频器
synthesizer ['sɪnθəsaɪzər]	*n.* 合成器；综合器
overtone ['oʊvərtoʊn]	*n.* 倍频；倍音；泛音
frequency multiplier	倍频器
Gunn diode	耿氏二极管；体效应管；甘恩二极管
dielectric resonator oscillator(DRO)	介质振荡器
bandpass [ˌbænd'pæs]	*n.* 带通
superheterodyne [ˌsupər'hetərəˌdaɪn]	*n.* 超外差式收音机；超外差式接收机 *adj.* 超外差的
IF: intermediate frequency	中间频率
low-noise amplifier(LNA)	低噪声放大器
phase-locked loop(PLL)	锁相环
VCO: voltage controlled oscillator	压控振荡器
local oscillator(LO)	本地振荡器
frequency divider	分频器
in-phase (I) signal	同相信号
quadrature (Q) signal	正交信号
low-pass filter(LPF)	低通滤波器
up-converting [ʌp kən'vɜːrtɪŋ]	*n.* 上变频；向上转换
bandpass filter(BPF)	带通滤波器
down-converting [daʊn kən'vɜːrtɪŋ]	*n.* 下变频；向下转换
spacer ['speɪsə]	*n.*垫片；隔板；衬垫；撑挡
washer ['wɑːʃər]	*n.*（螺母等的）垫圈；垫片；衬垫
nitrogen ['naɪtrədʒən]	*n.* 氮；氮气
half-wave dipole antenna	半波偶极天线（或称赫兹天线）
one-quarter wavelength vertical antenna	1/4 波垂直天线（或称马可尼天线）

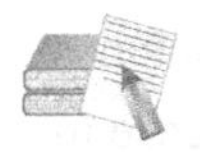

Notes

1. Instead of tuned circuits made of loops of wire for inductors and discrete capacitors, microstrip transmission lines are used as tuned circuits and as impedance-matching circuits.

微带传输线被用作调谐电路和阻抗匹配电路，而不是由电感线圈和离散电容器组成调谐电路。

2. A conventional crystal oscillator using fifth-overtone VHF crystals followed by a chain of frequency multipliers can be used to develop the local oscillator frequency.

使用五倍频 VHF 晶体的传统晶振，后接倍频器链，可用于产生本地振荡器频率。

3. However, coaxial cable has very high attenuation at microwave frequencies, and conventional cable is unsuitable for carrying microwave signals except for very short runs, usually several feet or less.

然而，同轴电缆在微波频率下具有非常高的衰减，传统电缆不适合传输微波信号，除非是非常短的距离，通常是几英尺或更短。

4. The insulation between the inner conductor and the outer tubing can be air; in some cases, a gas such as nitrogen is pumped into the cable to minimize moisture buildup, which causes excessive power loss.

内导体和外管之间的绝缘层可以是空气；在某些情况下，将氮气等气体泵入电缆中以最大限度地减少湿气积聚，这会导致过多的功率损耗。

Exercises

1. Match the terms (1)–(6) with the definitions A–F.

(1) frequency synthesizer	A. the time required for an electron or other charge carrier to travel between two electrodes in an electron tube or transistor
(2) BPF	B. a type of propagation that can transmit and receive data only where transmit and receive stations are in view of each other without any sort of an obstacle between them
(3) VCO	C. an electronic circuit that generates a range of frequencies from a single reference frequency
(4) transit time	D. an electronic oscillator whose output frequency is proportional to its input voltage
(5) LNA	E. a device that removes or attenuates frequencies both above and below the centre frequency at which it is set, and only passes a specific range of frequencies
(6) line-of-sight	F. a device that amplifies a very low-power signal without significantly degrading its signal-to-noise ratio

2. Translate into Chinese.

(1) The transmitter has two fundamental jobs: generating microwave energy at the required frequency and power level, and modulating it with the input signal so that it conveys meaningful information.

(2) The microstrip consists of a "conductor" strip, a "ground" and an intermediate dielectric substrate, and is commonly used in microwave integrated circuits for transmission of medium and high frequency microwave signals.

(3) Because of the line-of-sight transmission of microwave signals, highly directive antennas are preferred because they do not waste the radiated energy and because they provide an increase in gain, which helps offset the noise and distance problems at microwave frequencies.

3. Translate into English.

（1）从广义上讲，凡是能引导电磁波向一定方向传播的材料结构，都称为波导。

（2）波导通常用于高频区域，由单个金属管组成，根据其横截面的形状，会产生不同的波。

（3）顾名思义，点对点数字微波无线电（digital microwave radio，DMR）是一种数字传输技术，可在两点之间提供以微波频率运行的无线通信链路。

4. Read the following article and write a summary.

Marine communication between ships or with the shore was carried with the help of on board systems through shore stations and even satellites. While ship-to-ship communication was brought about by VHF radio, digital selective calling (DSC) came up with digitally remote control commands to transmit or receive distress alert, urgent or safety calls, or routine priority messages. DSC controllers can now be integrated with the VHF radio as per SOLAS (International Convention for the Safety of Life at Sea).

Satellite services, as opposed to terrestrial communication systems, need the help of geostationary satellites for transmitting and receiving signals, where the range of shore stations cannot reach. These marine communication services are provided by INMARSAT (a commercial company) and COSPAS-SARSAT (a multi-national government funded agency).

While INMARSAT gives the scope of two way communications, the COSPAS-SARSAT has a system that is limited to reception of signals from emergency position and places with no facilities of two way marine communications, indicating radio beacons (EPIRB).

For international operational requirements, the global maritime distress and safety system (GMDSS) has divided the world into four sub areas. These are four geographical divisions named as A1, A2, A3 and A4. Different radio communication systems are required by the vessel to be carried on board ships, depending on the area of operation of that particular vessel.

A1: it's about 20–30 nautical miles from the coast, which is under coverage of at least one VHF coast radio station in which continuous DSC alerting is available. Equipment used: A VHF, a DSC and a NAVTEX receiver (a navigational telex for receiving maritime and meteorological information).

A2: this area notionally should cover 400 nautical miles off shore but in practice it extends up to 100 nautical miles off shore but this should exclude A1 area. Equipment used: A DSC, and radio telephone (MF radio range) plus the equipment required for A1 area.

A3: this is the area excluding the A1 and A2 areas. But the coverage is within 70 degrees north and 70 degree south latitude and is within INMARSAT geostationary satellite range, where continuous alerting is available. Equipment used: A high frequency radio and/ or INMARSAT, a system of receiving MSI (maritime safety information) plus the other remaining systems for A1 and A2 areas.

A4: these are the areas outside sea areas of A1, A2 and A3. These are essentially the polar regions north and south of 70 degree of latitude. Equipment used: HF radio service plus those required for other areas.

All oceans are covered by HF marine communication services for which the IMO (International Maritime Organization) requires to have two coast stations per ocean region. Today almost all ships are fitted with satellite terminal for ship security alert system (SSAS) and for long range identification and tracking as per SOLAS requirements.

On distress, search and rescue operations from Maritime Rescue Co-ordination Centers are carried out among other methods, with the help of most of these marine navigation tools. Naturally, the sea has become a lot safer with these gadgets and other important navigation tools recommended by the IMO and as enshrined in GMDSS.

5. Language study: Reduced time clauses

Study these two actions:

(1) Ground waves pass over sand.

(2) Ground waves lose energy.

We can link these actions to make one sentence using a time clause:

When *ground waves pass over sand, they lose energy.*

Because the subject of both actions is the same—*ground waves*—there is a shorter method we can use to link the actions:

When passing *over sand, ground waves lose energy.*

When + ***-ing*** shows that Action (2) happens during the same period as Action (1).

Now study these two actions:

(1) The sky wave strikes the earth.

(2) The sky wave bounces back again.

Again we can link these actions to make one sentences, using a time clause:

When *the sky wave strikes the earth, it bounces back again.*

We can also link the actions in a shorter way:

On *strik**ing** the earch, the sky wave bounces back again.*

On + ***-ing*** shows that Action (2) follows immediately after Action (1).

Link these pairs of actions. Use short ways when it is possible.

(1) The switch is closed.

Current flows through the primary of the transformer.

(2) A cell discharges quickly.

A cell may become hot.

(3) The radar receiver receives the reflected signal.

The signal is compared with the transmitted signal.

(4) Microwave signals strike a high building.

Microwave signals are deflected.

(5) The remote control button is pressed.

The television set changes channel.

Unit 13 Basics of Computer Science

13.1 What Is a Computer?

A computer is an electronic device, operating under the control of instructions stored in its own memory, that can accept data, process the data according to specified rules, produce results, and store the results for future use.

Computers process data into information. Data is a collection of unprocessed items, which can include text, numbers, images, audios, and videos. Information conveys meaning and is useful to people.

Many daily activities either involve the use of or depend on information from a computer. As shown in Figure 13.1, for example, a computer processes several data items to print information in the form of a cash register receipt. In this simplified example, the item ordered, item price, quantity ordered, and amount received all represent data. The computer processes the data to produce the cash register receipt (information).

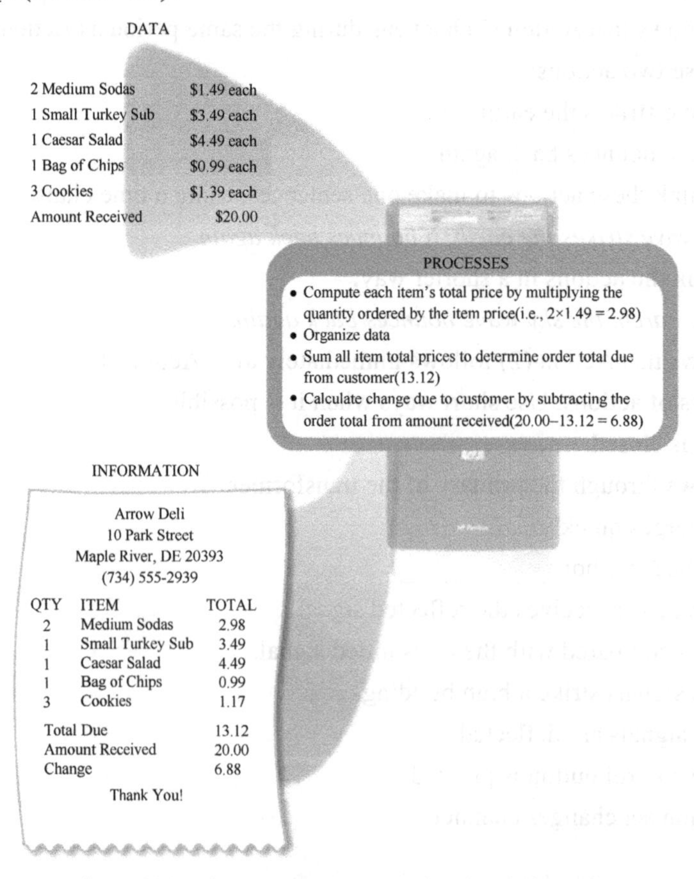

DATA
2 Medium Sodas $1.49 each
1 Small Turkey Sub $3.49 each
1 Caesar Salad $4.49 each
1 Bag of Chips $0.99 each
3 Cookies $1.39 each
Amount Received $20.00
PROCESSES
• Compute each item's total price by multiplying the quantity ordered by the item price(i.e., 2×1.49 = 2.98)
• Organize data
• Sum all item total prices to determine order total due from customer(13.12)
• Calculate change due to customer by subtracting the order total from amount received(20.00–13.12 = 6.88)
INFORMATION
Arrow Deli
10 Park Street
Maple River, DE 20393
(734) 555-2939
QTY ITEM TOTAL
2 Medium Sodas 2.98
1 Small Turkey Sub 3.49
1 Caesar Salad 4.49
1 Bag of Chips 0.99
3 Cookies 1.17
Total Due 13.12
Amount Received 20.00
Change 6.88
Thank You!

Figure 13.1 A computer processes data into information

Computers process data (input) into information (output). Computers carry out processes using instructions, which are the steps that tell the computer how to perform a particular task. A collection of related instructions organized for a common purpose is referred to as software. A computer often holds data, information, and instructions in storage for future use. Some people refer to the series of input, process, output, and storage activities as the information processing cycle. Most computers today communicate with other computers. As a result, communications also has become an essential element of the information processing cycle.

13.1.1 Computer Hardware

A computer contains many electric, electronic, and mechanical components known as hardware. These components include input devices, output devices, a system unit, storage devices, and communications devices(see Figure 13.2).

1. Input Devices

An input device is any hardware component that allows you to enter data and instructions into a computer. Five widely used input devices are keyboard, mouse, microphone, scanner, and webcam.

- A computer keyboard contains keys you press to enter data into the computer. For security purposes, some keyboards include a fingerprint reader, which allows you to work with the computer only if your fingerprint is recognized.
- A mouse is a small handheld device. With the mouse, you can control movement of a small symbol on the screen, called the pointer, and you make selections from the screen.
- A microphone allows you to speak into the computer.
- A scanner converts printed material (such as text and pictures) into a form the computer can use.
- A webcam is a digital video camera that allows you to create movies or take pictures and store them on the computer instead of on tape or film.

2. Output Devices

An output device is any hardware component that conveys information to one or more people. Three commonly used output devices are printer, monitor, and speakers.

- A printer produces text and graphics on a physical medium such as paper.
- A monitor displays text, graphics, and videos on a screen.
- Speakers allow you to hear music, voice, and other audio (sounds).

3. System Unit

The system unit is a case that contains the electronic components of the computer that are used to process data.

The circuitry of the system unit usually is part of or is connected to a circuit board called the motherboard. Two main components on the motherboard are processor and memory. The processor, also called a CPU (central processing unit), is the electronic component that interprets and carries out the basic instructions that operate the computer. Memory consists of electronic components that store instructions waiting to be executed and data needed by those instructions. **Although some forms of memory are permanent, most memory keeps data and instructions temporarily, which means its contents are erased when the computer is shut off.**

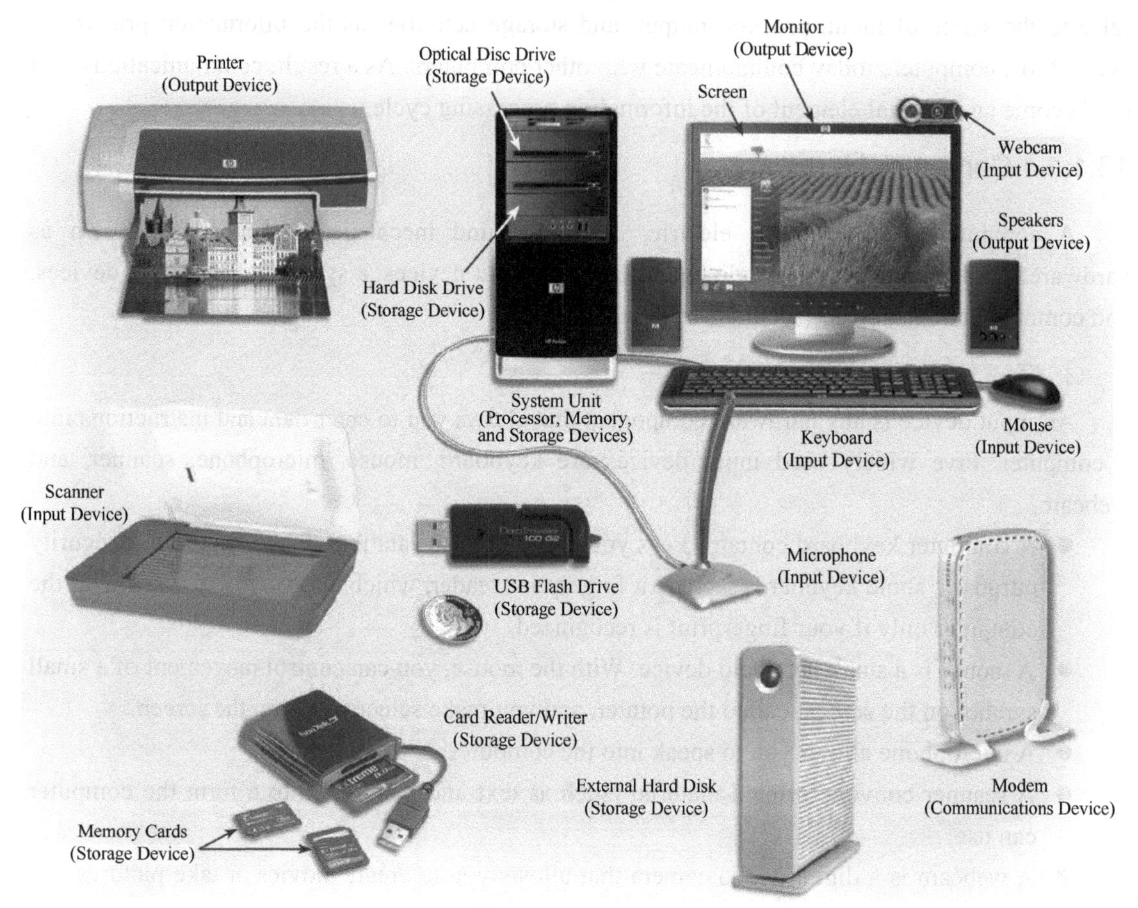

Figure 13.2 Common computer hardware components

4. Storage Devices

Storage holds data, instructions, and information for future use. For example, computers can store hundreds or millions of customer names and addresses. Storage holds these items permanently.

A computer keeps data, instructions, and information on storage media. Examples of storage media are USB flash drives, hard disks, optical discs, and memory cards. A storage device records (writes) and/or retrieves (reads) items to and from storage media. Drives and readers/writers, which are types of storage devices, accept a specific kind of storage media. For example, a DVD drive (storage device) accepts a DVD (storage media). Storage devices often function as a source of input because they transfer items from storage to memory.

- A USB flash drive is a portable storage device that is small and lightweight enough to be transported on a keychain or in a pocket. The average USB flash drive can hold about 4 billion characters. You plug a USB flash drive in a special, easily accessible opening on the computer.

- A hard disk provides much greater storage capacity than a USB flash drive. The average hard disk can hold more than 320 billion characters. Hard disks are enclosed in an airtight, sealed case. Although some are portable, most are housed inside the system unit. Portable hard disks are either external or removable. An external hard disk is a separate, freestanding unit, whereas you insert and remove a removable hard disk from the computer or a device connected to the computer.
- An optical disc is a flat, round, portable metal disc with a plastic coating. CDs, DVDs, and Blu-ray Discs are three types of optical discs. A CD can hold from 650 million to 1 billion characters. Some DVDs can store two full-length movies or 17 billion characters. Blu-ray Discs can store about 46 hours of standard video, or 100 billion characters.
- Memory cards are widely used in mobile devices, such as digital cameras, as the storage media. You can use a card reader/writer to transfer the stored items, such as digital photos, from the memory card to a computer or printer.

5. Communications Devices

A communications device is a hardware component that enables a computer to send (transmit) and receive data, instructions, and information to and from one or more computers or mobile devices. A widely used communications device is a modem.

Communications occur over cables, telephone lines, cellular radio networks, satellites, and other transmission media. Some transmission media, such as satellites and cellular radio networks, are wireless, which means they have no physical lines or wires.

13.1.2 Computer Software

Software, also called a program, consists of a series of related instructions, organized for a common purpose, that tells the computer what tasks to perform and how to perform them. You interact with a program through its user interface. The user interface controls how you enter data and instructions and how information is displayed on the screen. Software today often has a graphical user interface (GUI, pronounced gooey). With a graphical user interface, you interact with the software using text, graphics, and visual images such as icons. An icon is a miniature image that represents a program, an instruction, or some other object. You can use the mouse to select icons that perform operations such as starting a program.

The two categories of software are system software and application software. Figure 13.3 shows an example of each of these categories of software.

1. System Software

System software consists of the programs that control or maintain the operations of the computer and its devices. System software serves as the interface between the user, the application software, and the computer's hardware. Two types of system software are the operating system and utility programs.

Figure 13.3 Today's system software and application software usually have a GUI

- Operating system. An operating system is a set of programs that coordinates all the activities among computer hardware devices. It provides a means for users to communicate with the computer and other software. Many of today's computers use Microsoft's Windows, or Mac OS. When a user starts a computer, portions of the operating system are copied into memory from the computer's hard disk. These parts of the operating system remain in memory while the computer is on.
- Utility program. A utility program allows a user to perform maintenance-type tasks usually related to managing a computer, its devices, or its programs. For example, you can use a utility program to transfer digital photos to an optical disc. Most operating systems include several utility programs for managing disk drives, printers, and other devices and media. You also can buy utility programs that allow you to perform additional computer management functions.

2. Application Software

Application software consists of programs designed to make users more productive and/or assist them with personal tasks. A widely used type of application software related to communications is a Web browser, which allows users with an Internet connection to access and view Web pages or access programs. Other popular application software includes word processing software, spreadsheet software, database software, and presentation software.

Many other types of application software exist that enable users to perform a variety of tasks. **These include personal information management, note taking, project management, accounting, document management, computer-aided design (CAD), desktop publishing, paint/image editing, photo editing, audio and video editing, multimedia authoring, Web page authoring,**

personal finance, legal, tax preparation, home design/landscaping, travel and mapping, education, reference, and entertainment (e.g., games or simulations).

Software is available at stores that sell computer products and also online at many websites.

13.1.3 Categories of Computers

Industry experts typically classify computers in seven categories: personal computers (desktop), mobile computers and mobile devices, game consoles, servers, mainframes, supercomputers, and embedded computers (see Table 13.1). A computer's size, speed, processing power, and price determine the category it best fits. Due to rapidly changing technology, however, the distinction among categories is not always clear-cut. **This trend of computers and devices with technologies that overlap, called convergence, leads to computer manufacturers continually releasing newer models that include similar functionality and features.** For example, newer cell phones often include media player, camera, and Web browsing capabilities. As devices converge, users need fewer devices for the functionality that they require. When consumers replace outdated computers and devices, they should dispose of them properly.

Table 13.1 Categories of computers

Category	Physical Size	Number of Simultaneously Connected Users	General Price Range
Personal computers (desktop)	Fits on a desk	Usually one (can be more if networked)	Several hundred to several thousand dollars
Mobile computers and mobile devices	Fits on your lap or in your hand	Usually one	Less than a hundred dollars to several thousand dollars
Game consoles	Small box or handheld device	One to several	Several hundred dollars or less
Servers	Small cabinet	Two to thousands	Several hundred to a million dollars
Mainframes	Partial room to a full room of equipment	Hundreds to thousands	$300,000 to several million dollars
Supercomputers	Full room of equipment	Hundreds to thousands	$500,000 to several million dollars
Embedded computers	Miniature	Usually one	Embedded in the price of the product

1. Personal Computers (Desktop)

Two popular architectures of personal computers are the PC and the Apple. The term, PC-compatible, refers to any personal computer based on the original IBM personal computer design. Companies such as Dell, HP, and Toshiba sell PC-compatible computers. PC and PC-compatible computers usually use a Windows operating system. Apple computers usually use a Macintosh operating system (Mac OS).

Some desktop computers function as a server on a network. Others, such as a gaming desktop computer and home theater PC (HTPC), target a specific audience. Another expensive, powerful desktop computer is the workstation, which is geared for work that requires intense calculations and graphics capabilities. An architect uses a workstation to design buildings and homes. A graphic artist uses a workstation to create computer-animated special effects for full-length motion pictures and video games.

2. Mobile Computers and Mobile Devices

The most popular type of mobile computer is the notebook computer, also called a laptop computer. Netbook, ultra-thin, tablet PC are special types of notebook computers. Popular types of mobile devices are smartphones and (personal digital assistants, PDAs), e-book readers, handheld computers, portable media players, and digital cameras.

3. Game Consoles

A game console is a mobile computing device designed for single-player or multiplayer video games. Standard game consoles use a handheld controller(s) as an input device(s); a television screen as an output device; and hard disks, optical discs, and/or memory cards for storage. Three popular models are Microsoft's Xbox 360, Nintendo's Wii (pronounced wee), and Sony's PlayStation 3. Many handheld game consoles can communicate wirelessly with other similar consoles for multiplayer gaming. Two popular models are Nintendo DS Lite and Sony's PlayStation Portable (PSP). In addition to gaming, many game console models allow users to listen to music, watch movies, keep fit, and connect to the Internet. Game consoles can cost from a couple hundred dollars to more than $500.

4. Servers

A server controls access to the hardware, software, and other resources on a network and provides a centralized storage area for programs, data, and information. Servers can support from two to several thousand connected computers at the same time. In many cases, one server accesses data, information, and programs on another server. In other cases, people use personal computers or terminals to access data, information, and programs on a server. A terminal is a device with a monitor, keyboard, and memory.

5. Mainframes

A mainframe is a large, expensive, powerful computer that can handle hundreds or thousands of connected users simultaneously. Mainframes store tremendous amounts of data, instructions, and information. Most major corporations use mainframes for business activities. With mainframes, enterprises are able to bill millions of customers, prepare payroll for thousands of employees, and manage thousands of items in inventory. One study reported that mainframes process more than 83 percent of transactions around the world.

6. Supercomputers

A supercomputer is the fastest, most powerful computer — and the most expensive. The fastest supercomputers are capable of processing more than one quadrillion instructions in a single second. With weights that exceed 100 tons, these computers can store more than 20,000 times the data and information of an average desktop computer. Applications requiring complex, sophisticated mathematical calculations use supercomputers. **Large-scale simulations and applications in medicine,**

aerospace, automotive design, online banking, weather forecasting, nuclear energy research, and petroleum exploration use a supercomputer.

7. Embedded Computers

An embedded computer is a special-purpose computer that functions as a component in a larger product. Their functions depend on the requirements of the product in which they reside. Embedded computers in printers, for example, monitor the amount of paper in the tray, check the ink or toner level, signal if a paper jam has occurred, and so on.

Embedded computers are everywhere — at home, in your car, and at work. The following list identifies a variety of everyday products that contain embedded computers.

- Consumer electronics: mobile and digital telephones, digital televisions, cameras, video recorders, DVD players and recorders, answering machines.
- Home automation devices: thermostats, sprinkling systems, security monitoring systems, appliances, lights.
- Automobiles: antilock brakes, engine control modules, airbag controllers, cruise control.
- Process controllers and robotics: remote monitoring systems, power monitors, machine controllers, medical devices.
- Computer devices and office machines: keyboards, printers, fax and copy machines.

As technology continues to advance, computers have become a part of everyday life. Thus, many people believe that computer literacy is vital to success in today's world. Computer literacy, also known as digital literacy, involves having a current knowledge and understanding of computers and their uses. Because the requirements that determine computer literacy change as technology changes, you must keep up with these changes to remain computer literate.

Words and Expressions

cash register	收银机；现金出纳机
scanner ['skænər]	*n.* 扫描器；扫描仪
webcam ['webkæm]	*n.* 网络摄像头（=web camera）
recognize ['rekəgnaɪz]	*vt.* 识别；认出；承认；意识到
tape [teɪp]	*n.* 磁带；胶带
film [fɪlm]	*n.* 胶卷；薄膜；电影
motherboard ['mʌðərbɔːrd]	*n.* 主板；母板
USB flash drive	USB 闪存驱动器；优盘
memory card	存储卡
DVD: digital video disc	数字激光视盘
lightweight ['laɪtweɪt]	*adj.* 轻便的；轻量级的；比较轻的
house [haʊz]	*v.* 容纳；储存（某物）；藏有；给……房子住
external hard disk	外置硬盘；外部硬盘
CD: compact disc	激光唱盘
Blu-ray Disc	蓝光光碟

cellular radio network	蜂窝无线网络
graphical user interface(GUI)	图形用户界面
icon ['aɪkɑːn]	*n.* 图标
system software	系统软件
application software	应用软件
operating system	操作系统
utility program	应用程序；实用程序；公用程序
computer-aided design (CAD)	计算机辅助设计
game console	游戏机；游戏控制台
dispose [dɪ'spoʊz]	*vt.* 处置；处理；布置；安排
home theater PC(HTPC)	家庭影院 PC
workstation ['wɜːrksteɪʃn]	*n.* （计算机）工作站
gear [gɪr]	*v.* （使）搭配；（使）适合；换挡
computer-animated special effect	电脑动画特效
notebook ['noʊtbʊk]	*n.* 笔记本；笔记本电脑
netbook ['netbʊk]	*n.* 上网本
ultra-thin ['ʌltrə θɪn]	*n.* 超薄本 *adj.* 超薄的
tablet PC	平板电脑
personal digital assistant(PDA)	个人数字助理
PlayStation Portable(PSP)	掌上游戏机
centralized ['sentrəlaɪzd]	*adj.* 集中的
terminal ['tɜːrmɪnl]	*n.* 终端机；末端；终点站；航站楼
inventory ['ɪnvəntɔːri]	*n.* 详细目录；存货（清单）
transaction [træn'zækʃn]	*n.* 交易；办理；处理；事务
quadrillion [kwɑː'drɪljən]	*n.* 千的五次方；千万亿 *adj.* 千之五次方的
toner ['toʊnər]	*n.* 墨粉；护肤霜；爽肤水
thermostat ['θɜːrməstæt]	*n.* 恒温器；温控器
antilock brake	防抱死制动器
cruise control	巡行车速控制装置
robotics [roʊ'bɑːtɪks]	*n.* 机器人科学（或技术）
literacy ['lɪtərəsi]	*n.* 识字；读写能力；有文化

Notes

1. Although some forms of memory are permanent, most memory keeps data and instructions temporarily, which means its contents are erased when the computer is shut off.

尽管某些形式的内存是永久性的，但大多数内存会暂时保存数据和指令，这意味着当计算机关闭时其内容会被擦除。

2. Software, also called a program, consists of a series of related instructions, organized for a common purpose, that tells the computer what tasks to perform and how to perform them.

软件，也称为程序，由一系列相关指令组成，这些指令具有共同的目的，告诉计算机要执行哪些任务及如何执行这些任务。

3. These include personal information management, note taking, project management, accounting, document management, computer-aided design(CAD), desktop publishing, paint/image editing, photo editing, audio and video editing, multimedia authoring, Web page authoring, personal finance, legal, tax preparation, home design/landscaping, travel and mapping, education, reference, and entertainment (e.g., games or simulations).

这些（应用软件）包括个人信息管理、笔记记录、项目管理、会计、文档管理、计算机辅助设计（CAD）、桌面出版（排版）、绘画/图像编辑、照片编辑、音频和视频编辑、多媒体创作、网页创作、个人理财、法律、报税、家居设计/景观美化、旅行和测绘、教育、参考和娱乐（如游戏或模拟）。

4. This trend of computers and devices with technologies that overlap, called convergence, leads to computer manufacturers continually releasing newer models that include similar functionality and features.

计算机和设备技术重叠（称为融合）的趋势，导致计算机制造商不断发布包含类似功能和特性的新机型。

5. Large-scale simulations and applications in medicine, aerospace, automotive design, online banking, weather forecasting, nuclear energy research, and petroleum exploration use a supercomputer.

医学、航空航天、汽车设计、网上银行、天气预报、核能研究和石油勘探等领域的大规模模拟和应用都使用超级计算机。

13.2 Computer Networks

Computer networks have many uses, both for companies and for individuals, in the home and while on the move. Companies use networks of computers to share corporate information, typically using the client-server(C/S) model with employee desktops acting as clients accessing powerful servers in the machine room. For individuals, networks offer access to a variety of information and entertainment resources, as well as a way to buy and sell products and services. Individuals often access the Internet via their phone or cable providers at home, though increasingly wireless access is used for laptops and phones. The primary applications of the Internet are e-mail, file transfer, the world wide web(WWW), e-commerce, search engines, voice over Internet protocol, and video. Technology advances are enabling new kinds of mobile applications and networks with computers embedded in appliances and other consumer devices.

13.2.1 Network Hardware

Roughly speaking, networks can be divided into LANs, MANs (metropolitan area networks), WANs (wide area networks), and internetworks by scale.

LANs typical cover a building, like a home, office or factory, and operate at high speeds, connecting personal computers and consumer electronics. Ethernet is the predominant standard of wired LAN. MANs usually cover a city. An example is the cable television system, which is now used by many people to access the Internet. WANs may cover a country or a continent. Networks can be interconnected with routers to form internetworks, of which the Internet is the largest and best known example. Wireless networks, for example IEEE 802.11 LANs (WiFi) and 4G, 5G mobile telephony, are also becoming extremely popular.

1. Ethernet

One of the oldest and by far the most widely used of all LANs is Ethernet. Ethernet, which was developed by Xerox Corporation at Palo Alto Research Center in the 1970s, was based on the Aloha wide-area satellite network implemented at the University of Hawaii in the late 1960s.

In 1980, Xerox joined with Digital Equipment Corporation (now part of Hewlett-Packard) and Intel to sponsor a joint standard for Ethernet. The collaboration resulted in a definition that became the basis for the IEEE 802.3 standard. (The Institute of Electrical and Electronics Engineers [IEEE] establishes and maintains a wide range of electrical, electronic, and computing standards. The 802.X series relates to LANs.)

Today, there are numerous variants of Ethernet. More than 95 percent of all LANs use some form of Ethernet. Furthermore, Ethernet has grown more capable over the years and is now routinely used in both MANs and WANs.

The original versions of Ethernet used a bus topology. Today, most use a physical star configuration (see Figure 13.4). Each PC node with a network interface card (NIC) uses a twisted-pair cable to connect into a centrally located hub or switch. Switches have largely replaced hubs in most large LANs because switches greatly expand the number of possible PC nodes and improve performance. The switch is then connected to a router or gateway that provides access to the services needed by each user.

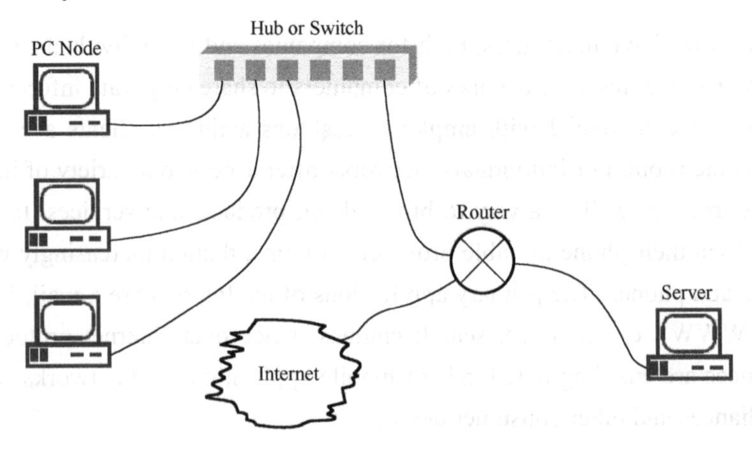

Figure 13.4 The Ethernet

The original transmission medium for Ethernet was coaxial cable. However, today twisted-pair versions of Ethernet are more popular. **By far the most popular version of 100-Mbps Ethernet is 100Base-T or 100Base-TX, also called Fast Ethernet, where the "Base" stands for baseband, indicating that Ethernet uses baseband data transmission methods, and the "T"**

stands for twisted-pair. It uses two unshielded twisted pairs (UTPs). One pair is used for transmitting, and the other is used for receiving, permitting full duplex operation, which is not possible with standard Ethernet. To achieve such high speeds on UTP, several important technical changes were implemented in Fast Ethernet.

First, the cable length is restricted to 100 m of category 5 UTP. This ensures minimal inductance, capacitance, and resistance, which distort and attenuate the digital data.

Second, a new type of encoding method is used. Called MLT-3, this encoding method is illustrated in Figure 13.5. The standard NRZ binary signal is shown at Figure 13.5(a), and the MLT-3 signal is illustrated at Figure 13.5(b). Note that three voltage levels are used: +1, 0, and –1 V. If the binary data is a binary 1, the MLT-3 signal changes from the current level to the next level. If the binary input is 0, no transition takes place. If the signal is currently at +1 and a 1111 bit sequence occurs, the MLT-3 signal changes from +1 to 0 to –1 and then to 0 and then to +1 again. What this encoding method does is to greatly reduce the frequency of the transmitted signal (to one-quarter of its original signal), making higher bit rates possible over UTP.

As semiconductor and optical technologies have progressed, it has been possible to push Ethernet speeds higher and higher. Over the years, speeds have increased by factors of 10 from 10 to 100 to 1,000 Mbps (1GE) and today 10 Gbps (10GE). Now 40 Gbps (40GE) and 100 Gbps (100GE) versions, standardized in 2010 as IEEE 802.3ba, are available and already deployed. With these speed levels, Ethernet can compete with MAN and WAN services as well as maintain its dominance in the LAN arena.

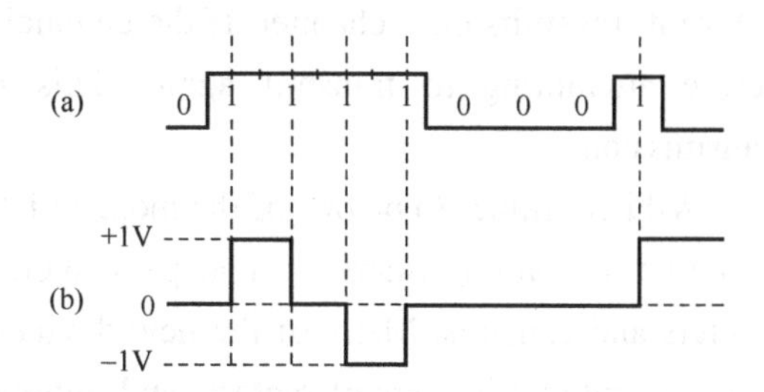

Figure 13.5 MLT-3 encoding used with 100Base-T Ethernet: (a) NRZ; (b) MLT-3

2. WiFi

Wireless LANs are very popular these days, especially in homes, older office buildings, cafeterias, and other places where it is too much trouble to install cables. In these systems, every computer has a radio modem and an antenna that it uses to communicate with other computers, see Figure 13.6.

Wireless router, also called an AP (access point), or base station, relays packets between the wireless computers and also between them and the Internet service provider. **This router uses a software approach called network address translation (NAT) to make it appear as if each networked PC has its own Internet address, when in reality only the one associated with the incoming broadband line is used.**

There is a standard for wireless LANs called IEEE 802.11, popularly known as WiFi, which has become very widespread. The base version of IEEE 802.11 was released in 1997 and has had subsequent amendments. **They use various frequencies including, but not limited to, 2.4 GHz, 5 GHz, 6 GHz, and 60 GHz frequency bands, running at speeds anywhere from 11 to hundreds of Mbps.**

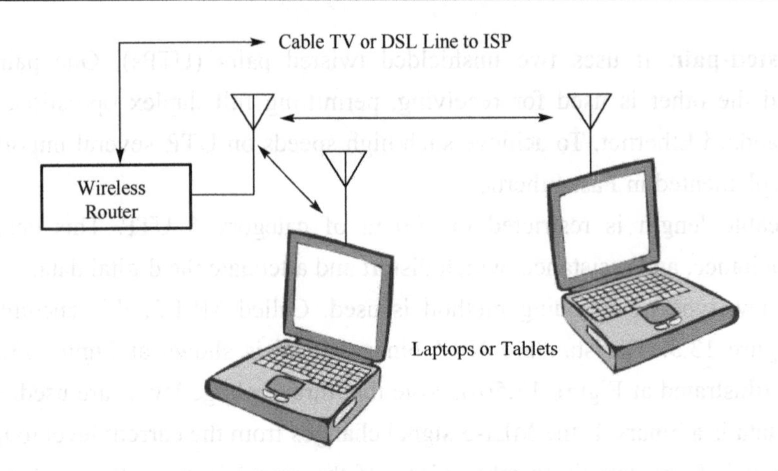

Figure 13.6 Home WiFi for Internet access

The carrier sense multiple access with collision avoidance (CSMA/CA) access method is used to minimize conflicts among those wireless nodes seeking access to the AP. Each transceiver listens before it transmits on a channel. If the channel is occupied, the transceiver waits a random period before attempting to transmit again. This process continues until the channel is free for transmission.

WiFi continues to be one of the most widely used wireless technologies in history. It is already available in many airliners, so that passengers can connect to the Internet in flight. It is used in printers and cameras. Many of the new devices, such as consumer media products, health/fitness/medical, automotive, smart meters, and automation products, are related to WiFi. Another WiFi target is the machine-to-machine (M2M) field and the internet of things (IoT).

3. Internet

The Internet is a worldwide interconnection of computers by means of many complex networks. The Internet was established in the late 1960s under the sponsorship of the United States Department of Defense and later through the National Science Foundation of the United States. It provided a way for universities and companies doing military and government research to communicate and to share computer files and software. In the early 1990s, the Internet was privatized and opened to anyone. Today it is a system with billions of users across the world.

The Internet is the ultimate data communication network. It uses virtually every conceivable type of data communication equipment and technique. Just keep in mind that the information is transmitted as serial binary pulses, usually grouped as bytes (8-bit chunks) of data within larger groups called packets. All the different types of communication media are used including twisted-pair cable, coaxial cable, fiber-optic cable, satellites, and other wireless connections.

Figure 13.7 shows a diagram of the main components of the Internet. On the left in Figure 13.7 are three ways in which a PC is connected for Internet access, to the telephone local office via a conventional or DSL modem or by way of a cable modem and cable TV company. The telephone company central office and Cable TV Company connect to an Internet service provider (ISP) by way of a local fiber-optic MAN. The ISP contains multiple servers that handle the traffic. Each connects to the Internet by way of a router that attaches to the Internet backbone at one of the many network access points. The primary equipment at network access points (NAPs) is routers that determine the destination of packets. The "cloud" represents the Internet backbone.

The Internet backbone is a collection of companies that install, service, and maintain large nationwide and even worldwide networks of high-speed fiber-optic cable. The companies own the equipment and operate it to provide universal access to the Internet. Although each of the backbone providers has its own hierarchical nationwide network, the providers are usually connected to one another to provide many different paths from one computer to another.

On the right in Figure 13.7 are additional connections. Here a company LAN accesses the backbone by way of a regional MAN. A company server contains a website that is regularly accessed by others. Also shown is a Web hosting company that stores the websites and Web pages of others.

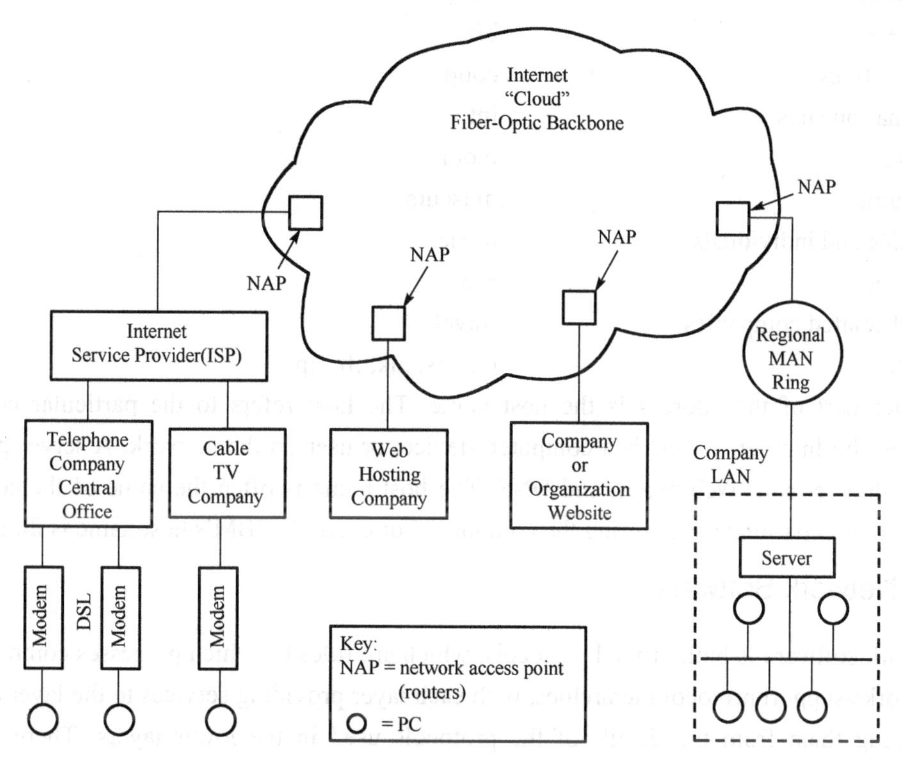

Figure 13.7 Simplified diagram of the Internet

The router is the single most important piece of equipment in the Internet. Routers interconnect the various segments of the WAN backbones. Routers are also used to connect MANs to WANs, LANs to other LAN, and LANs to MANs. The routers connect to one another and to the various servers to form a large mesh network connected usually by fiber-optic cables. A router is an intelligent computerlike device that looks at all packets transmitted to it and examines their Internet protocol (IP) destination addresses. Routing algorithms determine the best (closest, fastest) connection and then retransmit the packet.

Each individual or computer on the Internet must have some kind of identifier or address. An addressing system for the Internet uses a simplified name-address scheme that defines a particular hierarchy. The upper level of the hierarchy is called a top-level domain (TLD). A domain is a specific type of organization using the Internet. Such domains are assigned a part of the Internet address. The most common domains and their address segments are as follows.

Domain	**Address Segment**
Commercial companies	.com
Educational institutions	.edu
Nonprofit organizations	.org
Military	.mil
Government	.gov
Internet service providers	.net
Air transportation	.aero
Business	.biz
Cooperatives	.coop
Information sites	.int
Mobile	.mobi
Museums	.museum
Families and individuals	.name
Professions	.pro
Travel-related companies	.travel
Country	.cn, .us, .uk, .fr, .jp

Another part of the address is the host name. The host refers to the particular computer connected to the Internet. A host is a computer, device, or user on the network. A server provides services such as e-mail, Web pages, and DNS. The host name is often the name of the company, organization, or department sponsoring the computer. For example, IBM's host name is ibm.

13.2.2 Network Software

Network software is built around protocols, which are rules by which processes communicate. Most networks support protocol hierarchies, with each layer providing services to the layer above it and insulating them from the details of the protocols used in the lower layers. There are two important network architectures: the OSI reference model and the TCP/IP reference model(see Figure 13.8).

1. OSI (Open Systems Interconnection)

The OSI model is based on a proposal developed by the International Standards Organization (ISO) as a first step toward international standardization of the protocols. The OSI hierarchy is made up of seven layers.

Layer 1: Physical layer. The physical connections and electrical standards for the communication system are defined here. This layer specifies interface characteristics, such as binary voltage levels, encoding methods, data transfer rates, and the like.

Layer 2: Data link layer. This layer defines the framing information for the block of data. It identifies any error detection and correction methods as well as any synchronizing and control codes relevant to communication.

Layer 3: Network layer. This layer determines network configuration and the route the transmission can take. In some systems, there may be several paths for the data to traverse. The network layer determines the specific data routing and forwarding methods that can occur in the system, e.g., selection of a dial-up line, a private leased line, or some other dedicated path.

Layer 4: Transport layer. Included in this layer are multiplexing, if any; error recovery; partitioning of data into smaller units so that they can be handled more efficiently; and addressing and flow control operations.

Layer 5: Session layer. This layer handles such things as management and synchronization of the data transmission. It typically includes network logon and logoff procedures, as well as user authorization, and determines the availability of the network for processing and storing the data to be transmitted.

Layer 6: Presentation layer. This layer deals with the form and syntax of the message. It defines data formatting, encoding and decoding, encryption and decryption, synchronization, and other characteristics. It defines any code translations required, and sets the parameters for any graphics operations.

Layer 7: Application layer. This layer is the overall general manager of the network or the communication process. Its primary function is to format and transfer files between the communication message and the user's applications software.

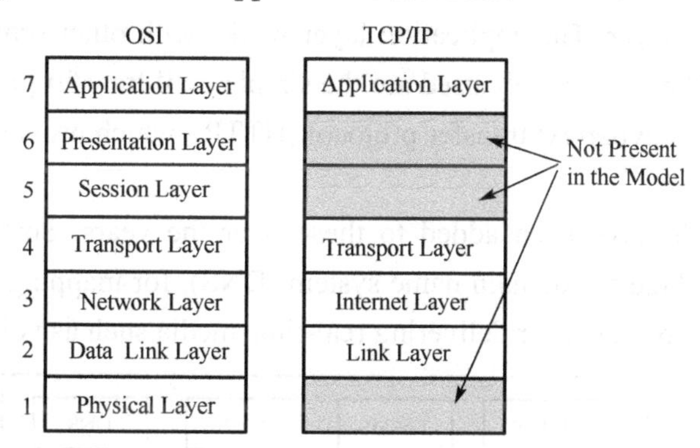

Figure 13.8 Comparing the OSI and TCP/IP layers

2. TCP/IP

TCP/IP was invented in the 1960s, when the Advanced Research Project Agency (ARPA) of the United States Department of Defense (DoD) tested and established the first packet network. Today, TCP/IP is by far the most widely implemented data communication protocol, and it is at the heart of the Internet.

As shown in Figure 13.8, TCP/IP does not implement all seven layers, although the effect is the same. Some operations of the individual seven layers are combined to form four layers.

Layer 1: Link layer. The link layer describes what links such as serial lines and classic Ethernet must do to meet the needs of this connectionless internet layer. It is not really a layer at all, in the normal sense of the term, but rather an interface between hosts and transmission links.

Layer 2: Internet layer. The internet layer is corresponding roughly to the OSI network layer. Its job is to permit hosts to inject packets into any network and have them travel independently to the destination. Every computer has one IP address assigned to it. The IP address is usually expressed in what is called the dotted decimal form. The 32-bit address is divided into four 8-bit segments. Each segment can represent 256 numbers from 0 through 255. A typical address would look like this: 125.63.208.7. Dots separate the four decimal segments. These addresses are assigned by an organization known as the Internet Assigned Numbers Authority (IANA). IP, plus a companion protocol called ICMP (Internet control message protocol) that helps it function, is responsible for delivering IP packets where they are supposed to go.

Layer 3: Transport layer. It is designed to allow peer entities on the source and destination hosts to carry on a conversation, just as in the OSI transport layer. Two end-to-end transport protocols have been defined here: TCP (transmission control protocol) and UDP (user datagram protocol).

TCP is a reliable connection-oriented protocol that allows a byte stream originating on one machine to be delivered without error on any other machine in the internet. While, UDP is an unreliable, connectionless protocol for applications that do not want TCP's sequencing or flow control. It is widely used for applications in which prompt delivery is more important than accurate delivery, such as transmitting speech or video.

Layer 4: Application layer. The application layer works with other protocols that implement the desired application. The most widely used are the simple mail transfer protocol (SMTP), which implements e-mail; and the hypertext transfer protocol (HTTP), which provides access to the world wide web.

Many other protocols have been added to these over the years. Some important ones are shown in Figure 13.9, include the domain name system (DNS), for mapping host names onto their IP addresses, and RTP, the protocol for delivering real-time media such as voice or movies.

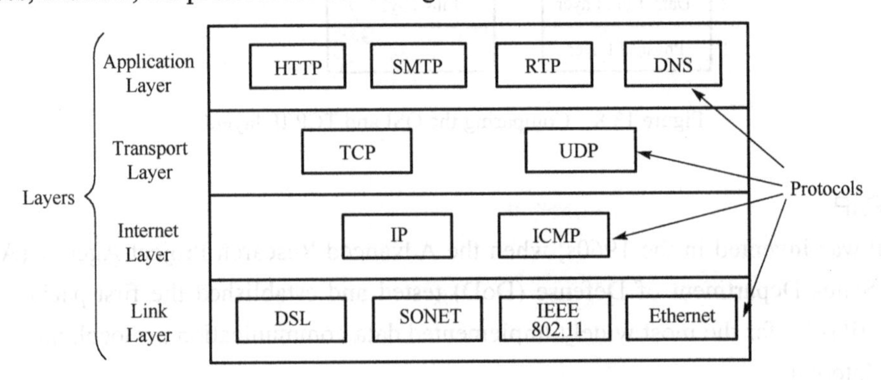

Figure 13.9 The TCP/IP protocol stack

3. OSI vs TCP/IP

The OSI reference model was devised before the corresponding protocols were invented. This ordering meant that the model was not biased toward one particular set of protocols, a fact that made it quite general. The downside of this ordering was that the designers did not have much experience with the subject and did not have a good idea of which functionality to put in which layer.

With TCP/IP the reverse was true: the protocols came first, and the model was really just a description of the existing protocols. There was no problem with the protocols fitting the model. They fit perfectly. The only trouble was that the model did not fit any other protocol stacks. Consequently, it was not especially useful for describing other, non-TCP/IP networks.

13.2.3 Internet Security

One of the most important aspects of the Internet is security of the data being transmitted. Security refers to protecting the data from interception and protecting the sending and receiving parties from unwanted threats such as viruses and spam. And it means protecting the equipment and software used in the networks. The Internet or any network connected computer is subject to threats by hackers, individuals who deliberately try to steal data or damage computer systems and software just for the challenge.

1. Types of Security Threats

Viruses. A virus is a small program designed to implement some nefarious action in a computer. A virus typically rides along with some other piece of information or program, so that it can be surreptitiously inserted into the computer and executed by the processor to do its damage. Viruses are usually transmitted by e-mails or by way of a Trojan horse, a seemingly useful and innocent program that hides the virus. Besides making the computer unusable, a virus can erase or corrupt files, cause unknown e-mails to be transmitted, or take other malicious actions. Like a real virus, computer viruses are designed to spread themselves within the computer or to be retransmitted to others in e-mails. These viruses are called worms as they automatically duplicate and transmit themselves from network to network and computer to computer.

Spam. A more recent threat, while not actually damaging, is unwanted ads and solicitations via e-mail called spam. Spam clogs up the e-mail system with huge quantities of unwanted data and uses transmission time and bandwidth that could be used in a more productive way. Spam is not illegal, but you must remove the spam yourself, thereby using up valuable time, not to mention memory space in your e-mail system.

Spyware. Spyware is a kind of software that monitors a computer and its user while he or she accesses the Internet or e-mail. It then collects data about how that user uses the Internet such as Internet website access, shopping, etc. It uses this information to send unsolicited ads and spam. Some examples of dangerous practices are the capture of credit card numbers, delivery of unsolicited pop-up ads, and capture of Web-browsing activity and transmission to a person or company for use in unauthorized promotions.

Denial-of-service (DoS) attack. This is a process that transmits errors in the communication protocol and causes the computer to crash or hang up. This type of vandalism doesn't steal information, but it does prevent the user from accessing the operating system, programs, data files, or communication links. It is the easiest form of attack that serves no useful purpose other than to hurt others.

One special type of DoS attack is called smurf attack. A smurf attack (see Figure 13.10) usually overwhelms ISP servers with a huge number of worthless packets, thereby preventing other ISP subscribers from using the system. Smurf attack makes use of a technique called ping. A ping is the transmission of an inquiry by way of the Internet control message protocol (ICMP) that is a part of TCP/IP to see if a particular computer is connected to the Internet and active. In response to the ping, the computer sends back a message confirming that it is connected. Hackers substitute the ISP's own address for the return message, so that it gets repeatedly transmitted, thus tying up the system.

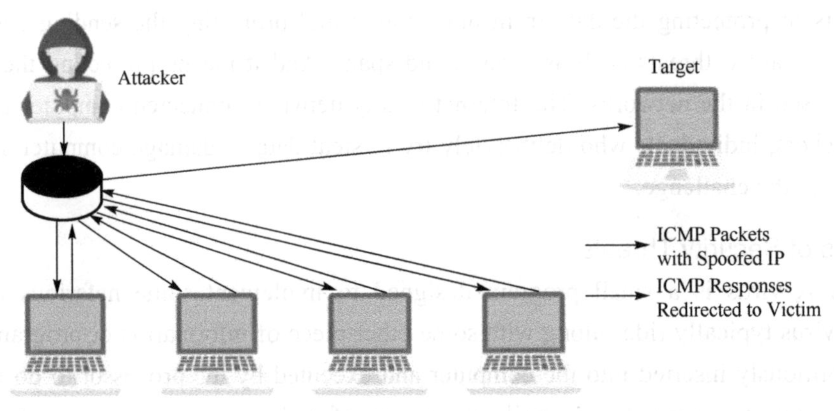

Figure 13.10 Smurf attack

2. Security Measures

To protect data and prevent the kinds of malicious hacking described, special software or hardware is used. Here is a brief summary of some of the techniques used to secure a computer system or network.

Encryption and decryption. Encryption is the process of obscuring information so that it cannot be read by someone else. It involves converting a message to some other form that makes it useless to the reader. Decryption is the reverse process that translates the encrypted message back to readable form.

Encryption has been used for centuries by the government and the military, mainly to protect sensitive material from enemies. Today it is still heavily used by the government and the military but also by companies and individuals as they strive to protect their private information. The Internet has made encryption more important than ever as individuals and organizations send information to one another. For example, encryption ensures that a customer's credit card number is protected in e-commerce transactions (buying items over the Internet). Other instances are automated teller machine (ATM) accesses and sending private financial information. Even digitized voice in a cell phone network can be protected by encryption.

Figure 13.11 shows the basic encryption/decryption process. The information or message to be transmitted is called plaintext. In binary form, the plaintext is encrypted by using some predetermined computer algorithm. The output of the algorithm is called ciphertext. The ciphertext is the secret code that is transmitted. At the receiving end, the reverse algorithm is performed on the ciphertext to generate the original plaintext.

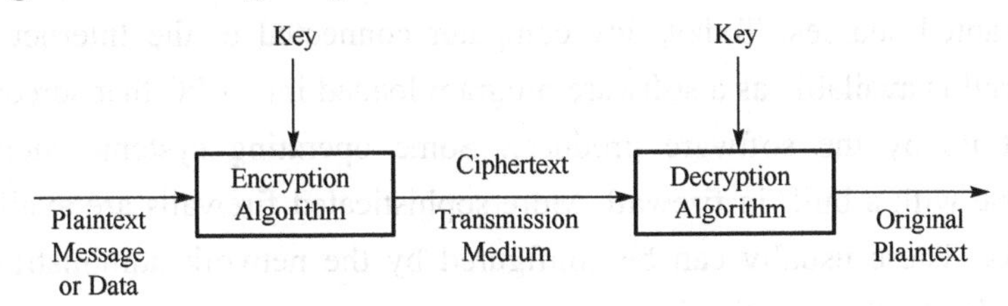

Figure 13.11 The basic encryption/decryption process

Most encryption processes combine the plaintext with another binary number called a secret key. The key is used as part of the algorithmic computing process. To translate the ciphertext back to plaintext, the receiving computer must also know the secret key. The strength of the encryption (meaning how secure the data is from deliberate attempts at decryption by brute force) is determined by the number of bits in the key. The greater the number of bits, the more difficult the key is to discover.

Authentication. Authentication ensures that the transmitting and receiving parties are really who they say they are and that their identities have not been stolen or simulated. **Digital authentication allows computer users to confidently access the Internet, other networks, computers, software, or other resources such as bank accounts if they can verify their identities.** Authentication is widely used in most Internet transactions such as e-commerce as it provides a way to control access, keep out unauthorized users, and keep track of those who are using the resources.

The most common methods of authentication are the use of passwords or personal identification numbers (PINs). Coded ID cards are another way. More recently, biometric methods of identification are being used as security tightens with more and more transactions. Some common biometric ID methods are fingerprint scans, retinal eye scans, voiceprints, or video facial recognition.

Secure socket layer (SSL). The processes of encryption/decryption and authentication are used together to ensure secure transactions over the Internet. All these processes are combined into a protocol known as the secure socket layer (SSL). E-commerce simply would not exist without the safeguards of SSL. SSL or a more advanced version called transport layer security (TLS) usually resides in layers 5, 6, or 7 of the OSI model. The SSL or TLS protocol is implemented in the browser software such as Microsoft Internet Explorer (IE) or Netscape.

Firewalls. A firewall is a piece of software that monitors transmissions on a network and inspects the incoming information to see if it conforms to a set of guidelines established by the software or the organization or person owning the network. The firewall controls the flow of traffic from the Internet to a LAN or PC or between LANs or other networks.

The most common type of firewall operates at the network layer in the OSI model. It examines TCP/IP packets and acts as a filter to block access from inputs that do not match a set of rules set up in the firewall. The firewall screens packets for specific IP sources or destinations, packet attributes, domain name, or other factors. Firewalls are the first line of defense against intrusions by unwanted sources. Today, any computer connected to the Internet should have a firewall. The firewall is available as a software program loaded into a PC that screens according to the guidelines set up by the software producer. Some operating systems such as Microsoft Windows now come with a built-in firewall. More sophisticated firewalls are available for LANs and other networks. These usually can be configured by the network administrator to filter on special rules as needed by the organization.

Antivirus software, antispam software, and antispyware. There are commercial programs designed to be installed on a computer to find and eliminate these security problems. The antivirus and antispyware programs scan all files on the hard drive either automatically or on command, to look for viruses. **The antivirus software looks for a pattern of code unique to each virus, and when it is identified, the software can remove the virus or in some cases quarantine and isolate the infected file so that it does no harm.** Antispyware works the same way by scanning all files, searching for patterns that designate a spyware program. It then removes the program.

Antispam software is typically set up to monitor incoming e-mail traffic and look for clues to whether it is legitimate e-mail or spam. It then blocks the spam from the e-mail inbox and places it in a special bulk e-mail file. You will never see the spam unless your e-mail system allows you to look in a special bulk file normally furnished by the e-mail provider. Antispam software is not perfect, and because of its rules for blocking spam it can also affect desired e-mail. It is worthwhile to examine the bulk files occasionally to be sure that legitimate e-mails are getting through. Most antispam programs allow you to change the filtering rules to ensure you get all desired mail while the real spam is rejected.

Virtual private network (VPN). One way to achieve security on a LAN is to use software measures to block off segments of a network or create a subnetwork using software to assign access only to authorized users. This is referred to as a virtual LAN or VLAN. Security can also be achieved when you are connecting two remote LANs by using a leased line. The leased line is totally dedicated to just the connection between the LANs. No one else has access. While this works well, it is very expensive. A popular alternative is to create a secure connection through the Internet by using a virtual private network (VPN). In a VPN, the data to be transmitted is encrypted, encapsulated in a special packet, and then sent over the Internet.

Wireless security. Security in wireless systems is important because it is relatively easy to capture a radio signal containing important information. A directional antenna and sensitive receiver designed for the specific wireless service, such as a wireless LAN and a computer, are all you need. Wireless data can be protected by encryption, and a number of special methods have been developed especially for wireless systems.

Most security measures are implemented in software. Some security techniques can also be implemented in hardware such as data encryption chips. Internet security is a very broad and complex subject that is far beyond the scope of this section. It is one of the most critical and fastest-growing segments of the networking industry.

Words and Expressions

client-server (C/S) model	客户-服务器模式
access ['ækses]	*vt.* 接入；进入；存取（计算机文件） *n.* 入口；通道；（使用或见到的）机会；使用权
world wide web (WWW)	万维网
e-commerce [iː'kɑːmɜːrs]	*n.* 电子商务
search engine	搜索引擎
metropolitan area network(MAN)	城域网
wide area network(WAN)	广域网
Ethernet ['iːθərnet]	*n.* 以太网
predominant [prɪ'dɑːmɪnənt]	*adj.* 主导的；占优势的；显著的
router ['ruːtər, 'raʊtər]	*n.* 路由器
WiFi: wireless fidelity	基于 IEEE 802.11 标准的无线局域网
Xerox Corporation	施乐公司
Digital Equipment Corporation	DEC，数字设备公司
Hewlett-Packard	HP，惠普公司
joint standard	联合标准
bus topology	总线拓扑
network interface card(NIC)	网络接口卡；网卡
hub [hʌb]	*n.* 集线器；（某地或活动的）中心
switch [swɪtʃ]	*n.* 交换机；开关 *v.* 交换；转换；改变
unshielded twisted pair(UTP)	非屏蔽双绞线
MLT-3: multi-level transmit -3	多电平传输码，三阶基带编码；多阶基带编码 3
access point(AP)	接入点
network address translation(NAT)	网络地址转换
amendment [ə'mendmənt]	*n.* 改进版本；改进；修正；修正案
carrier sense multiple access with collision avoidance(CSMA/CA)	带冲突避免的载波感应多路访问
smart meter	智能仪表；智能电表
privatize ['praɪvətaɪz]	*v.* 使私有化；将……私营化
conceivable [kən'siːvəbl]	*adj.* 想得到的；可想象的；可信的
Web hosting company	网络托管公司
identifier [aɪ'dentɪfaɪər]	*n.* 标识符；鉴定人

top-level domain(TLD)	顶级域
host name	主机名
open systems interconnection(OSI)	开放系统互连
International Standards Organization(ISO)	国际标准化组织
physical layer	物理层
data link layer	数据链路层
network layer	网络层
transport layer	传输层
session layer	会话层
presentation layer	表示层
application layer	应用层
error detection and correction	错误检测和纠正；误差探测与校正
logon ['lɔgɑːn]	*n.* 登录
logoff ['lɔːgɔːf]	*n.* 注销
encryption [ɪn'krɪpʃn]	*n.* 加密
decryption [diː'krɪpʃn]	*n.* 解密；译码
connectionless [kə'nekʃnles]	*adj.* 无连接的
dotted decimal	点分十进制
Internet Assigned Numbers Authority (IANA)	因特网编号分配机构
Internet control message protocol(ICMP)	因特网控制消息协议
peer [pɪr]	*n.* 对等端；同行
transmission control protocol(TCP)	传输控制协议
user datagram protocol(UDP)	用户数据报协议
connection-oriented [kə'nekʃn 'ɔːrientɪd]	*adj.* 面向连接的
simple mail transfer protocol(SMTP)	简单邮件传送协议
hypertext transfer protocol(HTTP)	超文本传送协议
domain name system(DNS)	域名系统
RTP: real-time transport protocol	实时传输协议
protocol stack	协议栈
devise [dɪ'vaɪz]	*vt.* 设计；发明
interception [ˌɪntər'sepʃn]	*n.* 截取；拦截；窃听
virus ['vaɪrəs]	*n.* 病毒
spam [spæm]	*n.* 垃圾电子邮件
hacker['hækər]	*n.* 黑客
nefarious [nɪ'feriəs]	*adj.* 违法的；邪恶的
surreptitiously [ˌsɜːrəp'tɪʃəsli]	*adv.* 偷偷摸摸地；暗中地
Trojan horse	特洛伊木马
malicious [mə'lɪʃəs]	*adj.* 怀有恶意的；恶毒的
worm [wɜːrm]	*n.* 蠕虫
solicitation [səˌlɪsɪ'teɪʃn]	*n.* 请求，（意见的）征求；引诱；教唆

spyware ['spaɪwer]	*n.* 间谍软件
unsolicited [ˌʌnsə'lɪsɪtɪd]	*adj.* 未经要求的；自发的；主动提供的
pop-up ads	弹出式广告；弹窗广告
denial-of-service (DoS) attack	拒绝服务攻击
vandalism ['vændəlɪzəm]	*n.* 故意破坏公物（艺术或文化等）的行为
automated teller machine (ATM)	自动柜员机；自动出纳机；自动取款机
plaintext [pleɪn'tekst]	*n.* 明文；纯文件
ciphertext ['saɪfərˌtekst]	*n.* 密码文本
secret key	密钥；密匙
brute force	暴力破解
authentication [ɔːˌθentɪ'keɪʃn]	*n.* 认证功能；验证方式；安全验证；报文鉴别
personal identification number (PIN)	个人识别号码；个人身份号
biometric [ˌbaɪoʊ'metrɪk]	*adj.* 生物测定的；生物统计的
retinal ['retɪnl]	*adj.* 视网膜的
voiceprint ['vɔɪsprɪnt]	*n.* 声纹
facial recognition	面部识别
secure socket layer (SSL)	安全套接字层
transport layer security (TLS)	传输层安全协议
firewall ['faɪərwɔːl]	*n.* 防火墙
built-in [ˌbɪlt 'ɪn]	*adj.* 嵌入式的；内置的
quarantine ['kwɔːrəntiːn]	*n.* 隔离
antivirus software	防病毒软件；杀毒软件
antispam software	反垃圾邮件软件
antispyware	反间谍软件
bulk file	批量文件
virtual private network (VPN)	虚拟专用网
VLAN: virtual local area network	虚拟局域网
leased line	专线；租用线路
directional antenna	定向天线

Notes

1. By far the most popular version of 100-Mbps Ethernet is 100Base-T or 100Base-TX, also called Fast Ethernet, where the "Base" stands for baseband, indicating that Ethernet uses baseband data transmission methods, and the "T" stands for twisted-pair.

目前最流行的 100 Mbps 以太网版本是 100Base-T 或 100Base-TX，也称为 Fast Ethernet（快速以太网），其中"Base"代表基带，表示以太网采用基带数据传输方式，"T"代表双绞线。

2. This router uses a software approach called network address translation (NAT) to make it appear as if each networked PC has its own Internet address, when in reality only the one associated with the incoming broadband line is used.

该路由器使用一种称为网络地址转换（NAT）的软件方法，使其看起来好像每台联网 PC 都有自己的因特网地址，而实际上只有一个与入户宽带线路关联的地址。

3. They use various frequencies including, but not limited to, 2.4 GHz, 5 GHz, 6 GHz, and 60 GHz frequency bands, running at speeds anywhere from 11 to hundreds of Mbps.

它们（IEEE 802.11 协议的各个版本）使用各种频率，包括但不限于 2.4 GHz、5 GHz、6 GHz 和 60 GHz 频段，运行速度从 11 Mbps 到数百 Mbps 不等。

4. Digital authentication allows computer users to confidently access the Internet, other networks, computers, software, or other resources such as bank accounts if they can verify their identities.

数字身份验证允许计算机用户自如地访问因特网、其他网络、计算机、软件或其他资源（如银行账户），前提是他们可以验证自己的身份。

5. The antivirus software looks for a pattern of code unique to each virus, and when it is identified, the software can remove the virus or in some cases quarantine and isolate the infected file so that it does no harm.

防病毒软件会查找每个病毒特有的代码模式，当识别出病毒后，该软件可以删除病毒，或者在某些情况下隔离被感染的文件，使其不会造成伤害。

Exercises

1. Match the terms (1)–(6) with the definitions A–F.

(1) VPN	A. a set of rules or procedures for transmitting data between electronic devices, such as computers
(2) AP	B. a type of malicious software designed to enter your computer device, gather data about you, and forward it to a third-party without your consent
(3) firewall	C. unsolicited usually commercial messages (such as e-mails, text messages, or Internet postings) sent to a large number of recipients or posted in a large number of places
(4) spyware	D. a private computer network that functions over a public network
(5) spam	E. a stand-alone device or computer that allows wireless devices (such as laptop computers) to connect to and communicate with a wired computer network
(6) protocol	F. a network security device that monitors incoming and outgoing network traffic and permits or blocks data packets based on a set of security rules

2. Translate into Chinese.

(1) All of the IEEE 802.11 techniques use short-range radios to transmit signals in either the 2.4 GHz or the 5 GHz frequency bands, freely available to any transmitter willing to meet some estrictions, such as radiated power of at most 1 W (though 50 mW is more typical for wireless LAN radios).

(2) A protocol defines the format and the order of messages exchanged between two or more communicating entities, as well as the actions taken on the transmission and/or receipt of a message or other event.

(3) Adware is a special type of malware which is used to either redirect the page to some advertising page or pop-up an additional page which promotes some product or event, financially

supported by the organizations whose products are advertised.

3. Translate into English.

（1）因为路由器的工作是在一条链路上接收数据报并在另一条链路上转发数据报，所以路由器必然有两条或更多条链路与之相连。

（2）网络应用程序是计算机网络存在的理由，如果我们不能构想出任何有用的应用，也就没有任何必要去设计支持它们的网络协议了。

（3）Internet DoS 攻击非常常见，Web 服务器、电子邮件服务器、DNS 服务器和机构网络都可能受到 DoS 攻击。

4. Read the following article and write a summary.

Switching is a process of determining of the logical path, how data would travel between sender and receiver in a network. Three main methods, which are commonly used in modern networks are: circuit switching, packet switching, and multiprotocol label switching.

Circuit switching. Main feature of the circuit switching is the connection that is being established before any data would be transmitted, and it would remain up until the transmission is over. All the data follows the same path, in exact order. Most common example of this connection in our everyday life is the traditional telephone. Since the connection is opened for the whole time data is being transmitted, the bandwidth in the path between sender and receiver is constantly reserved and is not available for any other application or service. That makes huge waste of resources in situations when the connection is up, but no data is actually being transmitted. Circuit switching is perfect option for video or audio streaming, where data need to be sent constantly, and there is no time for packet reorganization at the receiver end.

Packet switching. Circuit switching has its flaws, which are overcomed in packet switching, which makes it the most popular method of connecting nodes on the network. Unlike circuit switching, packet switching breaks data into packets and sends them separately to the receiver. Each packet is free to choose its own path to the destination, where they all will be assembled back to the data message. It doesn't matter in which order they arrive to the destination, since they have information about the sequence they should be assembled. At the destination node, data gets reassembled based on this information. Since reassembling of data takes time, this method of switching cannot be used for video or audio streaming, due to necessity of instant data availability on the arrival to destination. Packet switching is popular due to its ability not to waste bandwidth and to use it only for the packet transmission. These factors make packet switching the main solution for the Internet and Ethernet networks.

MPLS (multiprotocol label switching). MPLS is data forwarding technology that increases the speed and controls the flow of network traffic. With MPLS, data is directed through a path via labels instead of requiring complex lookups in a routing table at every stop. This technique works with Internet protocol (IP) and asynchronous transport mode (ATM).

When data enters a traditional packet switching IP network, it moves among network nodes based on long network addresses. With this method, each router on which a data packet lands must make its own decision, based on routing tables, about the packet's next stop on the network. MPLS, on the other hand, assigns a label to each packet to send it along a predetermined path. At the end of the path, the label is removed and the packet is delivered via normal IP routing.

It was designed to provide a unified data-carrying service for both circuit-based clients and packet-switching clients which provide a datagram service model. It can be used to carry many different kinds of traffic, including IP packets, as well as native asynchronous transfer mode (ATM), frame relay, synchronous optical network (SONET) or Ethernet.

5. Language study: *-ing* forms

Words which end in *–ing* and sometimes behave like nounus are called "*-ing* forms". They often refer to actions, processes and activites. Examples from the text are:

processing, computing, operating, transmitting, receiving, multiplexing, framing, addressing, coding, encoding, mapping, formatting, storing, synchronizing, routing, forwarding, hosting, using, connecting ...

They are often used when there are no ordinary nouns available. For example:

This layer defines the framing information for the block of data..

They are used after prepositions. For example:

*Besides **making** the computer unusable, a virus can erase or corrupt files.*

Use the correct form of the word in brackets in each of these sentences:

(1) A packet switch takes a packet (arrive) on one of its (income) communication links and forwards that packet on one of its (outgo) communication links.

(2) ISPs also provide Internet access to content providers, (connect) websites directly to the Internet.

(3) End systems, packet switches, and other pieces of the Internet run protocols that control the (send) and (receive) of information within the Internet.

(4) These applications include electronic mail, Web (surf), instant (message), VoIP, Internet radio, video (stream), distributed games, peer-to-peer file (share), television over the Internet, remote login, and much, much more.

(5) How the network makes the decision as to which path to use is called the (route) algorithm. How each router makes the decision as to where to send a packet next is called the (forward) algorithm.

Unit 14 New Technology Trends

14.1 Promising Opportunities for Artificial Intelligence

This section describes active areas of AI research and innovation poised to make beneficial impact in the near term.

We focus on two kinds of opportunities. The first involves AI that augments human capabilities. Such systems can be very valuable in situations where humans and AI have complementary strengths. **For example, an AI system may be able to synthesize large amounts of clinical data to identify a set of treatments for a particular patient along with likely side effects; a human clinician may be able to work with the patient to identify which option best fits his lifestyle and goals, and to explore creative ways of mitigating side effects that were not part of the AI's design space.** The second category involves situations in which AI software can function autonomously. For example, an AI system may automatically convert entries from handwritten forms into structured fields and text in a database.

14.1.1 AI for Augmentation

Whether it's finding patterns in chemical interactions that lead to a new drug discovery or helping public defenders identify the most appropriate strategies to pursue, there are many ways in which AI can augment the capabilities of people. Indeed, given that AI systems and humans have complementary strengths, one might hope that, combined, they can accomplish more than either alone. **An AI system might be better at synthesizing available data and making decisions in well-characterized parts of a problem, while a human may be better at understanding the implications of the data (say if missing data fields are actually a signal for important, unmeasured information for some subgroup represented in the data), working with difficult-to-fully-quantify objectives, and identifying creative actions beyond what the AI may be programmed to consider.**

Unfortunately, several recent studies have shown that human-AI teams often do not currently outperform AI-only teams. Still, there is a growing body of work on methods to create more effective human-AI collaboration in both the AI and human-computer interaction communities. As this work matures, we see several near-term opportunities for AI to improve human capabilities and vice versa. We describe three major categories of such opportunities below.

1. Drawing Insights

There are many applications in which AI-assisted insights are beginning to break new ground and have large potential for the future. In chemical informatics and drug discovery, AI assistance is helping identify molecules worth synthesizing in a wet lab. In the energy sector, patterns identified by AI algorithms are helping achieve greater efficiencies. By first training a model to be very good at making predictions, and then working to understand why those predictions are so good, we have deepened our scientific understanding of everything from disease to earthquake dynamics. AI-based tools will continue to help companies and governments identify bottlenecks in their operations.

AI can assist with discovery. **While human experts can always analyze an AI from the outside—for example, dissecting the innovative moves made by AlphaGo—new developments in interpretable AI and visualization of AI are making it much easier for humans to inspect AI programs more deeply and use them to explicitly organize information in a way that facilitates a human expert putting the pieces together and drawing insights.** For example, analysis of how an AI system internally organizes words (known as an embedding or a semantic representation) is helping us understand and visualize the way words like "awful" (formally "inspiring awe") undergo semantic shifts over time.

2. Assisting with Decision-Making

The second major area of opportunity for augmentation is for AI-based methods to assist with decision-making. For example, while a clinician may be well-equipped to talk through the side effects of different drug choices, they may be less well-equipped to identify a potentially dangerous interaction based on information deeply embedded in the patient's past history. A human driver may be well-equipped for making major route decisions and watching for certain hazards, while an AI driver might be better at keeping the vehicle in lane and watching for sudden changes in traffic flow. Ongoing research seeks to determine how to divide up tasks between the human user and the AI system, as well as how to manage the interaction between the human and the AI software. In particular, it is becoming increasingly clear that all stakeholders need to be involved in the design of such AI assistants to produce a human-AI team that outperforms either alone. Human users must understand the AI system and its limitations to trust and use it appropriately, and AI system designers must understand the context in which the system will be used (for example, a busy clinician may not have time to check whether a recommendation is safe or fair at the bedside).

There are several ways in which AI approaches can assist with decision-making. One is by summarizing data too complex for a person to easily absorb. In oncology and other medical fields, recent research in AI-assisted summarization promises to one day help clinicians see the most important information and patterns about a patient. Summarization is also now being used or actively considered in fields where large amounts of text must be read and analyzed—whether it is following news media, doing financial research, conducting search engine optimization, or analyzing contracts, patents, or legal documents. **Summarization and interactive chat technologies have great potential to help ensure that people get a healthy breadth of information on a topic, and to help break filter bubbles rather than make them—by providing a range of information, or at least an awareness of the biases in one's social-media**

or news feeds. Nascent progress in highly realistic (but currently not reliable or accurate) text generation, such as GPT-3, may also make these interactions more natural.

In addition to summarization, another aid for managing complex information is assisting with making predictions about future outcomes (sometimes also called forecasting or risk scoring). An AI system may be able to reason about the long-term effects of a decision, and so be able to recommend that a doctor ask for a particular set of tests, give a particular treatment, and so on, to improve long-term outcomes. AI-based early warning systems are becoming much more commonly used in health settings, agriculture, and more broadly conveying the likelihood of an unwanted outcome—be it a patient going into shock or an impending equipment failure—can help prevent a larger catastrophe. AI systems may also help predict the effects of different climate change-mitigation or pandemic-management strategies and search among possible options to highlight those that are most promising. These forecasting systems typically have limits and biases based on the data they were trained on, and there is also potential for misuse if people overtrust their predictions or if the decisions impact people directly.

AI systems increasingly have the capacity to help people work more efficiently. In the public sector, relatively small staffs must often process large numbers of public comments, complaints, potential cases for a public defender, requests for corruption investigations, and more, and AI methods can assist in triaging the incoming information. On education platforms, AI systems can provide initial hints to students and flag struggling students to educators. In medicine, smartphone-based pathology processing can allow for common diagnoses to be made without trained pathologists, which is especially crucial in low-resource settings. Language processing tools can help identify mental health concerns at both a population and individual scale and enable, for example, forum moderators to identify individuals in need of rapid intervention. AI systems can help assist both clinicians and patients in deciding when a clinic visit is needed and provide personalized prevention and wellness assistance in the meantime. More broadly, chatbots and other AI programs can help streamline business operations, from financial to legal. As always, while these efficiencies have the potential to expand the positive impact of low resourced, beneficial organizations, such systems can also result in harm when designed or integrated in ways that do not fully and ethically consider their sociotechnical context.

Finally, AI systems can help human decision-making by leveling the playing field of information and resources. Especially as AI becomes more applicable in lower-data regimes, predictions can increase economic efficiency of everyday users by helping people and businesses find relevant opportunities, goods, and services, matching producers and consumers. These uses go beyond major platforms and electronic marketplaces; kidney exchanges, for example, save many lives, combinatorial markets allow goods to be allocated fairly, and AI-based algorithms help select representative populations for citizen-based policy-making meetings.

3. AI as Assistant

A final major area of opportunity for augmentation is for AI to provide basic assistance during a task. For example, we are already starting to see AI programs that can process and

translate text from a photograph, allowing travelers to read signage and menus. Improved translation tools will facilitate human interactions across cultures. Projects that once required a person to have highly specialized knowledge or copious amounts of time—from fixing your sink to creating a diabetes-friendly meal— may become accessible to more people by allowing them to search for task- and context-specific expertise (such as adapting a tutorial video to apply to unique sink configuration).

Basic AI assistance has the potential to allow individuals to make more and better decisions for themselves. In the area of health, the combination of sensor data and AI analysis is poised to help promote a range of behavior changes, including exercise, weight loss, stress management, and dental hygiene. Automated systems are already in use for blood-glucose control and providing ways to monitor and coordinate care at home. AI-based tools can allow people with various disabilities—such as limitations in vision, hearing, fine and gross mobility, and memory—to live more independently and participate in more activities. Many of these programs can run on smartphones, further improving accessibility.

Simple AI assistance can also help with safety and security. We are starting to see lane-keeping assistance and other reaction-support features in cars. It is interesting that self-driving cars have been slow in development and adoption, but the level of automation and assistance in "normal" cars is increasing—perhaps because drivers value their (shared) autonomy with cars, and also because AI-based assistance takes certain loads off drivers while letting them do more nuanced tasks (such as waving or making eye contact with pedestrians to signal they can cross). AI-assisted surgery tools are helping make movements in surgical operations more precise. AI-assisted systems flag potential e-mail based phishing attacks to be checked by the user, and others monitor transactions to identify everything from fraud to cyberattacks.

14.1.2 AI Agents on Their Own

Finally, there is a range of opportunities for AI agents acting largely autonomously or not in close connection with humans. AlphaFold recently made significant progress toward solving the protein-folding problem, and we can expect to see significantly more AI-based automation in chemistry and biology. AI systems now help convert handwritten forms into structured fields, are starting to automate medical billing, and have been used recently to scale efforts to monitor habitat biodiversity. They may also help monitor and adjust operations in fields like clean energy, logistics, and communications; track and communicate health information to the public; and create smart cities that make more efficient use of public services, better manage traffic, and reduce climate impacts. The pandemic saw a rise in fully AI-based education tools that attempt to teach without a human educator in the loop, and there is a great deal of potential for AI to assist with virtual reality scenarios for training, such as practicing how to perform a surgery or carry out disaster relief. We expect many mundane and potentially dangerous tasks to be taken over by AI systems in the near future.

In most cases, the main factors holding back these applications are not in the algorithms themselves, but in the collection and organization of appropriate data and the effective integration

of these algorithms into their broader sociotechnical systems. For example, without significant human-engineered knowledge, existing machine-learning algorithms struggle to generalize to "out of sample" examples that differ significantly from the data on which they were trained. Thus, if AlphaFold trained on natural proteins fails on synthetic proteins, or if a handwriting-recognition system trained on printed letters fails on cursive letters, these failures are due to the way the algorithms were trained, not the algorithms per se. (Consider the willingness of big tech companies like Facebook, Google, and Microsoft to share their deep learning algorithms and their reluctance to share the data they use in-house.)

Similarly, most AI-based decision-making systems require a formal specification of a reward or cost function, and eliciting and translating such preferences from multiple stakeholders remains a challenging task. For example, an AI controller managing a wind farm has to manage "standard" objectives such as maximizing energy produced and minimizing maintenance costs, but also harder-to-quantify preferences such as reducing ecological impact and noise to neighbors. As with the issue of insufficient relevant data, a failure of the AI in these cases is due to the way it was trained—on incorrect goals—rather than the algorithm itself.

In some cases, further challenges to the integration of AI systems come in the form of legal or economic incentives; for example, malpractice and compliance concerns have limited the penetration of AI in the health sector. Regulatory frameworks for safe, responsible innovation will be needed to achieve these possible near-term beneficial impacts.

Words and Expressions

augmentation [ˌɔːgmenˈteɪʃən] *n.* 增加；加强；提高
human-computer interaction 人机交互
molecule [ˈmɑːlɪkjuːl] *n.* 分子
visualization [ˌvɪʒuələˈzeɪʃn] *n.* 可视化
embed [ɪmˈbed] *v.* 嵌入
semantic representation 语义表示
visualize [ˈvɪʒuəlaɪz] *v.* 可视化；设想
decision-making [dɪˈsɪʒn meɪkɪŋ] *n.* 决策
recommendation [ˌrekəmenˈdeɪʃn] *n.* 推荐
forecast [ˈfɔːrkæst] *v.* 预测；预报
likelihood [ˈlaɪklihʊd] *n.* 可能；可能性
facilitate [fəˈsɪlɪteɪt] *v.* 促进；使便利
accessibility [əkˌsesəˈbɪləti] *n.* 可访问性；可达性

Notes

1. For example, an AI system may be able to synthesize large amounts of clinical data to identify a set of treatments for a particular patient along with likely side effects; a human clinician may be able to work with the patient to identify which option best fits his lifestyle and goals, and to explore creative ways of mitigating side effects that were not part of the AI's design space.

例如，人工智能系统可合成大量的临床数据，以确定特定患者的一整套治疗方案及可能的副作用；人类临床医生可与患者合作，确定哪种选择最适合他的生活方式和目标，并探索减轻副作用的创造性方法，而这些并不属于人工智能设计的领域。

2. An AI system might be better at synthesizing available data and making decisions in well-characterized parts of a problem, while a human may be better at understanding the implications of the data (say if missing data fields are actually a signal for important, unmeasured information for some subgroup represented in the data), working with difficult-to-fully-quantify objectives, and identifying creative actions beyond what the AI may be programmed to consider.

人工智能系统可能更善于综合可用数据，并对问题特征明确的部分做出决策，而人类可能更善于理解数据的含义（例如，如果缺失数据字段实际上是数据中表示的某些子群的重要且未测量信息的信号），并在目标难以完全量化的情况下进行决策，以及识别人工智能编程考虑之外的创造性行为。

3. While human experts can always analyze an AI from the outside—for example, dissecting the innovative moves made by AlphaGo—new developments in interpretable AI and visualization of AI are making it much easier for humans to inspect AI programs more deeply and use them to explicitly organize information in a way that facilitates a human expert putting the pieces together and drawing insights.

虽然人类专家总是可以从外部分析人工智能（例如，剖析 AlphaGo 的创新行为），但可解释人工智能和人工智能可视化方面的新发展使人类更容易深入地审视人工智能程序，并使用它们明确地组织信息，从而有助于人类专家将各个部分组合起来并得出见解。

4. Summarization and interactive chat technologies have great potential to help ensure that people get a healthy breadth of information on a topic, and to help break filter bubbles rather than make them—by providing a range of information, or at least an awareness of the biases in one's social-media or news feeds.

摘要和交互式聊天技术具有巨大的潜力，有助于确保人们在某个主题上获得健康的信息广度，并通过提供一系列信息或至少意识到社交媒体或新闻源中的偏见，帮助打破过滤泡沫，而不是制造泡沫。

5. Finally, AI systems can help human decision-making by leveling the playing field of information and resources. Especially as AI becomes more applicable in lower-data regimes, predictions can increase economic efficiency of everyday users by helping people and businesses find relevant opportunities, goods, and services, matching producers and consumers.

最后，人工智能系统可以通过平衡信息和资源的竞争环境来帮助人类决策。特别是随着人工智能越来越适用于较低的数据体系，预测可以帮助人们和企业找到相关的机会、商品和服务，匹配生产者和消费者，从而提高日常用户的经济效率。

6. A final major area of opportunity for augmentation is for AI to provide basic assistance during a task. For example, we are already starting to see AI programs that can process and translate text from a photograph, allowing travelers to read signage and menus. Improved translation tools will facilitate human interactions across cultures. Projects that once required a person to have highly specialized knowledge or copious amounts of time—from fixing your sink to creating a diabetes-friendly meal—may become accessible to more people by allowing them to

search for task- and context-specific expertise (such as adapting a tutorial video to apply to unique sink configuration).

增强的最后一个主要机会领域是人工智能在任务中提供基本帮助。例如，我们已经开始看到可以处理和翻译照片中文本的人工智能程序，让旅行者可以阅读标牌和菜单。改进的翻译工具将促进人类跨文化交流。曾经需要一个人拥有高度专业知识或大量时间的项目（从修理水槽到制作糖尿病友好餐）可能会通过允许更多人搜索特定任务和特定环境的专业知识(如调整教程视频以应用于独特的水槽配置）而被更多人使用。

7. In most cases, the main factors holding back these applications are not in the algorithms themselves, but in the collection and organization of appropriate data and the effective integration of these algorithms into their broader sociotechnical systems.

在大多数情况下，阻碍这些应用的主要因素不在于算法本身，而在于收集和组织适当的数据，并将这些算法有效地集成到更广泛的社会技术系统中。

14.2 Ten Predictions on the Future of Machine Learning

Machine learning is emerging as one of the most important developments in the software industry. While this advanced technology has been around for decades, it is now becoming commercially viable. We're moving into an era where machine learning techniques are essential tools to create value for businesses that want to understand the hidden value of their data. What does the future hold for machine learning? In this section, you explore our top ten predictions.

1. Machine Learning Will Be Embedded in Most Applications

Today, machine learning techniques are beginning to become popular in a variety of specialized environments. Businesses are looking to machine learning techniques to help them anticipate the future and create competitive differentiation.

In the next several years, you'll begin to see machine learning models embedded in nearly every application and on a variety of devices, including mobile devices and IoT hubs. In many cases, users will not know that they're interacting with machine learning models. Two examples where machine learning models are already embedded into everyday applications are retail websites and online advertisements. In both cases, machine learning models are often used to provide a more customized experience for users.

The impact of machine learning on a variety of industries will be dramatic and disruptive. Therefore, machine learning will significantly change how you do things. For example, hospitals can use machine learning models to anticipate the rate of admission based on conditions within their communities. Admissions can be related to weather conditions, the outbreak of a communicable illness, and other situations such as large events taking place in the city.

We are just beginning to see more and more machine learning models embedded into packaged solutions, such as customer management solutions and factory management systems. With the addition of machine learning models, these same systems become smarter and are able to provide predictive capability to enhance the value for the organization.

2. Trained Data as a Service Will Become a Prerequisite

One of the major obstacles in developing cognitive and machine learning models is training the data. Traditionally, data scientists have had to assume the jobs of gathering, labeling, and training the data. Another approach is to use publicly available data sets or crowdsourcing tools to collect and label data. While both of these approaches work, they are time consuming and complicated to execute.

To overcome these difficulties, a number of vendors offer pretrained data models. For example, a company may provide hundreds of thousands of pre-labeled medical images to help customers create an application that can help screen medical images and spot potential health issues.

3. Continuous Retraining of Models

Currently, the majority of machine learning models are offline. These offline models are trained using trained data and then deployed. After an offline model is deployed, the underlying model doesn't change as it is exposed to more data. The problem with offline models is that they presume the incoming data will remain fairly consistent.

Over the next few years, you will see more machine learning models available for use. **As these models are constantly updated with new data, the better the models will be at predictive analytics. However, preferences and trends change, and offline models can't adapt as the incoming data changes.** For example, take the situation where a machine learning model makes predictions on the likelihood that customers will churn. The model could have been very accurate when it was deployed, but as new, more flexible competitors emerge, and once customers have more options, their likelihood to churn will increase. Because the original model was trained on older data before new market entrants emerged, it will no longer give the organization accurate predictions. On the other hand, if the model is online and continuously adapting based on incoming data, the predictions on churn will be relevant even as preferences evolve and the market landscape changes.

4. Machine Learning as a Service Will Grow

As the models and algorithms that support machine learning mature, you'll see the growing popularity of machine learning as a service (MLaaS). MLaaS describes a variety of machine learning capabilities that are delivered via the cloud. Vendors in the MLaaS market offer tools like image recognition, voice recognition, data visualization, and deep learning. A user typically uploads data to a vendor's cloud, and then the machine learning computation is processed on the cloud.

Some of the challenges of moving large data sets to the cloud include networking costs, compliance and governance risks, and performance. However, by using a cloud service, organizations can use machine learning without the upfront time and costs associated with procuring hardware.

In addition, MLaaS abstracts much of the complexity involved with machine learning. For example, a team can use natural language processing (NLP) — a tool used to interpret text — to create dialogs between humans and machines. Both NLP and image recognition are well suited for the application of cloud services that has been designed to process specific compute intensive tasks. The performance differences are especially important when training and iterating many models. Large graphic processing units (GPUs) are designed to speed the rendering of images so that they can significantly reduce the cycle time.

5. The Maturation of NLP

We expect that in the coming decade, NLP will mature enough to be the norm for users to communicate with systems via a written or spoken interface. **NLP is the technology that allows machines to understand the structure and meaning of the spoken and written languages of humans. In addition, NLP technology allows machines to output information in spoken language understood by humans.** Researchers have been working on NLP technology for decades, and machine learning is helping to accelerate the implementation of NLP systems. Currently, it is very difficult for machines to understand the context of words and sentences. By applying machine learning to NLP, systems are able to learn the context and meaning of words and sentences. Take for example the sentence "A bat flew toward the crowd.". The sentence could be referring to a baseball bat that a hitter inadvertently let go of or a flying mammal that was heading toward a crowd of people. To understand the meaning of the sentence, a system would need to ingest the context around that sentence.

6. More Automation Will Streamline Machine Learning Pipelines

Automating the machine learning process will give less-technical employees access to machine learning capabilities. Additionally, by adding automation, technical users will be able to focus on more challenging work rather than simply automating repetitive tasks. There are many tedious details involved with machine learning that are important but ripe for automation (for example, data cleaning). **Data visualization is another area where automation is helping to streamline the machine learning process. Systems can be designed to select the most appropriate visualization for a given data set, making it easy to understand the relationship between data points.**

7. Specialized Hardware Will Improve the Performance of Machine Learning

We are approaching an era where sophisticated hardware is now affordable. Therefore, many organizations can procure hardware that is powerful enough to quickly process machine learning algorithms. In addition, this powerful hardware removes the processing bottleneck of machine learning, thus allowing machine learning to be embedded in more applications.

Traditionally, CPUs have been used to support the deep learning training process with mixed results. These CPUs are problematic because of the cumbersome way that they process steps in a neural network. In contrast, GPUs have hundreds of simpler cores that allow thousands of concurrent hardware threads. Because of the importance of GPUs in deep learning applications, there has been considerable research going into the technology in order to offer more powerful chips. Cloud computing vendors also recognize the value of GPUs, and more of them are offering

GPU environments on the cloud.

In addition to GPUs, researchers are using field programmable gate arrays (FPGAs) to successfully run machine learning workloads. Sometimes FPGAs outperform GPUs when running neural network and deep learning operations.

8. Automate Algorithm Selection and Testing Algorithms

Data scientists typically need to understand how to use dozens of specific machine learning algorithms. A variety of algorithms are used for different types of data or different types of questions you're trying to answer.

Choosing the right algorithm to create a machine learning model is not always easy. A data scientist may try several different algorithms until he finds the one that creates the best model. This process takes time and requires a high degree of expertise. Automation is being applied to help speed the task of algorithm selection. By using automation, data scientists are able to quickly focus on just one or two algorithms rather than manually testing many more. In addition, this automation helps developers and analysts with less machine learning experience work with machine learning algorithms.

9. Transparency and Trust Become a Requirement

Understanding not just how but why a machine learning model recommends a specific outcome will be essential in order to trust the results. A deep learning model used for medical image scanning may flag an image for a potential cancerous growth. However, simply identifying the image isn't enough. The physician will need to understand why the machine model thought the growth was cancerous. What information was analyzed to lead the model to conclude the diagnosis? The physician must be convinced that the results are confirmed by the data.

10. Machine Learning as an End-to-End Process

Now that we are moving into an era of commercialization of machine learning, we will begin to see machine learning as an end-to-end process from a development and operations perspective. This means that the process includes identifying the right data to solve a complex problem, ensuring that the data is properly trained, modeled, and managed on an ongoing basis. This life cycle of machine learning is critical because there is so much at stake. Machine learning models can be a powerful tool for predicting the future.

Words and Expressions

machine learning	机器学习
IoT: internet of things	物联网
prerequisite [ˌpriː'rekwəzɪt]	*n.* 前提；先决条件
cognitive ['kɑːgnətɪv]	*adj.* 认知的；感知的
label ['leɪbl]	*v.* 标注；贴标签于；用标签说明
offline model	离线模型
image recognition	图像识别

data visualization	数据可视化
data set	数据集
cloud service	云服务
complexity [kəm'pleksəti]	*n.* 复杂性
natural language processing(NLP)	自然语言处理
graphic processing unit (GPU)	图形处理单元
streamline ['stri:mlaɪn]	*vt.* 使成流线型；使（系统、机构等）效率更高
pipeline ['paɪplaɪn]	*n.* 管道
data cleaning	数据清洗
concurrent [kən'kɜ:rənt]	*adj.* 并存的；同时发生的
cloud computing	云计算
field programmable gate array (FPGA)	现场可编程门阵列
transparency [træns'pærənsi]	*n.* 透明；显而易见
end-to-end process	端到端的过程

Notes

1. As these models are constantly updated with new data, the better the models will be at predictive analytics. However, preferences and trends change, and offline models can't adapt as the incoming data changes.

随着这些模型不断使用新数据进行更新，这些模型在预测分析方面会做得越来越好。然而，随着偏好和趋势的变化，离线模型无法适应输入数据的变化。

2. NLP is the technology that allows machines to understand the structure and meaning of the spoken and written languages of humans. In addition, NLP technology allows machines to output information in spoken language understood by humans.

自然语言处理是一种让机器理解人类口语和书面语言的结构和意义的技术。此外，自然语言处理技术可使机器输出人类可理解的口语信息。

3. Data visualization is another area where automation is helping to streamline the machine learning process. Systems can be designed to select the most appropriate visualization for a given data set, making it easy to understand the relationship between data points.

数据可视化是自动化帮助简化机器学习过程的另一个领域。系统的设计可以为给定的数据集选择最合适的可视化方式，从而便于理解数据点之间的关系。

4. In addition to GPUs, researchers are using field programmable gate arrays (FPGAs) to successfully run machine learning workloads. Sometimes FPGAs outperform GPUs when running neural network and deep learning operations.

除了图形处理单元（GPU），研究人员还使用现场可编程门阵列（FPGA）成功运行机器学习的任务。在运行神经网络和深度学习任务时，FPGA 的性能在有些情况下会优于 GPU。

5. Now that we are moving into an era of commercialization of machine learning, we will begin to see machine learning as an end-to-end process from a development and operations perspective. This means that the process includes identifying the right data to solve a complex problem, ensuring that the data is properly trained, modeled, and managed on an ongoing basis.

现在，我们正进入机器学习商业化的时代，我们将开始从开发和运营的角度将机器学习视为一个端到端的过程。这意味着该过程包括确定解决复杂问题的正确数据，确保数据在持续的基础上得到正确的训练、建模和管理。

14.3 Big Data

Managing and analyzing data have always offered the greatest benefits and the greatest challenges for organizations of all sizes and across all industries. Businesses have long struggled with finding a pragmatic approach to capturing information about their customers, products, and services. When a company only had a handful of customers who all bought the same product in the same way, things were pretty straightforward and simple. But over time, companies and the markets they participate in have grown more complicated. To survive or gain a competitive advantage with customers, these companies added more product lines and diversified how they deliver their products. Data struggles are not limited to business. Research and development (R&D) organizations, for example, have struggled to get enough computing power to run sophisticated models or to process images and other sources of scientific data.

Indeed, we are dealing with a lot of complexity when it comes to data. Some data is structured and stored in a traditional relational database, while other data, including documents, customer service records, and even pictures and videos, is unstructured. Companies also have to consider new sources of data generated by machines such as sensors. Other new information sources are human generated, such as data from social media and the click-stream data generated from website interactions. In addition, the availability and adoption of newer, more powerful mobile devices, coupled with ubiquitous access to global networks will drive the creation of new sources for data. Although each data source can be independently managed and searched, the challenge today is how companies can make sense of the intersection of all these different types of data. When you are dealing with so much information in so many different forms, it is impossible to think about data management in traditional ways. Although we have always had a lot of data, the difference today is that significantly more of it exists, and it varies in type and timeliness. Organizations are also finding more ways to make use of this information than ever before. Therefore, you have to think about managing data differently. That is the opportunity and challenge of big data.

14.3.1 Definitions of Big Data

Although the term "big data" has become popular, there is no general consensus about what it really means. Often, many professional data analysts would imply the process of extraction, transformation, and load (ETL) for large datasets as the connotation of big data. A popular description of big data is based on three main attributes of data: volume, velocity, and variety (or 3Vs). Nevertheless, it does not capture all the aspects of big data accurately. **In order to provide a comprehensive meaning of big data, we will investigate this term from a historical perspective and see how it has been evolving from yesterday's meaning to today's connotation.**

Historically, the term "big data" is quite vague and ill defined. It is not a precise term and does not carry a particular meaning other than the notion of its size. The word "big" is too generic; the question how "big" is big and how "small" is small is relative to time, space, and circumstance. **From an evolutionary perspective, the size of "big data" is always evolving. If we use the current global Internet traffic capacity as a measuring stick, the meaning of big data volume would lie between the terabyte (TB) and zettabyte (ZB) range.** Based on the historical data traffic growth rate, Cisco claimed that humans have entered the ZB era in 2015. To understand the significance of the data volume's impact, let us glance at the average size of different data files shown in Table 14.1.

Table 14.1 Average size of different data files

Media	Average Size of Data File	Notes (2014)
Web page	1.6–2 MB	Average 100 objects
eBook	1–5 MB	200–350 pages
Song	3.5–5.8 MB	Average 1.9 MB per minute (MP3), 256 kbps rate (3 minutes)
Movie	100–120 GB	60 frames per second (MPEG-4 format, full high definition, 2 hours)

Intuitively, neither yesterday's data volume (absolute size) nor that of today can be defined as "big". Moreover, today's "big" may become tomorrow's "small". In order to clarify the term "big data" precisely and settle the debate, we will first investigate the historical definition from an evolutionary perspective.

1. Gartner's 3Vs Definition

Since 1997, many attributes have been added to big data. Among these attributes, three of them are the most popular and have been widely cited and adopted. The first one is so called Gartner's interpretation or 3Vs; the root of this term can be traced back to Feb. 2001. It was casted by Douglas Laney in his white paper published by Meta Group, which Gartner subsequently acquired in 2004. Douglas noticed that due to surging of e-commerce activities, data had grown along three dimensions, namely:

(1) Volume, which means the incoming data stream and cumulative volume of data.

(2) Velocity, which represents the pace of data used to support interactions and generated by interactions.

(3) Variety, which signifies the variety of incompatible and inconsistent data formats and data structures.

Douglas Laney's 3Vs definition has been widely regarded as the "common" attributes of big data but he stopped short of assigning these attributes to the term "big data".

2. IBM's 4Vs Definition

IBM added another attribute or "V" for "veracity" on the top of Douglas Laney's 3Vs notation, which is known as the 4Vs of big data. It defines each "V" as follows.

(1) Volume stands for the scale of data.

(2) Velocity denotes the analysis of streaming data.

(3) Variety indicates different forms of data.

(4) Veracity implies the uncertainty of data.

Zikopoulos et al. explained the reason behind the additional "V" or veracity dimension, which was "in response to the quality and source issues our clients began facing with their big data initiatives". They were also aware of some analysts including other V-based descriptors for big data, such as variability and visibility.

3. Microsoft's 6Vs Definition

For the sake of maximizing the business value, Microsoft extended Douglas Laney's 3Vs attributes to 6Vs, in which it added veracity, variability, and visibility.

(1) Volume stands for the scale of data.

(2) Velocity denotes the analysis of streaming data.

(3) Variety indicates different forms of data.

(4) Veracity focuses on the trustworthiness of data source.

(5) Variability refers to the complexity of data set. In comparison with "variety" (or different data formats), it means the number of variables in data set.

(6) Visibility emphasizes that you need to have a full picture of data in order to make informative decision.

4. More Vs for Big Data

A 5Vs big data definition was also proposed by Yuri Demchenko in 2013. He added the value dimension along with the IBM's 4Vs definition. Since Douglas Laney published 3Vs in 2001, there have been additional "Vs," even as many as 11.

All these definitions, such as 3Vs, 4Vs, 5Vs, or even 11Vs, are primarily trying to articulate the aspects of data. Most of them are data-oriented definitions, but they fail to articulate big data clearly in a relationship to the essence of big data analysis (BDA). In order to understand the essential meaning, we have to clarify what data is.

Data is everything within the universe. This means that data is within the existing limitation of technological capacity. If the technological capacity is allowed, there is no boundary or limitation for data. The question is why we should capture it in the first place. **Clearly, the main reason of capturing data is not because we have enough capacity to capture high volume, high velocity, and high variety data rather than to find a better solution for our research or business problem, which is to search for actionable intelligence.** Pure data-driven analysis may add little value for a decision maker; sometimes, it may only add the burden for the costs or resources of BDA. Perhaps this is why Harper believes big data is really hard.

Table 14.2 shows seven popular big data definitions, summarized by Timo Elliott and based on more than 33 big data definitions. Each of the definitions intends to describe a particular issue from one aspect of big data only and is very restrictive. However, a comprehensive definition can become complex and very long. **A solution for this issue is to use "rational reconstruction" offered by Karl Popper, which intends to make the reasons behind practice, decision, and process explicit and easier to understand.**

Table 14.2 Seven popular big data definitions

No	Type	Description
1	The original big data (3Vs)	The original type of definition is referred to Douglas Laney's volume, velocity, and variety, or 3Vs. It has been widely cited since 2001. Many have tried to extend the number of Vs, such as 4Vs, 5Vs, 6Vs… up to 11Vs.
2	Big data as technology	This type of definition is oriented by new technology development, such as MapReduce, bulk synchronous parallel (BSP-Hama), resilient distributed dataset (RDD, Spark), and Lambda architecture (Flink).
3	Big data as application	This kind of definition emphasizes different applications based on different types of big data. Barry Devlin defined it as applications of process-mediated data, human-sourced information, and machine-generated data. Shaun Connolly focused on analyzing transactions, interactions, and observation of data. It looks for hindsight of data.
4	Big data as signal	This is another type of application-oriented definition, but it focuses on timing rather than the type of data. It looks for foresight of data or new "signal" pattern in dataset.
5	Big data as opportunity	Matt Aslett: "Big data as analyzing data was previously ignored because of technology limitations." It highlights many potential opportunities by revisiting the collected or archived datasets when new technologies are variable.
6	Big data as metaphor	It defines big data as a human thinking process. It elevates BDA to the new level, which means BDS is not a type of analytic tool rather it is an extension of human brain.
7	Big data as new term for old stuff	This definition simply means the new bottle (ralabel the new term "big data") for old wine (BI, data mining, or other traditional data analytic activities). It is one of the most cynical ways to define big data.

14.3.2 Big Data Management Architecture

As data has become the fuel of growth and innovation, it is more important than ever to have an underlying architecture to support growing requirements. It is important to take into account the functional requirements for big data. Figure 14.1 illustrates that data must first be captured, and then organized and integrated. After this phase is successfully implemented, data can be analyzed based on the problem being addressed. Finally, management takes action based on the outcome of that analysis. For example, Amazon.com might recommend a book based on a past purchase or a customer might receive a coupon for a discount for a future purchase of a related product to one that was just purchased. Although this sounds straightforward, certain nuances of these functions are complicated. Validation is a particularly important issue. If your organization is combining data sources, it is critical that you have the ability to validate that these sources make sense when combined. Also, certain data sources may contain sensitive information, so you must implement sufficient levels of security and governance.

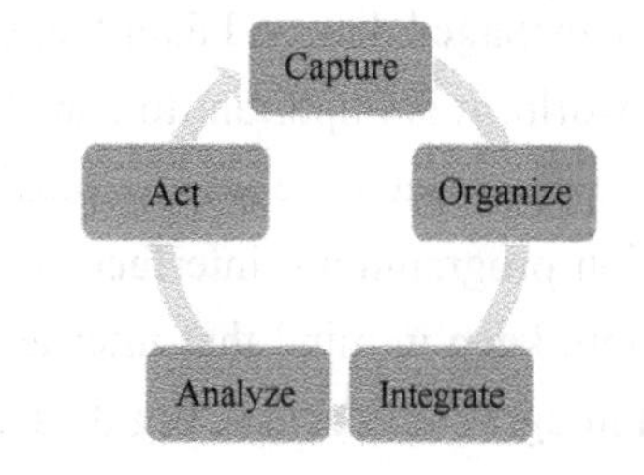

Figure 14.1 The cycle of big data management

In addition to supporting the functional requirements, it is important to support the required performance. Your needs will depend on the nature of the analysis you are supporting. You will need the right amount of computational power and speed. While some of the analysis you will do will be performed in real time, you will inevitably be storing some amount of data as well. Your architecture also has to have the right amount of redundancy so that you are protected from unanticipated latency and downtime.

To understand big data, it helps to lay out the components of the architecture. A big data management architecture must include a variety of services that enable companies to make use of myriad data sources in a fast and effective manner. To help you make sense of this, we put the components into a diagram (see Figure 14.2) that will help you see what's there and the relationship between the components.

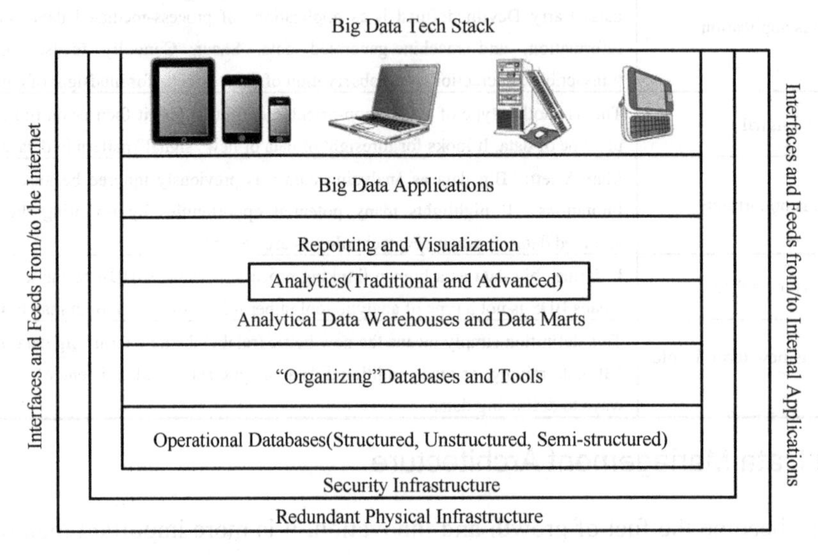

Figure 14.2 The big data management architecture

1. Interfaces and Feeds

Before we get into the nitty-gritty of the big data technology stack itself, we'd like you to notice that on either side of the diagram are indications of interfaces and feeds into and out of both internally managed data and data feeds from external sources. To understand how big data works in the real world, it is important to start by understanding this necessity. In fact, what makes big data big is the fact that it relies on picking up lots of data from lots of sources. Therefore, open application programming interfaces (APIs) will be core to any big data management architecture. In addition, keep in mind that interfaces exist at every level and between every layer of the stack. Without integration services, big data can't happen.

2. Redundant Physical Infrastructure

The supporting physical infrastructure is fundamental to the operation and scalability of a big data management architecture. In fact, without the availability of robust physical infrastructure, big data would probably not have emerged as such an important trend. To support an unanticipated or unpredictable volume of data, the physical infrastructure for big data has to be different than that for traditional data. The physical infrastructure is based on a distributed computing model. This means that data may be physically stored in many different locations and can be linked together through networks, the use of a distributed file system, and various big data analytic tools and applications.

Redundancy is important because we are dealing with so much data from so many different sources. Redundancy comes in many forms. If your company has created a private cloud, you will want to have redundancy built within the private environment so that it can scale out to support changing workloads. If your company wants to contain internal IT growth, it may use external cloud services to augment its internal resources. In some cases, this redundancy may come in the form of a software as a service (SaaS) offering that allows companies to do sophisticated data analysis as a service. The SaaS approach offers lower costs, quicker startup, and seamless evolution of the underlying technology.

3. Security Infrastructure

The more important big data analysis becomes to companies, the more important it will be to secure that data. For example, if you are a healthcare company, you will probably want to use big data applications to determine changes in demographics or shifts in patient needs. This data about your constituents needs to be protected both to meet compliance requirements and to protect the patients' privacy. You will need to take into account who is allowed to see the data and under what circumstances they are allowed to do so. You will need to be able to verify the identity of users as well as protect the identity of patients. These types of security requirements need to be part of the big data fabric from the outset and not an afterthought.

4. Operational Data Sources

When you think about big data, it is important to understand that you have to incorporate all the data sources that will give you a complete picture of your business and see how the data impacts the way you operate your business. **Traditionally, an operational data source consisted of highly structured data managed by the line of business in a relational database. But as the world changes, it is important to understand that operational data now has to encompass a broader set of data sources, including unstructured sources such as customer and social media data in all its forms.**

You find new emerging approaches to data management in the big data world, including document, graph, columnar, and geospatial database architectures. Collectively, these are referred to as NoSQL, or not only SQL, databases. In essence, you need to map the data architectures to the types of transactions. Doing so will help to ensure the right data is available when you need it. You also need data architectures that support complex unstructured content. You need to include both relational databases and nonrelational databases in your approach to harnessing big data. It is also necessary to include unstructured data sources, such as content management systems, so that you can get closer to that 360-degree business view.

14.3.3 Performance Matters

Your data architecture also needs to perform in concert with your organization's supporting infrastructure. For example, you might be interested in running models to determine whether it is safe to drill for oil in an offshore area given real-time data of temperature, salinity, sediment resuspension, and a host of other biological, chemical, and physical properties of the water column. It might take days to run this model using a traditional server configuration. However, using a distributed computing model, what took days might now take minutes.

Performance might also determine the kind of database you would use. For example, in some situations, you may want to understand how two very distinct data elements are related. What is the relationship between buzz on a social network and the growth in sales? This is not the typical query you could ask of a structured, relational database. A graph database might be a better choice, as it is specifically designed to separate the "nodes" or entities from its "properties" or the information that defines that entity, and the "edge" or relationship between nodes and properties. Using the right database will also improve performance. Typically the graph database will be used in scientific and technical applications.

Other important operational database approaches include columnar databases that store information efficiently in columns rather than rows. This approach leads to faster performance because input/output is extremely fast. When geographic data storage is part of the equation, a spatial database is optimized to store and query data based on how objects are related in space.

1. Organizing Data Services and Tools

Not all the data that organizations use is operational. A growing amount of data comes from a variety of sources that aren't quite as organized or straightforward, including data that comes from machines or sensors, and massive public and private data sources. In the past, most companies weren't able to either capture or store these vast amounts of data. It was simply too expensive or too overwhelming. Even if companies were able to capture the data, they did not have the tools to do anything about it. Very few tools could make sense of these vast amounts of data. The tools that did exist were complex to use and did not produce results in a reasonable time frame. In the end, those who really wanted to go to the enormous effort of analyzing this data were forced to work with snapshots of data. This has the undesirable effect of missing important events because they were not in a particular snapshot.

2. MapReduce, Big Table, and Hadoop

With the evolution of computing technology, it is now possible to manage the immense volumes of data that previously could have only been handled by supercomputers at great expense. Prices of systems have dropped, and as a result, new techniques for distributed computing are mainstream. The real breakthrough in big data happened as companies like Yahoo!, Google, and Facebook came to the realization that they needed help in monetizing the massive amounts of data their offerings were creating.

These emerging companies needed to find new technologies that would allow them to store, access, and analyze huge amounts of data in near real time, so that they could monetize the benefits of owning this much data about participants in their networks. Their resulting solutions are transforming the data management market. In particular, the innovations MapReduce, Big Table, and Hadoop proved to be the sparks that led to a new generation of data management. These technologies address one of the most fundamental problems — the capability to process massive amounts of data efficiently, cost effectively, and in a timely fashion.

- MapReduce. MapReduce was designed by Google as a way of efficiently executing a set of functions against a large amount of data in batch mode. The "map" component distributes the programming problem or tasks across a large number of systems and handles the placement of the tasks in a way that balances the load and manages recovery from failures. After the distributed computation is completed, another function called "reduce" aggregates all the elements back together to provide a result. An example of MapReduce usage would be to determine how many pages of a book are written in each of 50 different languages.
- Big Table. Big Table was developed by Google to be a distributed storage system intended to manage highly scalable structured data. Data is organized into tables with rows and columns. Unlike a traditional relational database model, Big Table is a sparse, distributed, persistent multidimensional sorted map. It is intended to store huge volumes of data across commodity servers.
- Hadoop. Hadoop is an Apache-managed software framework derived from MapReduce and Big Table. Hadoop allows applications based on MapReduce to run on large clusters of commodity hardware. The project is the foundation for the computing architecture supporting Yahoo!'s business. Hadoop is designed to parallelize data processing across computing nodes to speed computations and hide latency. Two major components of Hadoop exist: a massively scalable distributed file system that can support petabytes of data and a massively scalable MapReduce engine that computes results in batch.

14.3.4 Traditional and Advanced Analytics

What does your business now do with all the data in all its forms to try to make sense of it for the business? It requires many different approaches to analysis, depending on the problem being solved. Some analyses will use a traditional data warehouse, while other analyses will take advantage of advanced predictive analytics. Managing big data holistically requires many different approaches to help the business to successfully plan for the future.

1. Analytical Data Warehouses and Data Marts

After a company sorts through the massive amounts of data available, it is often pragmatic to take the subset of data that reveals patterns and put it into a form that's available to the business. These warehouses and marts provide compression, multilevel partitioning, and a massively parallel processing architecture.

2. Big Data Analytics

The capability to manage and analyze petabytes of data enables companies to deal with clusters of information that could have an impact on the business. This requires analytical engines that can manage this highly distributed data and provide results that can be optimized to solve a business problem. Analytics can get quite complex with big data. For example, some organizations are using predictive models that couple structured and unstructured data together to predict fraud. Social media analytics, text analytics, and new kinds of analytics are being utilized by organizations looking to gain insight into big data.

3. Reporting and Visualization

Organizations have always relied on the capability to create reports to give them an understanding of what the data tells them about everything from monthly sales figures to projections of growth. Big data changes the way that data is managed and used. If a company can collect, manage, and analyze enough data, it can use a new generation of tools to help management truly understand the impact not just of a collection of data elements but also how these data elements offer context based on the business problem being addressed. With big data, reporting and data visualization become tools for looking at the context of how data is related and the impact of those relationships on the future.

4. Big Data Applications

Traditionally, the business expected that data would be used to answer questions about what to do and when to do it. Data was often integrated as fields into general-purpose business applications. With the advent of big data, this is changing. Now, we are seeing the development of applications that are designed specifically to take advantage of the unique characteristics of big data. Some of the emerging applications are in areas such as healthcare, manufacturing management, traffic management, and so on. What do all these big data applications have in common? They rely on huge volumes, velocities, and varieties of data to transform the behavior of a market. In healthcare, a big data application might be able to monitor premature infants to determine when intervention is needed. In manufacturing, a big data application can be used to prevent a machine from shutting down during a production run. A big data traffic management application can reduce the number of traffic jams on busy city highways to decrease accidents, save fuel, and reduce pollution.

Words and Expressions

big data	大数据
attribute ['ætrɪbjuːt]	*n.* 属性；特征
volume ['vɑːljuːm, 'vɑːljəm]	*n.* 量；容量
velocity [və'lɑːsəti]	*n.* 速度；高速；快速
variety [və'raɪəti]	*n.* 不同种类；多变性；多样化
terabyte ['terəbaɪt]	*n.* 太字节；万亿字节
zettabyte ['zetəbaɪt]	*n.* 泽字节；十万亿亿字节
dimension [daɪ'menʃn, dɪ'menʃn]	*n.* 维；规模；程度；范围
assign [ə'saɪn]	*v.* 分配；指定
veracity [və'ræsəti]	*n.* 真实性
variability [ˌveriə'bɪləti, væriə'bɪləti]	*n.* 可变性
visibility [ˌvɪzə'bɪləti]	*n.* 可见性
data-oriented definition	面向数据的定义

boundary ['baʊndri] *n.* 边界；界限
rational ['ræʃnəl] *adj.* 理性的；合理的；有理数的
nitty-gritty [ˌnɪti'grɪti] *n.* 重要细节；基本事实；本质
application programming interface 应用编程接口
relational database 关系数据库
nonrelational database 非关系数据库
salinity [sə'lɪnəti] *n.* 盐分；盐浓度；含盐量
sediment ['sedɪmənt] *n.* 沉积物；沉淀物
resuspension [riːzəs'penʃn] *n.* 再悬浮；重新悬浮

Notes

1. Although the term "big data" has become popular, there is no general consensus about what it really means. Often, many professional data analysts would imply the process of extraction, transformation, and load (ETL) for large datasets as the connotation of big data.

尽管"大数据"一词已经很流行，但对于它的真正含义还没有达成普遍共识。通常，许多专业数据分析师会将大型数据集的提取、转换和加载（ETL）过程作为大数据的内涵。

2. In order to provide a comprehensive meaning of big data, we will investigate this term from a historical perspective and see how it has been evolving from yesterday's meaning to today's connotation.

为提供大数据的全面含义，我们将从历史的角度来研究这个术语，看看它是如何从过去的含义演变为现在的含义的。

3. From an evolutionary perspective, the size of "big data" is always evolving. If we use the current global Internet traffic capacity as a measuring stick, the meaning of big data volume would lie between the terabyte (TB) and zettabyte (ZB) range.

从进化的角度来看，"大数据"的规模总是在不断变化的。如果我们以当前全球因特网流量作为衡量标准，大数据量的含义将介于太字节（TB）和泽字节（ZB）之间。

4. Clearly, the main reason of capturing data is not because we have enough capacity to capture high volume, high velocity, and high variety data rather than to find a better solution for our research or business problem, which is to search for actionable intelligence.

显然，捕获数据的主要原因并不是我们有足够的能力捕获高容量、高速度和多样化的数据，而是为我们的研究或业务问题找到更好的解决方案，即搜索可操作的智能。

5. A solution for this issue is to use "rational reconstruction" offered by Karl Popper, which intends to make the reasons behind practice, decision, and process explicit and easier to understand.

解决这个问题的一个方法是使用卡尔•波普尔提供的"理性重建"，它旨在使实践、决策和过程背后的原因明确和更易于理解。

6. Traditionally, an operational data source consisted of highly structured data managed by the line of business in a relational database. But as the world changes, it is important to understand that operational data now has to encompass a broader set of data sources, including unstructured sources such as customer and social media data in all its forms.

传统上，运营数据源由业务线在关系数据库中管理的高度结构化数据组成。但随着世界的变化，重要的是要理解，运营数据现在必须包含更广泛的数据源，包括非结构化源，如各种形式的客户和社交媒体数据。

14.4 Quantum Internet

14.4.1 Introduction

Quantum internet (QI) is trending internet technology that facilitates quantum communication through quantum bits (qubits) among remote quantum devices or nodes. Such a technology will work in synergy with classical internet to overcome the limitations posed by traditional interconnect technologies. This novel technology stems from the laws of quantum mechanics, where one of the laws states that it is impossible to measure a property of a system without changing its state. Consequently, qubits cannot be copied and any attempt to do so will be detected, thus making the communication more secure and private. The properties of qubits give an edge to QI over the traditional internet in many ways.

Qubits also exhibit quantum entanglement, where qubits at remote nodes are correlated with each other. This correlation is stronger than ever possible in the classical domain. Entanglement is inherently private, as it is not possible, because of no-cloning, for a third qubit to be entangled with either of the two entangled qubits. Therefore, this quantum effect could open up a different galaxy of applications with world-changing potential. As qubits are prone to environmental losses, transmitting qubits over long distances is a challenging task due to decoherence. Therefore, extensive research is going on to achieve long-haul quantum communication. To mitigate this noise, various techniques are employed, such as quantum error-correcting (QEC) codes and fault-tolerant techniques.

In the modern era, where the internet plays a vital role in everyday life, providing secure communication and data privacy is of prime importance, ensuring no eavesdropping between two communicating parties. On the technical side, some of the contemporary cryptographic techniques might be broken in the future, with the advent of quantum computers. **New potential in cryptography emerges with advances in quantum cryptography that exploits quantum mechanical principle of no-cloning. These features play a vital role in providing secrecy and integrity to the data to be communicated, thus making messages unintelligible to any unauthorized party.** If a malevolent third party eavesdrops on this key distribution, privacy of the communication will be compromised. This problem is addressed by the best-known application of quantum networking, i.e., quantum key distribution (QKD), which provides secure access to computers on the cloud utilizing various protocols. QKD networks are commercially available, and are studied and deployed covering metropolitan distances. Long distance QKD networks with trusted nodes are also currently possible. Ultra-secure QKD protocols are developed based on well-accepted laws that govern quantum physics for sharing secret keys among two parties. Further advances in the field ask for the significant interdisciplinary effort by research groups working on this novel technology to make it widespread.

It is envisioned that a future QI will somewhat look like as seen in Figure 14.3. It will have a network of remote nodes interconnected with each other with the help of multiparty entanglement and quantum teleportation using fiber-based or free space channels. The quantum effects of quantum mechanics will make this network secure using various QKD protocols. QEC codes will be required to encode the qubits carrying the information that will protect them from decoherence and environmental effects. With advancements in multi qubit processors, full scale quantum computers will come into force and, along with existing classical infrastructure, will revolutionize the world with inherently secure global QI.

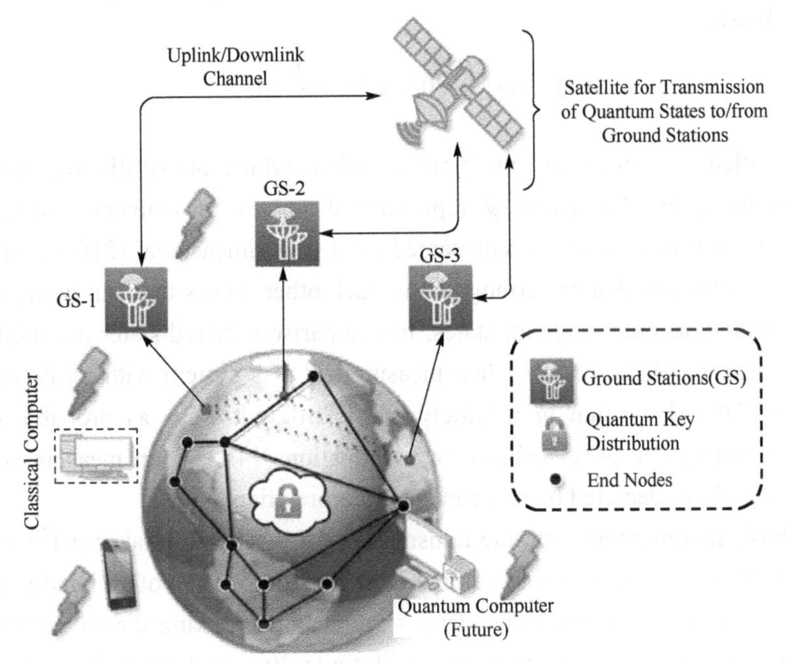

Figure 14.3 Vision of future quantum internet working in synergy with classical internet

14.4.2 Preliminaries

In this section, several fundamental concepts of quantum internet are introduced.

1. Quantum Mechanics

Quantum mechanics is a discipline with a continuously increasing importance in today's world applied with colossal success in many fields of science and technology including physics, chemistry, AI, life sciences, and military applications. It is formulated as a mathematical framework, or set of rules, for the development of physical hypotheses by exploiting the properties of single quantum systems involving qubits. For harnessing the power of quantum mechanics in various applications, the study of the qubit and of the single quantum system is of utter importance. It has the potential to offer secure communications that can make them cryptographically as secure as the one-time-pad (OTP), unlike today's classical communication and cryptography with its vulnerabilities. For better understanding the quantum effects of quantum mechanics, certain theorems were postulated by physicists to get a clear view of properties of qubits.

1) State Space and Superposition Principle

An isolated or closed system of qubits is associated with complex Hilbert space, also known as the state space of the system. A unit vector completely describes the state vector of a system in its state space. A qubit is the simplest quantum mechanical system that can be represented geometrically by a Bloch sphere, as depicted in Figure 14.4. Any pure quantum state $|\psi\rangle$ can be represented by a point on the surface of the Bloch sphere with spherical coordinates of θ as a polar angle, i.e., angle that a line makes with Z-axis, and ϕ as an azimuthal angle, i.e., angle that a line makes with X-axis:

$$|\psi\rangle = \cos\left(\frac{\theta}{2}\right)|0\rangle + \mathrm{e}^{\mathrm{i}\phi}\sin\left(\frac{\theta}{2}\right)|1\rangle \tag{14.1}$$

where "$|\bullet\rangle$" is called a "ket-vector" in Dirac notation, which physically represents the pure quantum state of the qubit. The angle ϕ represents the phase of a quantum state $|\psi\rangle$. Every individual state in the Bloch sphere is represented by a two-dimensional (2D) vector, where $|0\rangle$ and $|1\rangle$ are its basis vectors that are orthogonal to each other. States that can be represented by a "ket-vector" are known as pure quantum states. In comparison, mixed states are mixtures of pure states, resulting in probabilistic results when measured. It is a system with weak state or whose state is not fully defined, written in infinitely many different ways as probable outcomes of well-defined pure states. The pure state, given in Equation (14.1), upon measurement results in either state $|0\rangle$ or $|1\rangle$, as depicted by Copenhagen interpretation.

While in classical communication, the transmission of information takes place in the form of bits taking either "0" or "1" value, the quantum communication on the other hand uses qubits that can be in a superposition of the two basis states simultaneously taking the value of "0" and "1". The result of measurement of such states is not definite $|0\rangle$ or definite $|1\rangle$. The principle of superposition can be best understood by visualizing the polarization state of a photon. At 45 degree of polarization, a photon is simultaneously vertically and horizontally polarized, representing both states at the same time. A famous thought experiment paradigm of superposition principle is Schrödinger's cat, where the cat is, as colloquially interpreted, simultaneously dead and alive as a result of random event that may or may not occur.

2) Quantum No-Cloning

This theorem, which was first formulated by Wooters, Zurek and Deiks in 1982, states that the quantum state of any particle carrying a qubit in a quantum channel cannot be either copied, amplified or cloned, thus making it a reliable and secure form of communication. Any attempt to clone it will be detected, further activating the QKD protocols. **However, at the same time, no-cloning prevents qubits to be sent over long distances by amplification or sending copies of qubits, as it is done in classical communication, for efficient detection of bits without any loss of information. For this purpose, quantum repeaters are studied to solve the problem of long distance communication of qubits, which is the core functionality required for full-scale QI.**

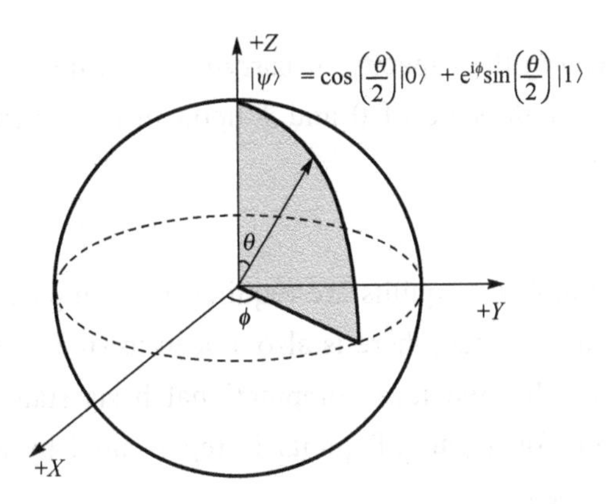

Figure 14.4 Bloch sphere: geometrical representation of pure quantum state of a qubit. Here any state $|\psi\rangle = \alpha|0\rangle + \beta|1\rangle$ can be shown by a point on the surface of the sphere, where $\alpha = \cos\left(\frac{\theta}{2}\right)$ and $\beta = e^{i\phi}\sin\left(\frac{\theta}{2}\right)$

2. Single Qubit System

In classical digital communication systems, the information is encoded in the form of binary bits. The bit represents a logical state with one of two possible values, i.e., "1" or "0". **On the other hand, in quantum communication, instead of bits, qubits are employed to carry quantum information from source to destination. A qubit can be imagined as a two-way system: direction of polarization of a photon, the up or down spin of an electron, or two energy levels of an electron orbiting an atom.** For instance, the electron can exist in either "ground" or "excited" state in an atom, represented by a state $|\psi\rangle = \alpha|0\rangle + \beta|1\rangle$, where $|0\rangle$ and $|1\rangle$ corresponds to ground state and excited state, respectively. By directing light on the atom at the right time and the right intensity, it is possible to move the electron from the $|0\rangle$ state to the $|1\rangle$ state and the other way around. Manipulating the time interval of the light, an electron, at first, in the state $|0\rangle$ can now be moved somewhere between $|0\rangle$ and $|1\rangle$, into $|\psi\rangle = \frac{1}{\sqrt{2}}|0\rangle + \frac{1}{\sqrt{2}}|1\rangle$ state, i.e., superposition of basis states. The variables α and β are complex numbers representing probability amplitudes, which means that $|\alpha|^2$ is the probability of getting $|\psi\rangle = 0$ as a result of the measurement on qubit $|\psi\rangle$ and $|\beta|^2$ is the probability of getting $|\psi\rangle = 1$ as a result of the measurement on qubit $|\psi\rangle$. It must be satisfied that

$$|\alpha|^2 + |\beta|^2 = 1 \tag{14.2}$$

A classical bit is like a coin resulting in definite outputs of heads or tails. Supposing, a qubit with $\alpha = \beta = \frac{1}{\sqrt{2}}$, will be in the state:

$$|\pm\rangle = \frac{1}{\sqrt{2}}|0\rangle \pm \frac{1}{\sqrt{2}}|1\rangle \tag{14.3}$$

which gives result 0 with 0.5 probability and result 1 with 0.5 probability, when measured over the time. This state often plays a vital role in quantum communication and is denoted by $|+\rangle$ and $|-\rangle$.

Despite qubit theoretically having the capacity of carrying an infinite amount of information, all quantum states collapse to a single state of 0 and 1 upon measurement and the reason for this characteristic is still unknown.

3. Multiple Qubits

In the real world, more number of qubits are required for complex computations to gain full advantage of quantum postulates. Hence, there is also a need to study multiple-qubit systems. For instance, a two-qubit system will have four computational basis states denoted by $|00\rangle$, $|01\rangle$, $|10\rangle$, $|11\rangle$. The quantum state for a pair of qubits is represented by state vectors having their respective amplitude coefficients as

$$|\psi\rangle = a_{00}|00\rangle + a_{01}|01\rangle + a_{10}|10\rangle + a_{11}|11\rangle \tag{14.4}$$

Similar to the condition that has been discussed for a single qubit, the measurement result of Equation (14.4) occurs with probability a_{00}^2 for $|00\rangle$, a_{01}^2 for $|01\rangle$, a_{10}^2 for $|10\rangle$, and a_{11}^2 for $|11\rangle$. If $x = 00, 01, 10, 11$, then normalization condition is $\sum_x |a_x|^2 = 1$. Thus, it can be seen that with the addition of single qubit, the computational basis states are increased exponentially thereby giving space for complex computations that were beyond reach with classical resources.

1) Bell States or EPR Pairs

EPR is the name which came from famous paradox raised by three scientists Einstein, Podolsky, and Rosen in their famous paper "Can a quantum mechanical description of physical reality be considered complete?", where unintuitive properties of Bell states were pointed out. In a two-qubit mechanical system, Bell states are mathematically expressed as

$$|\Phi^+\rangle = \frac{1}{\sqrt{2}}|00\rangle + \frac{1}{\sqrt{2}}|11\rangle \tag{14.5}$$

$$|\Phi^-\rangle = \frac{1}{\sqrt{2}}|00\rangle - \frac{1}{\sqrt{2}}|11\rangle \tag{14.6}$$

$$|\Psi^+\rangle = \frac{1}{\sqrt{2}}|01\rangle + \frac{1}{\sqrt{2}}|10\rangle \tag{14.7}$$

$$|\Psi^-\rangle = \frac{1}{\sqrt{2}}|01\rangle - \frac{1}{\sqrt{2}}|10\rangle \tag{14.8}$$

The two particles in these Bell states are termed EPR pairs and are responsible for entanglement generation and distribution in the case of quantum teleportation, which is the key functionality of QI. A Bell state has the property that when the first qubit state is measured, it results in two outcomes, i.e., 0 with $\frac{1}{2}$ probability and 1 with $\frac{1}{2}$ probability leaving the post-measurement state to be $|00\rangle$ and $|11\rangle$, respectively. Due to this, if the second qubit is measured, it results in the same outcome as the first qubit. This explains the interesting correlation between these two qubits. Even if some local operations are applied to the first or second qubit, these correlations still exited because of entanglement. EPR's paradox was improvised by John Bell, who stated that these measurement correlations in the Bell state are much stronger than anything possible in classical systems. This enables processing of the information that goes well beyond what is possible in the classical world.

2) Entanglement

Quantum entanglement, shown in Figure 14.5, is the phenomenon of correlation between two pairs of qubits that are separated by a physical distance, such that any random measurement on entangled qubits will generate the same set of random outcomes. The concept of quantum entanglement, which is governed by laws of quantum mechanics, is the core functionality for many applications and does not have any counterpart in classical communication. Independently measuring these qubits results in a random distribution of "1" and "0" in equal probability. This means that the state of entangled qubits is instantaneously fixed when the other pair of entangled qubits are measured. **Entanglement does not allow any exchange of information between remote nodes, rather it allows obtaining only mutual information. For information exchange, qubits can be transmitted between remote nodes without actually sending them from source to destination by the phenomenon of quantum teleportation, which is fundamental for quantum networks.** An unknown quantum state can also be transmitted through quantum teleportation from source to destination over long distances with the aid of EPR pair and local operations.

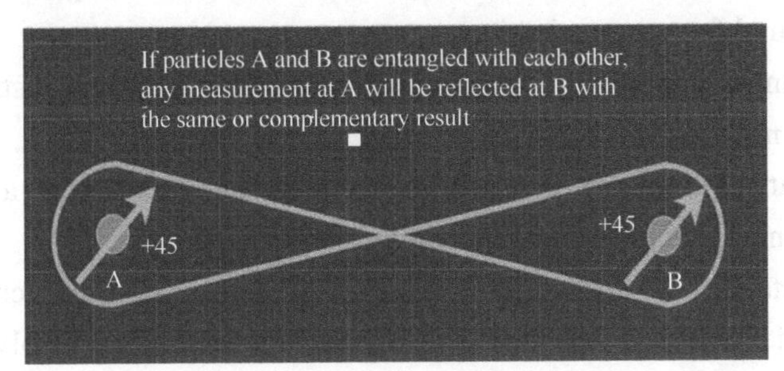

Figure 14.5 Quantum entanglement: two particles A and B are correlated with each other even if they are a long distance apart. If any measurement is done at A having spin +45, then the same is reflected at B. This happens due to the weird phenomena of quantum entanglement

14.4.3 Quantum Internet Functionalities

Quantum internet is a system that consists of various functionalities required for communicating information among remote nodes of a network. In this section, various functionalities of QI like quantum teleportation, quantum repeaters, quantum channels, quantum memories, QKD are briefly introduced.

1. Quantum Teleportation

Quantum teleportation comprises a system for transmitting qubits inside a network of quantum communication without physically transferring the particle storing the qubit. The concept of teleporting qubits has started with the seminal work presented in "Teleporting an unknown quantum state via dual classical and Einstein-Podolsky-Rosen channels". It was proved experimentally by Anton Zeilinger group in 1997 for the first time through the successful transmission of quantum optical states. At the same time, group of Sandu Popescu successfully experimented with the photonic teleportation in 1998. The teleportation distance of 55 m was achieved under laboratory conditions by Swiss researcher Nicolas Gisin and his team in 2003 and later verified it in 2007 by using Swisscom's commercial telecommunication network based on fiber-optics. Later in 2004, researchers from Innsbruck and USA succeeded in teleporting atomic states for the first time followed by an Austrian group led by Anton Zeilinger who achieved a distance of 600 m through a fiber-optic across Danube. This distance was improved to 1.3 km in following research. Significant experimental progresses in quantum teleportation networks improved the beeline distance to 8.2 km and 100 km. A larger distance—144 km between the islands of La Palma and Tenerife—of teleportation was further achieved by a team of Anton Zeilinger in 2012 using a free space channel. With the advancements in satellite technology, teleportation over a distance of more than 1,200 km has been successfully realized with the launch of the Micius satellite by Jian-Wei Pan and his team from China.

A large portion of the examinations has been confined to the teleportation of single-body quantum states, i.e., quantum teleportation of two-level states, multidimensional states, continuous variable teleportation, and discrete variable teleportation. Quantum teleportation is realized with two parallel links required for communication.

(1) Quantum channel link, for generating entangled EPR pair and distributing it between source and destination nodes.

(2) Classical channel link, for transmitting two classical bits from a source node to a destination node for sending the measurement result of source qubit.

Quantum teleportation is not straightforward for application, which renders efficient communication and fidelity of the system. However, at the same time, it involves imperfections in its process imposed by photon loss in environment decoherence or the result of a sequence of operations that are applied for the processing of teleportation. Thus, the technology opted for quantum teleportation plays a strong part in improving the system fidelity and reducing decoherence.

2. Quantum Repeaters

A quantum repeater is another functionality of QI that is employed in between the remote nodes, to transmit quantum information to long distance as shown in Figure 14.6. Quantum repeaters generate maximally entangled pairs between two far away end nodes by bifurcating the network into segments, establishing the long-distance entanglement between end points. In long distance communication, most of the photons carrying information get lost because of channel attenuation or other environmental factors. This exponential problem of photon loss was addressed by Jurgen Briegel in 1998 by introducing the model of a quantum repeater. In the classical domain, the issue is resolved by employing classical repeaters that copy the information after regular intervals for amplification. However, in the case of quantum communication, due to the no-cloning theorem, information carried in qubits cannot be copied or amplified since classical repeaters are not suitable for this communication. Hence, the quantum repeater model proposed by Briegel is employed to transmit qubits without any loss of information.

The evolution of quantum repeaters is classified into three distinct generations. In the first generation of quantum repeaters, intermediate stations are placed in such a way that links are short distance apart and probability for establishing entanglement among repeater stations is high. The second generation of quantum repeaters employ QEC (one-way classical communication) for the purpose of entanglement purification. In the third generation of quantum repeaters, two-way signalling is replaced by one-way signalling using certain loss-tolerant codes.

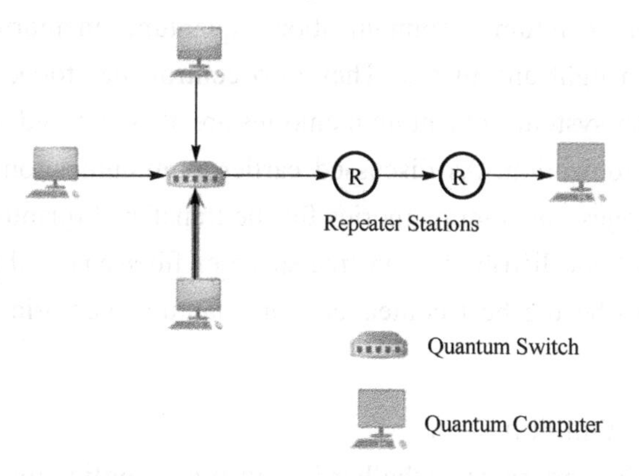

Figure 14.6 Repeater stations: these stations are employed at short distances so that entanglement can be established between intermediary nodes and ultimately between end nodes which are long distance apart. Here, quantum switch is also employed to transmit information between end nodes, traversing channels in superposition of different orders

3. Quantum Channels

The QI is an inherently secure network that will be implemented along with classical internet, where end nodes will be connected via classical and quantum channels. A continuous quantum channel is created until all the end nodes are connected. The classical information as well as quantum information can be sent through quantum channels. This channel can be a free space channel for sending quantum information to long distances or can be a fiber-based channel, which can be used to send quantum information to relatively shorter distances. **How efficiently an entanglement can be distributed depends upon the capacity of a channel and entanglement distribution process, which is mainly limited by the photon loss in channel that scales exponentially in a usual scenario.** The communication channel capacity defines the ability of the channel to deliver the information in a faithful and recoverable way.

The steps of quantum communication through quantum channels differ from the communication through classical channels in many ways. In quantum channel, communication of quantum information takes place in three different phases.

- The first phase is channel encoding phase.
- The next phase is quantum evolution phase.
- The final phase is decoding phase, which includes quantum information to be decoded by destination.

4. Quantum Memories

Quantum memories are an integral part of quantum communication system, which are required by intermediary devices to store and process the qubits for sending and receiving photons with high fidelity. Quantum computing includes a tool for the synchronization of the processes within a quantum computer. In addition, quantum memories are employed with quantum repeaters for extending the range of quantum communication. For achieving the scalability required for large-scale long-distance quantum communication, quantum memories are employed for converting qubits between light and matter. They also control the storage of qubits and retrieve them with stationary matter systems. Quantum memories are also required for timely operations for perfect timing and synchronization. As discussed earlier, communication between distant nodes storing matter qubits, engages quantum memories for the transfer of quantum information between these nodes with entanglement distribution, in free space or fiber-based channels. For this, atomic ensembles were proved to be the best contenders for quantum memories, with good amount of experimental progress.

5. Quantum Key Distribution

Every field in today's modern era, whether it is military applications, financial institutions, private multinational companies or government agencies, etc., relies upon a full-fledged internet to accomplish the day to day tasks and share gigabytes of data over the internet. The security of this data is fundamental, as any eavesdropping attack on private/public channels, will be a serious attack on the privacy. The security of systems using traditional cryptographic protocols is based upon the complexity in solving mathematical problems such as factorization of large integers or discrete logarithmic problem, and believing that the hackers computing performance is inadequate. Cryptography is classified into symmetric, asymmetric and hybrid cryptography where some type of keys are used for encryption and decryption of messages. If the same key is used for encryption and decryption of messages, such a system is known as private key cryptosystem. On the other hand, public key cryptosystem uses public key for encryption and private/secret key for decryption. The idea of public key cryptosystem was put forth by Diffie and Hellman in 1976. The first implementation was done by Rivest, Shamir, and Adleman in 1978. Since public key cryptosystems are slow, hybrid schemes are employed that combine private key and public key methods offering considerable speed advantages. In public key cryptosystems, without the knowledge of the secret key a message cannot be decoded, thus enhancing the security of the systems. Unfortunately, private key cryptosystems suffer from key distribution problem, i.e., distributing the keys in a secure manner. QKD provides excellent privacy against hacks, and provides solution to key distribution problem using quantum cryptography.

14.4.4 Possible Applications

Quantum internet is a diverse field with a possibility of a whole bunch of applications. As this technology is still in its infancy, it is not possible to predict the exact area of applications that it can offer. But, as this technology is developing, many applications are already predicted and proved with relevant literature. Some of them are discussed below.

1. Secure Communication

Secure communication is the first and foremost requirement in modern everyday life. There are numerous applications which depend on transfer of data from one node to another using internet like online banking, stock exchange trading, online shopping, military applications, national security, IoT (internet of things), etc. QI with various QKD protocols will maintain the security and privacy of the information exchanged over the internet. **With QI and security protocols such as QKD, quantum secret sharing (QSS), quantum secure direct communication (QSDC), quantum teleportation, and quantum dense coding, any attempt to hack the system will be detected on-site preserving the privacy of systems.** QKD has been researched widely, and successfully applied to a distance covering from few hundreds of kilometers to thousands of kilometers with the launch of Micius satellite by scientists from China. As IoT plays an important role in our lives due to wide area of applications, thus protecting data involved in wide IoT applications is the most important aspect which can be possible with inherently secure QI.

2. Blind Quantum Computation

Blind quantum computation (BQC) is another application of QI that is related to the security of a user. In this way, the privacy is not leaked to a quantum computer, which is involved in computation. BQC was originally proposed by Chaum in 1983 and was first formulated by Childs in 2001. QI in future will deploy costly and powerful quantum computers on the cloud, majorly owned by big tech companies. However, in addition to providing security and privacy to the user data, BQC would also check whether the computation is executed by a quantum processor.

3. Clock Synchronization

An important issue in many practical applications of modern technologies is accurate timekeeping and synchronization of clocks all over the world. Due to accurate synchronization of clocks, all the scientific applications dependant upon time synchronization, show minimal errors. These applications include quantum positioning system, synchronization of electric power generators feeding into national power grids, synchronous data transfers and financial transactions, long baseline of telescopes and distributed quantum computation. But, these navigation systems, such as GPS, suffer from security threats known as spoofing. Therefore, a scheme of distributing high precision time data through quantum Micius satellite securely is reported. It is based on two-way QKD using free space channel.

4. Longer Baseline of Telescopes

In modern world, astronomy totally depends on very long baseline interferometry between telescopes having greater resolving power than a single telescope. Quantum communication using quantum repeaters can reliably send quantum states of qubits over noisy communication channels, that allows the teleportation of quantum states over long distances with the least amount of error.

Using this technology, the baseline of telescope arrays can be extended as compared to conventional telescope arrays.

14.4.5 Future Directions

In the section, some future perspectives of QI technology are presented to make this vision a reality.

1. Evolution of Quantum Processors

Quantum computers with great computation capabilities will be required in future QI technologies, for performing complex computations. They can solve the problems that are practically impossible with state-of-the-art supercomputers. The quantum network is becoming powerful and efficient because of the increase in the number of qubits at the end nodes. To meet these requirements, quantum processors with double digit qubits are evolving, that will shape the future of quantum networking. This technological development would help boost the economies of the world, as various cloud operators will make fortunes, with BQC protocols maintaining user privacy. Also, quantum computers will be linked to each other through pure entanglement, and communication will take place using quantum teleportation. The QI might also assist quantum computer networking to reach even greater performance capacities.

2. Implementation of Trusted Repeater Network

The repeater network plays an important role in modern networking and therefore, a lot of theoretical and practical research has been established, even if it is still at a developing stage. Implementing repeater network requires number of intermediary nodes that are entangled to each other so that entanglement is generated between two far away end nodes with the phenomena of entanglement swapping. Each pair of adjacent repeater nodes exchange the encrypted keys using QKD which allows the far away end nodes to generate their own keys with an assumption that nodes should be trusted. But, until recently in Bristol work, trust-free nodes are also proved scalable in a city wide quantum network utilizing entanglement-based QKD. Thus, trusted nodes assumption can be avoided in four ways: by use of entanglement-based QKD, by use of measurement-device independent QKD, by use of multi-path communication techniques such as quantum network coding and by use of quantum repeaters. But still, trusted repeater is a hot topic of research that will be implemented along with trust-free nodes to overcome distance limitations between QKD nodes.

3. Quantum Error-Correcting Codes

The future QI technologies will require advanced QEC codes that work on the basis of encoding qubits carrying quantum information in a unique way that mitigates the influence of noise, decoherence and other environmental factors. After that, quantum information is decoded to retrieve the original quantum state. The decoherence free subspace (DFS) stores data in multiple qubits instead of storing data in a single qubit, which selects specific aspects of the system that are less affected by one or more important environmental factors. In this way, DFS stores data in a subspace of the Hilbert space related to quantum systems that are least influenced by the interaction of its system with the surrounding environment. **But, identifying DFS for complex quantum systems is extremely difficult, which requires more advanced DFS techniques such as dynamic decoupling (DD), where radio-frequency pulses are applied as external interaction to manipulate the non-unitary component of quantum system.** Nevertheless, the robust design of the sequence provided by the DD technique suppresses the imperfections in the experiments upto a greater extent. This implies that more advanced techniques are required and further research is needed in QEC codes.

4. Exploring Interface Between DV and CV

Discrete variable (DV) and continuous variable (CV) describes the states where quantum states can be represented in quantum communication. In DV states, data is represented by discrete features such as the polarization of single photons, which can be detected by single-photon detectors. Also, the information is represented by finite number of basis states, such as "qubit" which is the standard unit of DV quantum states. Alternative to this approach is the CV quantum states. Here the information is encoded onto the optical field defined by quadrature variables that constitute infinite dimensional Hilbert space and are useful for quantum information carriers such as lasers. CV states can be detected by highly efficient homodyne or heterodyne detectors having faster transmission rates than single-photon detectors. Therefore, an extensive research in theoretical and experimental domains is needed at the interface of CV and DV states, that will extract the best features from both these technologies to exploit the best of both.

5. Integrating Qubits with Other Technologies

Many powerful computers that are built use different synergies of systems such as bits, neurons, and qubits for computation. Summit supercomputer that has been built by IBM, has a peak performance of 200,000 teraflops, and is the world's most powerful machine that is built on the "bits + neurons" platform. This supercomputer can solve complex computation problems and tasks related to AI. Conclusively, supercomputers based upon "bits + neurons" scheme expedite technologically suitable workloads and deliver novel scientific insights. Neurons can also be combined with qubits, and together they can build a quantum computer with neuron inspired machine learning algorithms to derive a quantum advantage over the classical computation. Qubits along with bits can be another combination based on which quantum processors can be developed, that has capabilities well beyond the reach of classical computations. Therefore, exploring possible synergies of "bits + qubits + neurons" will shape the future of computing. With this, the capabilities of "bits + qubits + neurons" must be extensively researched and will remain a topic of active research in coming years. However, some advanced computing systems and use cases are already operating at the intersection between the pair of these computing methods.

6. Exploration of Solid-State Memories

Solid-state qubits are formed by electron spins followed by superconducting circuits, comprising of Josephson junctions, interconnects, and passive elements, that are designed to behave as a quantum mechanical two-stage system with good isolation. Solid-state quantum memories are more advantageous and attractive as they can maintain their coherence even at cryogenic temperatures. For example, crystalline-solid spin ensembles formed by implanting defects in lattice, such as implanting nitrogen-vacancy (NV) centers into diamonds, or by rare-earth-doped crystals, can be proved coherent for hours at cryogenic temperatures. Hence, comprehensive efforts are needed to explore the union of superconducting processors and solid-state quantum memories to enhance the performance of transmission and generation of microwave photons. Accordingly, steps must be taken to investigate on-chip teleportation between a superconducting qubit and NV in a local quantum memory. If these integrations are successful, this hybrid technology would become the most encouraging design to be scaled up into a comprehensive quantum network. Therefore, with employing such hybrid quantum computers as end nodes in quantum networking, the implementation of a hybrid-technology QI globally could be seen in the next 5–10 years.

7. Quantum-Classical Synergy

A future QI will soon be developed owing to the launch of various quantum satellites paving the way for long-distance entanglement and QKD with eavesdropping detection. Quantum communication links will require already mature fiber channels for the exchange of classical data required for the quantum teleportation process. QKD also requires classical channels for information reconciliation and privacy amplification, to achieve higher key rates and reduced latencies over noisy communication channels. A fully secure QKD based channels will ultimately connect classical devices such as mobiles, tablets, smart wearables, satellites, smart vehicles with utmost security. A secure quantum cloud will exist, that will take care of user privacy and its data through BQC. Thus, future QI will work in synergy with the classical internet giving rise to a universe of applications with excellent security.

Words and Expressions

quantum internet (QI)	量子互联网
quantum communication	量子通信
quantum bit (qubit)	量子比特
synergy ['sɪnərdʒi]	*n.* 协同作用
quantum mechanics	量子力学
quantum entanglement	量子纠缠
decoherence [diːkoʊ'hɪrəns]	*n.* 退相干；消相干
quantum error-correcting (QEC) codes	量子纠错码
secure communication	保密通信
data privacy	数据保密；数据隐私
cryptographic [ˌkrɪptə'græfɪk]	*adj.* 密码的；关于暗号的；用密码写的
cryptography [krɪp'tɑːgrəfi]	*n.* 密码学
quantum key distribution (QKD)	量子密钥分发

quantum teleportation 量子隐形传态；量子隐形传输
quantum repeater 量子中继器；量子中继
bifurcate ['baɪfərkeɪt] *v.* 分叉；分支
no-cloning theorem 不可克隆原理
encryption [ɪn'krɪpʃn] *n.* 加密；加密技术
decryption [diː'krɪpʃn] *n.* 解密；译码
quantum secret sharing (QSS) 量子密钥共享
quantum secure direct communication (QSDC) 量子安全直接通信
blind quantum computation (BQC) 盲量子计算
clock synchronization 时钟同步
quantum positioning system 量子定位系统
baseline ['beɪslaɪn] *n.* 基础；底线；起点
homodyne ['hɑːmoʊdaɪn] *n.* 零差（拍）；同步检波；零差法；自差法
heterodyne ['hetərəˌdaɪn] *n.* 外差法；外差作用
cryogenic [ˌkraɪə'dʒenɪk] *adj.* 低温的；致冷的
reconciliation [ˌrekənsɪli'eɪʃn] *n.* 协调；调解

Notes

1. Quantum internet (QI) is trending internet technology that facilitates quantum communication through quantum bits (qubits) among remote quantum devices or nodes. Such a technology will work in synergy with classical internet to overcome the limitations posed by traditional interconnect technologies. This novel technology stems from the laws of quantum mechanics, where one of the laws states that it is impossible to measure a property of a system without changing its state. Consequently, qubits cannot be copied and any attempt to do so will be detected, thus making the communication more secure and private.

量子互联网（QI）是互联网技术的趋势，是通过推动远程量子设备或节点之间利用量子比特（qubit）进行量子通信的技术。这种技术将与传统互联网协同工作，以克服传统互联技术带来的限制。这项新技术源于量子力学定律，其中一条定律指出，不改变系统的状态就不可能测量系统的特性。因此，无法复制量子比特，任何这样做的尝试都将被检测到，从而使通信更加安全和私密。

2. New potential in cryptography emerges with advances in quantum cryptography that exploits quantum mechanical principle of no-cloning. These features play a vital role in providing secrecy and integrity to the data to be communicated, thus making messages unintelligible to any unauthorized party.

量子密码学利用了不可克隆的量子力学原理，伴随其发展，密码学出现了新的潜力。这些特征在为要传输的数据提供保密性和完整性方面起着至关重要的作用，从而使任何未经授权的一方都无法理解消息。

3. However, at the same time, no-cloning prevents qubits to be sent over long distances by amplification or sending copies of qubits, as it is done in classical communication, for efficient detection of bits without any loss of information. For this purpose, quantum repeaters are studied to

solve the problem of long distance communication of qubits, which is the core functionality required for full-scale QI.

然而，与此同时，不可克隆技术可防止像经典通信中那样通过放大或发送量子比特副本长距离发送量子比特，以便在不丢失任何信息的情况下有效地检测比特。为此，研究量子中继器可解决量子比特的远距离通信问题，这是全面量子互联网所需的核心功能。

4. On the other hand, in quantum communication, instead of bits, qubits are employed to carry quantum information from source to destination. A qubit can be imagined as a two-way system: direction of polarization of a photon, the up or down spin of an electron, or two energy levels of an electron orbiting an atom.

另一方面，在量子通信中，量子比特代替比特，用于将量子信息从信源传输到信宿。量子比特可以被想象为一个双向系统：光子的极化方向、一个电子的上下自旋或一个电子绕原子旋转的两个能级。

5. Entanglement does not allow any exchange of information between remote nodes, rather it allows obtaining only mutual information. For information exchange, qubits can be transmitted between remote nodes without actually sending them from source to destination by the phenomenon of quantum teleportation, which is fundamental for quantum networks.

量子纠缠不允许远程节点之间进行任何信息交换，而只允许获取相互信息。为了进行信息交换，量子比特可在远程节点之间传输，而无须通过量子隐形传态现象将它们从信源发送到信宿，这是量子网络的基础。

6. quantum Internet is a system that consists of various functionalities required for communicating information among remote nodes of a network. In this section, various functionalities of QI like quantum teleportation, quantum repeaters, quantum channels, quantum memories, QKD are briefly introduced.

量子互联网是一个由网络远程节点之间通信信息所需的各种功能组成的系统。在本节中，简要介绍量子互联网的各种功能，如量子隐形传态、量子中继器、量子信道、量子存储器、量子密钥分发。

7. How efficiently an entanglement can be distributed depends upon the capacity of a channel and entanglement distribution process, which is mainly limited by the photon loss in channel that scales exponentially in a usual scenario.

纠缠分布的效率取决于信道容量和纠缠分布过程，这主要受到信道中光子损失的限制，在通常情况下，信道中的光子损失呈指数级扩展。

8. With QI and security protocols such as QKD, quantum secret sharing (QSS), quantum secure direct communication (QSDC), quantum teleportation, and quantum dense coding, any attempt to hack the system will be detected on-site preserving the privacy of systems.

通过量子互联网和安全协议，如量子密钥分发、量子秘密共享（QSS）、量子安全直接通信（QSDC）、量子隐形传态和量子密集编码，任何入侵系统的企图都将在现场被检测出来，以保护系统的隐私。

9. But, identifying DFS for complex quantum systems is extremely difficult, which requires more advanced DFS techniques such as dynamic decoupling (DD), where radio-frequency pulses are applied as external interaction to manipulate the non-unitary component of quantum system.

但是，识别复杂量子系统的无消相干子空间极其困难，这需要更先进的无消相干子空间技术，如动态解耦，其中射频脉冲作为外部相互作用来操纵量子系统的非幺正分量。

Exercises

1. Match the terms (1)–(8) with the definitions A–H.

(1) cloud computing	A. a subclass of information filtering system that seeks to predict the "rating" or "preference" a user would give to an item
(2) GPU	B. an interdisciplinary field that deals with the graphic representation of data and information
(3) recommendation system	C. a subfield of linguistics, computer science, and artificial intelligence concerned with the interactions between computers and human language
(4) deep learning	D. the on-demand availability of computer system resources, especially data storage (cloud storage) and computing power, without direct active management by the user
(5) data visualization	E. a specialized electronic circuit designed to rapidly manipulate and alter memory to accelerate the creation of images in a frame buffer intended for output to a display device
(6) natural language processing	F. part of a broader family of machine learning methods based on artificial neural networks with representation learning
(7) quantum teleportation	G. secure communication method which implements a cryptographic protocol involving components of quantum mechanics
(8) quantum key distribution	H. technique for transferring quantum information from a sender at one location to a receiver some distance away

2. Translate into Chinese.

(1) AI applications include advanced web search engines, recommendation systems, understanding human speech, self-driving cars, automated decision-making and competing at the highest level in strategic game systems.

(2) Knowing the different families of machine learning algorithms and their strengths and weaknesses is fundamental to making wise decisions about developing and using AI systems. Supervised learning systems can deliver incredible performance, but acquiring labeled data may be a challenge. Unsupervised learning systems don't require labeled data, but their performance for some applications will generally be far more limited than supervised systems. Reinforcement learning systems can generate their own data, but can generally only be used for applications that offer access to simulators that closely resemble the operational environment. Thus far, this includes fewer applications.

(3) Reinforcement learning is an area of machine learning concerned with how software agents ought to take actions in an environment so as to maximize some notion of cumulative reward. Due to its generality, the field is studied in many other disciplines, such as game theory, control theory, operations research, information theory, simulation-based optimization, multi-agent systems, swarm intelligence, statistics and genetic algorithms.

(4) Quantum communication is the field of study related to the transmission of quantum states between two or more parties. The most widely accepted applications for quantum communication fall in the field of cryptography, where the laws of quantum mechanics are exploited to secure data, and specifically, to share secret keys between symmetric parties, a technique named quantum key distribution.

3. Translate into English.

（1）开发一个可操作的机器学习 AI 系统有多个步骤。通常，最大的挑战是获取足够的高质量训练数据。系统性能与数据量、质量和代表性直接相关。

（2）虚拟现实（VR）是一种模拟体验，可以与真实世界相似，也可以完全不同。虚拟现实的应用包括娱乐（尤其是视频游戏）、教育（如医疗或军事训练）和商业（如虚拟会议）。其他不同类型的 VR 风格技术包括增强现实和混合现实（有时被称为扩展现实或 XR）。

（3）尽管有许多优势和应用，但为了提高服务质量，大数据中仍有许多挑战需要解决，如大数据分析、大数据管理及大数据隐私和安全。区块链以其分散性和安全性，在改善大数据服务和应用方面具有巨大潜力。

（4）量子通信是经典信息论和量子力学相结合的一门新兴交叉学科，它利用量子态来携带信息。量子通信有望在通信安全、计算能力、信息传输、信道容量和测量精度等方面突破经典通信技术的限制，成为通信和信息领域的一个新方向，将成为主流。

4. Read the following article and write a summary.

Cloud computing is a rapidly growing technology. Cloud services are cheap, provide flexibility in accessing data anywhere anytime just with the help of good internet connection, ensure security of the data and hava many more uses. Cloud networking is the method of sourcing or utilization of one or more network resources (shared networks) and services from the cloud. Cloud networking basically means sharing of network resources in a shared network, whereas cloud computing involves just sharing of computer resources. Examples of some network resources include virtual routers, firewalls, bandwidth, and network management software, to name a few. There are different types of cloud computing services and they are categorized into 3 types.

- Infrastructure as a service (IaaS). This service provides the infrastructure like servers, operating systems, virtual machines, networks, and storage on rent basis. In this service, an organization uses its own platform and applications within a service provider's infrastructure. The user has to just pay on rental basis to the vendor for accessing the computing resources such as servers, storage and networking. The largest cloud service providers Amazon Web Services and Microsoft Azure use this service.
- Platform as a service (PaaS). This service is used in developing, testing and maintaining of software. PaaS is same as IaaS, but also provides additional tools like DBMS and BI service. As the name suggests, here the service provider offers access to a cloud-based environment in which users can build and deliver applications. The provider supplies underlying infrastructure. The advantage of this service over IaaS is that it not only provides access to computer resources but also it comes with a suite of prebuilt tools to develop, customize and test their own applications. Apprenda and Red Hat OpenShift are the cloud service providers which use this model.
- Software as a service (SaaS). This service makes the users connect to the applications through the Internet on a subscription basis and accessing is done via the Web or vendor APIs. In this service, users do not have to manage, install or upgrade software, they can simply access the applications through remote cloud network. Google Apps and Salesforce use SaaS.

There are various cloud service providers throughout the world such as Kamatera, phoenixNAP, Amazon Web Services, Microsoft Azure, Google Cloud Platform, and Adobe to name a few top services. Recent studies suggest that AWS (Amazon Web Services) has emerged the largest cloud service provider. It has become more popular because of the features like mobile friendly features, security, serverless cloud functions which are believed to improve the efficiency of any organization.

5. Language study: Cause and effect

Study this sentence:

Dust on records causes crackle.

It contains a cause and an effect. Identify them.

We can link a cause and an effect as follows.

Cause		Effect
Dust on records	causes leads to results in is the cause of	crackle

We can also put the effect first:

Effect		Cause
Crackle	is caused by results from is the effect of is due to	dust on records

Items in List 1 can be causes or effects of items in List 2. Match the pairs. Compare your answers with our partner. For example:

mains frequency interference *hum*

List 1	List 2
(1) distortion	A. interference on radios
(2) noise generated within components	B. too high a recording level
(3) overheating a transistor	C. the tape rubbing against the head
(4) dirty heads	D. scratches on records
(5) a build-up of oxide on the head	E. hiss
(6) jumping	F. damage
(7) unwanted signals	G. poor recordings

Write sentences to show the relationship between the pairs you linked in above table. For example:

Mains frequency interference results in hum.

Part II Practical Guidance

Unit 15 Learning Skills

Electronics and communication engineering (ECE) is a discipline which uses the scientific knowledge of the behavior and effects of electrons to develop components, devices, systems, and equipment. It is the backbone of the modern technology, supporting almost everything from rocket science to home appliances. This engineering field deals with analog and digital circuits, solid-state devices, integrated circuits, signal processing, various communication systems, control systems, internet of things (IoT), robotics and so on.

The basic subjects include Engineering Mathematics, Engineering Physics, Programming in C++, Electronic Devices, Circuits, Microcontrollers, Principle of Communication, Electromagnetic Field Theory, Signals and Systems, Digital Signal Processing, Optical Fiber Communication, Mobile Communication, IC Design...

So, it seems that ECE is a hard major. You'll need to dedicate hours to studying, researching, and learning in order to become an expert in your field. Yet, with the practical learning skills, it will be a rewarding experience. If implemented, the learning skills outlined below will not only help you become a more successful engineering student, they'll make the study of engineering much more enjoyable!

1. Become a Problem Solver

In high school, all you had to do was show up for class, listen to the wisdom and truth spewing forth from your teacher and soak it up. If you did this, you'd be able to complete your assignments with relative ease and pass your exams. That approach may have worked for you in high school, but it starts to fall apart in college—especially if you choose to major in a field of engineering.

In the real world, there aren't any professor spoon feeding you information, giving you homework, providing lectures full of useful information, or end of semester exams for you to prove yourself. In the real world, there are simply problems—usually poorly defined problems. These problems require solutions that are either acceptable or unacceptable. There is no partial credit for solutions that don't work—or that sort of work. If you design landing gear for twenty airplanes and one set of gear fails, you're not going to get a 95 percent score and a pat on the back.

In order to excel in engineering, either academically or professionally, you need to change your mindset. You need to learn to not count on someone else to tell you everything you need to know to solve problems. You need to learn how to discover what you need to know and then where to go to find it.

2. Discover Your Learning Style

Every student has a different learning style. The most widely accepted model of learning styles is called the VARK model, which stands for visual, aural/auditory, reading/writing, and kinesthetic. In brief:

- Visual (spacial) learners learn best by seeing.
- Auditory (aural) learners learn best by hearing.
- Reading/writing learners learn best by reading and writing.
- Kinesthetic (physical) learners learn best by moving and doing.

You may have a dominant learning style or a mix of learning styles. As an engineering student, identifying and understanding your learning style can be helpful, especially when your instructor's teaching style does not match your learning style.

Some engineering instructors are guilty of using a lot of words and formulas as they lecture but neglect to employ visual imagery (pictures, diagrams, flow charts, sketches, etc.). For a student that is a visual learner, this can be frustrating and problematic. Other instructors are big into mathematical theory and formulas but provide little in the way of real-world examples and applications. For a student that is a physical learner, this makes learning challenging. Your first job as an aspiring engineer is to discover your learning style, your professor's teaching style, and figure out how to fill in the gaps.

3. Seek Help from Your Instructor

Contrary to popular belief, instructors are hired to teach, not to lecture. Lecturing is just one form of teaching — and not always the most effective. Instructors are there to help you learn, and in most cases they really want you to succeed. If there is something you're not understanding during a lecture, raise your hand and ask a "clarifying" question. Although it's acceptable to say "I don't understand", it is far better to ask a clarifying question that allows the instructor to provide you specific information to help you gain clarity and understanding. Asking clarifying questions also tells the instructor that you were paying attention. Examples of clarifying questions include:

- How is this theory applied in the real world?
- Could you provide an example of when this formula might be used?
- Could you sketch what that (solution, device, etc.) might look like?
- How is this equation applied in practice?
- Where did that formula come from?
- I still don't understand when that formula is used.

In order to ask clarifying questions, it's important that you come to class prepared and pay attention to the lecture. Asking clarifying questions not only helps you learn, but also helps your entire class.

Most engineering instructors are happy to answer any question you have during class. However, there are a few who don't handle questions very well—especially if they have a lecture they're trying to get through. If you happen to get an instructor who is hostile towards questions during class, make an appointment to meet with your instructor during office hours to get your questions answered.

Don't ask questions that you could have answered yourself with a little study or research. No one likes to have his time wasted. Never ask your instructor for help with a problem until after you've spent ample time and effort trying to figure it out on your own. When you ask your instructor for help figuring out the solution to a problem, be prepared to present all the work you've performed in your attempt to solve it.

When at all possible, go as a group during your instructor's office hours to seek the solution to a problem you're unable to solve on your own. Going as a group shows your instructor that you've made a legitimate attempt to solve the problem and he'll likely feel like his time is being better spent by helping several of his students instead of just one.

4. Read Your Textbook with Purpose

There are many reasons why students read textbooks. Often, it's to find answers to homework problems they're trying to solve. In an effort to find specific information, they skim through the text, ignoring much of what's presented, in order to find clues and examples that will help them solve their homework problems. After completing their homework, they ignore their text until another set of problems is assigned.

Many engineering texts cover important theoretical material, providing real-world examples of how engineering theories are applied in practice. When texts are only used to answer assigned homework problems, students miss out on valuable learning opportunities that will help them get on track in other engineering courses and later in their careers.

5. Form a Study Group

Working with a study group can be beneficial for any student. For engineering students, workings with a study group is particularly advantageous. The benefits of working with a study group for engineering students include:

- Engineering is a changing field of study. It's not uncommon for students to get stuck on a problem and want to give up. When working as a group, students are able to find solutions to challenging engineering problems that they may not have been able to solve on their own.
- Study groups allow for various perspectives and expose alternative ways to solve problems. Even when you're able to solve a problem, someone in your group may come up with a solution that is more effective and efficient than your solution.
- Study groups create environments where teaching occurs. As engineering students share with one another their knowledge, insights and understanding on engineering theories, formulas, equations, they reinforce their own understanding. Most instructors will attest that teaching a subject is the most effective way to learn it.
- Study groups foster a collaborative learning environment. Research suggests that collaborative learning is very effective. Studies show that students who regularly participate in study groups retain what they learn longer, gain better understanding of concepts and theories, enjoy coursework more, gain more self-confidence, and perform better in class than students who work on their own.
- Study groups engender teamwork. Once you enter the workforce, you'll find that almost all engineering projects are performed by teams of engineers. Working with a study group

will help you develop team building skills and prepare you to be a team player.

Some study groups are more effective than others. Effective study groups for engineering students share the following characteristics.

- Groups of 3 to 5. Study groups should include at least three engineering students but no more than five. With fewer than three students, study groups tend to be ineffective because a sufficient variety of approaches, insights, knowledge and ideas are not offered. In groups greater than five, there is a tendency for some students to do most of the work while others are left out of the active problem-solving process.
- Do the work by yourself first. The most challenging aspect of solving engineering problems is figuring out how to get started. It's important that every student figures how to solve engineering problems on their own. It's not uncommon for one student in a group to be quicker than the rest in initiating the problem-solving process. If the same student initiates every solution, then the other students in the group will never gain the confidence or ability to set up and tackle engineering problems on their own. Before working on engineering problems with your group, outline the solutions to problems yourself.
- Every group member must understand every solution. One of the challenges of study groups (especially in engineering) is that one or two group members often develop a solution to a problem while the others just sort of follow along, not really participating in the problem-solving process. Effective study groups ensure that all group members participate in finding solutions and that all solutions are thoroughly understood by each group member. Before a group study session is ended, each group member should explain to the rest of the group how each solution was obtained in order to ensure understanding.

6. Time Management

Time management is not something you do just once and never worry about again (e.g., creating a schedule). Time management is an ongoing process of awareness and control. It is also closely tied to achieving the goals you have set for yourself.

- Treat the school day like a work day. Regardless of when your classes are, get to school at 8:00 AM, and stay until 5:00 PM. If you have an exam or something the next day, then stay longer. Work hard during that time to finish all your homework, studying, office hours, e-mails, etc.
- Start working on assignments based on the assigned date, not the due date. Instead of planning assignments from the due date backwards, start doing work from the assigned date forwards. For example, if a homework is assigned on 11/5 but isn't due until 11/20, don't plan on doing the homework on 11/18. Instead, plan on doing the homework on 11/6. You could even trick yourself by writing down that the homework is due a few days before it actually is. That way you are done early and aren't stressed out.
- Start studying for exams a few (2–3) weeks before they come. If you start studying the week before, you will be stressed, not cover enough material, and struggle. Help yourself out and start early. It feels amazing when the exam is a day or two away and you have

already finished studying for it. Now any more studying is just an easy review, which makes you feel more confident, which makes you do better on the exam.

"Awesome with no effort" doesn't happen in college. You will have to make a deliberate effort to learn. You will have to make time to study and do homework. Remember, this is a rewarding major. ECE will pay off! Jobs are often exciting and interesting, and pay well. Go to your career fairs on campus each semester if they happen that frequently. Try talking with each of the representatives and find out what kind of classes and what kind of work they are doing. You get to practice your communication skills and also learn more about the industry.

Exercises

1. Are you a procrastinator?

Read each statement below and choose the word that best describes your behavior. Write the corresponding number you choose on your paper.

Never – 1 Occasionally – 2 Often – 3 Always – 4

(1) I feel I have to "cram" before an exam.

(2) My homework is turned in on time.

(3) I think I get enough sleep.

(4) I pull all-nighters before mid-terms and finals.

(5) I plan activities with friends or family for a couple of nights a week and spend the amount of time with them that I planned.

(6) When I'm working on a paper, I put off writing until a few days before it's due.

(7) I cancel social activities because I feel I don't have enough time.

(8) I get my papers in on time.

(9) I find myself making a lot of excuses to my instructors about why my work isn't done.

(10) I feel comfortable about how I use time now.

(11) I feel that something is hanging over my head, that I'll never have enough time to do the work assigned.

(12) I feel tired.

Score A: Add up the numbers for questions 1,4,6,7,9,11, and 12. _________

Score B: Add up the numbers for questions 2,3,5,8, and 10. _________

Result: If Score A is greater than Score B, you are probably a procrastinator. If Score A is less than Score B, you manage your time well. If the scores are equal, you may procrastinate at times, but procrastination is not a habit.

2. How do you spend your time?

Use your completed 24 hour circle to calculate how much time you spend on each activity listed in the Activities Breakdown below each week. The blank lines are for any additional situations that take up your time. After you have totaled up all the items you can think of, figure out how much free time you have.

Activities Breakdown(Hours per Week)

(1) Class time _____
(2) Study, reviewing, projects, papers _____
(3) Commuting _____
(4) Dressing and eating _____
(5) Hours of employment _____
(6) Responsibilities at home _____
(7) Athletics requirements _____
(8) Telephone and computer_____
(9) Television _____
(10) Dating, outings, sports, movies, "going out", etc. (entertainment) _____
(11) Sleeping _____
(12) ______________________________ _____
(13) ______________________________ _____
(14) Wasted hours _____

Total: _____

Total number of hours per week = 168.

Subtract your total time spent on the above items to get your total free hours per week: _____

Now that you know how you are currently spending your time, it is good to reflect on your life's priorities and goals. What is most important to you? What are your life priorities?

List your top 10 life priorities in order from most to least important:

(1) ____________________________	(6) ____________________________
(2) ____________________________	(7) ____________________________
(3) ____________________________	(8) ____________________________
(4) ____________________________	(9) ____________________________
(5) ____________________________	(10) ____________________________

How do your priorities match up to how you spend your time each week?

____

What do you need to adjust in your weekly schedule to better match your life priorities?

____

____

List any additions you want to add to your weekly schedule: ____________________

____

What will you remove or reduce in your weekly schedule? _______

____

Create your new ideal schedule: You can start with a day or construct a whole week. Be sure to include the changes you wish to make for yourself and don't forget to include your top priorities. You will need time to take care of yourself, for instance, when will you sleep, eat, etc.?

Unit 16 Research Skills

From the time it is assigned until the day it is due, a research paper can occupy your mind like any other type of assignment. While the ability to do good research may in fact require personal traits and characteristics that are beyond the capability to teach, such as creativity and persistence, there are some fundamental skills that are necessary (although not sufficient). These skills include the ability to decide what to investigate, gather information, devise a framework, and write the paper.

1. Decide What to Investigate

Finding a suitable topic is often the biggest stumbling block in research. It's essential that you know how to choose a topic easily and efficiently. There are three steps in the process of selecting a topic: begin with a general subject that interests you, narrow it down, and then sharpen it even further by finding a focus.

Suppose you are fascinated by image processing technologies and want to learn more about them. But the subject "image processing" includes scores of topics: image acquisition, image enhancement, image restoration, image compression, image segmentation, and object recognition, to name just a few. How can you do justice to them all? Obviously, you can't. You must narrow your topic.

General topic: image processing.

First narrowing: image enhancement.

Second narrowing: traffic image enhancement.

Third narrowing: traffic image enhancement on haze days.

Once you've narrowed your topic, give your research direction and purpose by developing a compelling question about that topic. The information you gather from your research can then be used to develop an answer. For the topic "The use of haze removal algorithms on traffic images", you might ask, "How helpful is the use of haze removal algorithms on traffic images?" Whether you actually arrive at a definitive answer to your research question isn't crucial. The important thing is to focus your research efforts on answering the broad question.

2. Gather Information

Although you can always consult with experts and do a general search by surfing the Web, the best place to do your preliminary research is still the library, where you'll have access to a variety of reference sources. Most libraries in universities provide access to scholarly databases like CNKI, Web of Science, ScienceDirect, or a database specific to one engineering field, such as IEEE Xplore.

Using keywords searching enables you to receive a list of relevant articles, as shown in Figure

16.1. The refinement options in the left-pane can help narrow your search results. For example, limiting the search to results published in the last few years can help show only the most current results. Also, you can use Boolean searches to refine your searching results, by combining two or more keywords with basic logical operators (and, not, or), as shown in Figure 16.2.

As you incorporate your sources into a document, you will need to record your use of those sources in a bibliography and in in-text citations. Technical journal articles typically include many references, especially in the introductions, where the authors ground their researches in previous work.

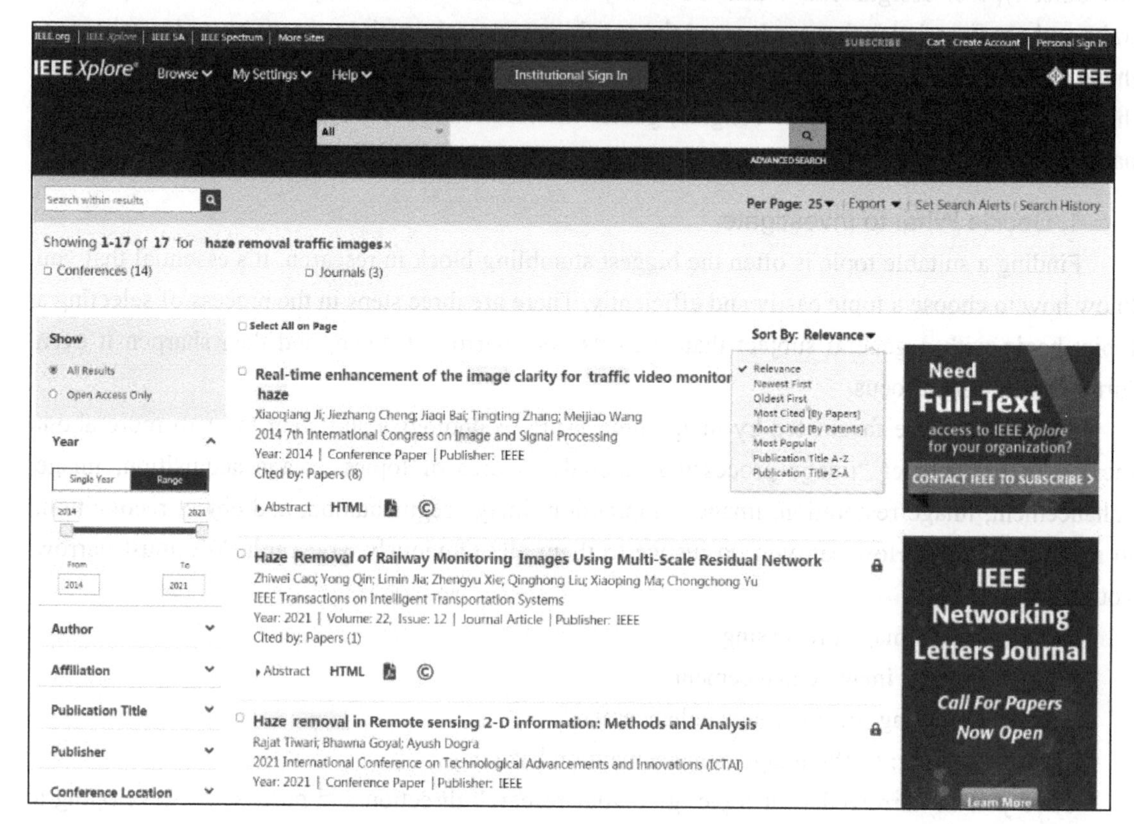

Figure 16.1 IEEE Xplore search results

Usually, technical journals require authors to use a certain citation style. Some library databases provide properly formatted citations, but you should make sure the output format is consistent with the requirements.

3. Devise a Framework

In the same way that choosing a focus helps provide direction for your research, deciding on a basic premise from the references you now have lays the foundation for your paper's organization. If a premise doesn't become obvious to you, go back over the information and ask yourself some hard questions. For example:

- Where is this paper heading?
- What is the impact of the information I've assembled?

- What point is most important?
- What am I saying?
- What do I want to say?
- If there's a choice of viewpoint—for or against a question, for example—which view has the most evidence to support it?

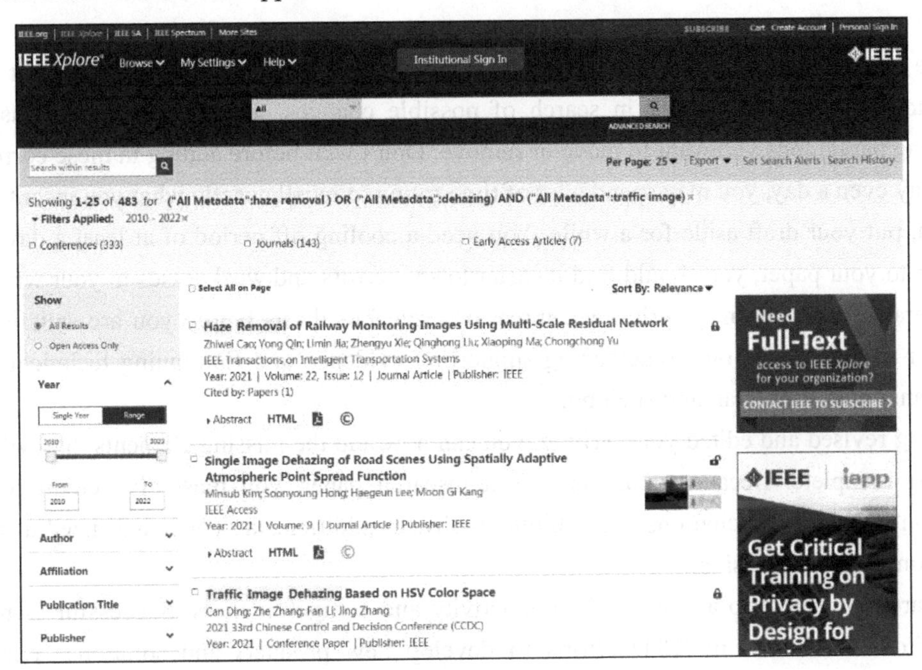

Figure 16.2 IEEE Xplore Boolean search results

If you've done a good job of research, you should be able to decide now what you want to say in your paper, and you should have the evidence to support that view.

Your basic premise and personal choice largely determine the pattern your paper will follow. The time pattern (events presented in order), the spatial pattern (info presented based on location or arrangement), or the process pattern (steps or events lead to desired situation or product) is appropriate for most college papers. For some papers, however, you may be required to develop an argument. A good pattern for such papers is to begin with a statement of your premise and then support it with logical examples that build to a conclusion. This kind of organization affords more flexibility than the others.

You may need to experiment with several patterns before you arrive at a framework that adequately accommodates the information you want to include in your paper. If you find gaps in your organization, you may need to return to the library or the Internet to search for more information.

4. Write the Paper

You already have most of your paper worked out—information, sources, and organization. Now all you have to do is to put your data into sentences and paragraphs and work up a first draft of your paper. Once that is accomplished, allow yourself plenty of time to go back and revise and edit what you've written, add the missing elements, and type the final copy.

The best way to start writing is simply to write. Write your first draft as rapidly and spontaneously as possible. To ensure continuity, record your thoughts as they go through your mind. Don't stop to ponder alternatives. Although you will probably write too much, don't be concerned; it's easier to cut than it is to add.

Once you've completed your first draft, you should step back and take a look at what you've written, and go over it carefully in search of possible changes, adding words or phrases and highlighting paragraphs you want to move or remove. Don't wait before adding in these corrections. If you delay even a day, you may spend a lot of time trying to recall exactly what you meant.

Then, put your draft aside for a while. You need a cooling off period of at least a day. When you return to your paper, you should find it easier to spot errors and weaknesses in your writing. In the next drafts—and you may write two, three, or even four drafts before you are satisfied with your paper—you'll focus on strengthening supporting evidence and fine-tuning technical details such as transitions, grammar, and spelling.

Having revised and edited your writing, you can now add the missing elements that will make your paper complete. Because your paper is a research paper, you must give credit for your information by including citations and a bibliography. In addition, the paper will need a title, an introduction, and a conclusion.

Research is critical to a company's productivity and competitiveness. Successful companies invest billions of dollars in R&D efforts to develop new products and solutions. Therefore, engineering students should be aware of the importance of research in their education and careers.

Exercises

1. Work in groups to choose a topic related to your major and complete a research report.

2. Writing the abstract for your report.

An abstract is a 150- to 250-word paragraph that provides readers with a quick overview of your report. The abstract should include:

- Why the work was done (the basic problem), the specific purpose or objective, and the scope of the work if that is relevant.
- How the work was done, the test methods or means of investigation.
- What was found—the results, conclusions, and recommendations.

A good abstract should be complete, concise, specific, and self-sufficient.

Unit 17 Careers in ECE

The applications of electronics and communication engineering is an endless list with a lot of scope in each and every subfield. It is a versatile branch that opens path for core electronics and communication jobs. As compared to other branches, there are more interdisciplinary specializations in higher studies for ECE students like embedded systems, VLSI, wireless communication, signal processing. An ECE engineer is indispensable at every front. Opportunity in this highly sophisticated and advanced branch of engineering is just immense.

Therefore, ECE students have a wider choice. ECE students usually get jobs in companies/ organizations having operations in hardware, i.e., electronics (electronics industry), communication (telecom industry), as well as software (IT industry). However, the options are much broader than this, both within the engineering industry and outside it. Electronics and communication engineering jobs represent over 50% of all available jobs in engineering.

1. Types of Jobs

Broadly speaking, there are four different types of jobs for ECE students in the future: product development, electronics manufacturing, teaching, and research.

- Product development. ECE covers a wide range of products and tools, ranging from televisions to cellular phones. Product development is central to creating new devices, improving the efficiency of existing devices and making the best possible use of new power sources and processing chips as they become available. This type of work is typically completed by people with doctoral degrees in electronics engineering or related fields.
- Electronics manufacturing. There is a broad range of jobs available in the electronics manufacturing sector. This industry supports careers in design, implementation, production, support, and repair. There are electronics manufacturing and design companies located in countries around the world.
- Teaching. ECE instructors or teachers may work in colleges or technical schools, teaching courses to engineers or technicians. People with doctoral degrees can find instructor positions at the university level. In order to become an effective instructor, many professionals complete a certificate program in adult education. Learning the most effective way to teach adults can be a huge help when making this career transition.
- Research. Research is a huge area of exploration. Grants are available from government agencies, private industry, and dedicated research groups who seek to expand the field of knowledge in this area. Many engineers who want to explore this career option go on to become university professors. These positions require teaching, but provide the opportunities

for extensive research.

Besides, there are other non technical positions requiring professional knowledge background. For example, there are many outstanding jobs in technical sales. Selling complex electronic communication equipment often requires a strong technical education and background. The work may involve determining customer needs and related equipment specifications, writing technical proposals, making sales presentations to customers, and attending shows and exhibits where equipment is sold. Another position is that of technical writer. Technical writers generate the technical documentation for communication equipment and systems, producing installation and service manuals, maintenance procedures, and customer operations manuals. This important task requires considerable depth of education and experience.

If you're keen to explore career paths outside engineering, most engineering employers recruit graduates for roles in HR, finance and supply chain, to name just a few. Or if you'd like to take a completely new career direction, there are plenty of options. In particular, you might like to consider options such as consulting, law (e.g., intellectual property law), financial services, sales and pre-sales, education, which are areas in which you can put your analytical skills and high level of numeracy to good use. You could also put your background to good use in careers such as science journalism or technical publishing.

The level of the position available varies by industry and level of eduction. Most ECE jobs have a broad range of career advancement opportunities available. Regardless of the industry where initial experience is obtained, all skills are transferable to other areas.

2. Jobs in Different Industries

Electronic engineering graduates are typically sought by the following industries.

- Aerospace industry. Electronic engineers in the aerospace industry would be working on cutting edge technology, introducing or enhancing power-dense electrical controllers and electronics. Also, an increasing focus is on emerging technologies for hybrid/electric propulsion for aerospace platforms.
- Automotive industry. Electronics is now an important part of the automotive industry and there is a big call for power electronics skills. Electronic engineers will work on a variety of systems including engine control units, dashboard indicators, air conditioning, safety systems, braking systems and infotainment systems. Their skill sets are also needed for the development of autonomous, connected and electrified (ACE) vehicles.
- Defence industry. Electronic engineers in the defence industry optimise hardware and software design concepts, develop sophisticated design processes and test complex products to ensure the equipment is fit for air, sea or land operating environments. Activities could include: circuit design, assessment of equipment behaviour, fault diagnosis, assessment of new technologies and components, simulation and modelling, and data analysis.
- Electronics industry. Electronic engineers in the electronics industry could work in roles such as design engineer (designing a product or component prior to launch) or applications

engineer (supporting a product for its entire life). They may work with chips, integrated circuits, components such as capacitors and resistors, and devices that use electricity.

- Consumer goods industry. Most of the graduate roles in the fast-moving consumer goods industry are in one of two areas: manufacturing/engineering or supply networking operations/logistics. For both of these areas, the work is not defined in nice separate buckets of mechanical, electrical, chemical, etc. but is normally a mixture of different engineering disciplines as a general manufacturing or logistics engineer. Graduates will pick up skills from other disciplines as they go through their training and career.
- Marine industry. Engineers in the marine industry usually either operate and maintain vessels or design and build them. Electronic engineers could be working on radar systems for warships or complex automation systems, reducing manning requirements at sea and tackling demands to reduce pollution and lower the cost of operation.
- Materials and metals industry. Electronic engineers in the materials and metals industry will be maintaining the control and instrumentation in place and optimising hardware and software design concepts. Activities could include: fault diagnosis, simulation and modeling, and data analysis. They could well be designing and running control systems for power stations, for example.
- Power generation industry. Electronic engineers in the power generation industry will often be involved in designing, building and maintaining control and instrumentation plant items such as SCADA (supervisory control and data acquisition), DCS (distributed control systems), instruments, telephony and data networks.
- Rail industry. Electronic engineers in the rail industry could work in a number of areas, including signalling power, point heating and lighting. Their jobs will involve writing specifications for power distribution systems, reviewing designs and answering technical queries. On the maintenance side, they will be going out onto the rail network to test equipment or replace components.
- Utilities industry. Electronic engineers in the utilities industry can work in telecoms and energy, e.g., designing and running control systems for nuclear power stations.
- IT industry. Electronic engineering graduates are often welcome to apply for technical roles in the IT industry. Don't assume that only computer scientists or software engineers are sought.

However, different employers will have different requirements, so do check out companies individually.

3. Major Employers

The overall structure of the communication electronics industry is shown in Figure 17.1. The four major segments of the industry are manufacturers, resellers, service organizations, and end users.

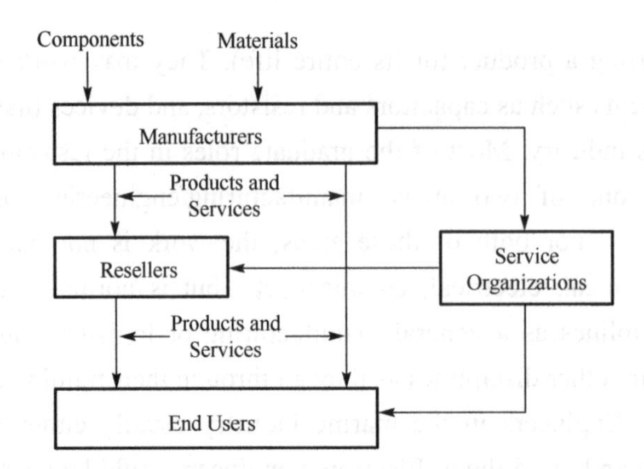

Figure 17.1 Structure of the communication electronics industry

- Manufacturers. It all begins, of course, with customer needs. Manufacturers translate customer needs into products, purchasing components and materials from other electronics companies to use in creating the products. Engineers design the products, and manufacturers produce them. There are jobs for engineers, technicians, salespeople, field service personnel, technical writers, and trainers.
- Resellers. Manufacturers who do not sell products directly to end users sell the products to reselling organizations, which in turn sell them to end users. For example, a manufacturer of marine communication equipment may not sell directly to a boat owner but instead to a regional distributor or a marine electronics store or shop. This shop not only sells the equipment but also takes care of service. A cellular telephone or fax machine manufacturer also typically sells to a distributor or dealer who takes care of sales and service. Most of the jobs available in the reselling segment of the industry are in sales, service, and training.
- Service organizations. These companies usually perform some kind of service, such as repair, installation, or maintenance. One example is an avionics company that does installation or service work on electronic equipment for private planes. Another is a systems integrator, a company that designs and assembles a piece of communication equipment or more often an entire system by using the products of other companies. Systems integrators put together systems to meet special needs and customize existing systems for particular jobs. Other types of service organizations are the communication service providers like cellular network carriers (e.g., China Mobile, China Telecom, and China Unicom), internet providers, cable TV companies, and internet web companies (e.g., Baidu, Tencent, and Alibaba).
- End users. The end user is the ultimate customer—and a major employer. Today, almost every person and organization is an end user of electronic and communication equipment. The major categories of end users in the communication field are:
  - ✓ Telephone companies.
  - ✓ Radio users (mobile, marine, aircraft, etc.).
  - ✓ Radio and TV broadcast stations and cable TV companies.

- ✓ Business and industry users of satellites, networks, etc.
- ✓ Transportation companies (airlines, shipping, railroads).
- ✓ Government and military.
- ✓ Internet companies.
- ✓ Personal and hobby.
- ✓ Consumers.

There are an enormous number of jobs with end users. Most are of the service type: installation, repair, maintenance, and operation of equipment.

4. Internships for ECE Students

Lots of students feel nervous when they are looking for internships during their graduation and it is a natural phenomenon for every fresher to feel anxious when they step into the corporate world. Every student will go through this phase and there is nothing to worry about if you have the required skill sets. By skill sets we mean about your practical exposure towards latest technologies and project building experience.

If you have good practical skills on latest technologies, it would be very easy for you to get into a internship. Also most of the companies today focus on latest technologies and they prefer candidates who have hands-on project building experience. So in addition to your theory exams, you should also spend time learning and building projects on latest technologies.

Here are some tips for an ECE student looking for some internship opportunities.

- How to apply for internships?

With the start-up culture spreading across the country contagiously, new internship opportunities are arising for engineering students. These start-ups generally post their internship requirements on recruitment websites that connect students to them.

There are many recruitment websites used by these companies to publish their internship opportunities. Some of them are Zhaopin, 51job, Liepin, etc. First you need to register yourself as a student by updating your skills, profile/resume. You can then subscribe to their alerts to get updates about the internships that best suit you.

There are also other organizations and companies that offer internships, but you cannot find those on recruitment websites. You need to visit their websites and upload your details on the career/ job opportunities pages to get shortlisted for the interview processes.

- What are the documents that you should prepare?

Before you apply for an internship, you need to prepare the following documents that best describe your objectives and goals.

- ✓ Resume. It should have all the necessary details like your educational credentials, skills you possess, responsibilities that you have taken in the past, project experience, and contact info.
- ✓ Cover letter. You should furnish the details about your availability (duration, preferred city), what is your intention of applying for the internship and referrals if any, in your cover letter.

✓ Transcript. This document should best describe your learning experience and the skills that you gained.

✓ Bonafide certificate. You need to submit this to the company. It is a kind of no objection certificate that you need to get from your college to get approval for the internship that you are joining.

✓ Other documents. You can keep other documents like state/national level certifications, value-added courses certificates, diploma certificates, etc., which will come in handy during your interview.

To increase your likehood of getting shortlisted for an interview, edit your resume/cover letter in a way to meet the company's objectives and goals. Don't be in a hurry and send a single resume to 100 companies. Make your profile precise to each company and apply, as companies prefer candidates who are more resilient.

● What skill sets do the companies expect from ECE students?

If you are looking to apply for internships in companies that have their research areas in several domains that are the emerging out of electronics and communication engineering, they look for specific technical skill sets. Some of them are:

✓ Analog design: strong knowledge on circuit design, concepts of amplifiers, regulators, oscillators, data converters, etc.

✓ Digital design: C, C++, Verilog, VHDL, SystemC, knowledge on signal processing, digital circuit concepts.

✓ Signal processing, control design: MATLAB, Simulink, C, C++.

✓ IC design: familiarity with tools like Cadence, ADS, Xilinx, Altera.

✓ Embedded programming: ability to program various microcontrollers.

Since there are a lot of engineering students looking to do internships in top reputed companies, the available opportunities are very limited. If you want to maximize your chance of securing those internships, you need to start working on improving your profile as early as possible.

● What can you do other than internships?

If for some reason you are unable to get an internship, don't worry. There are plenty of other opportunities that you can utilize to develop similar skills and experience you get by doing an internship. For example, you can try the following activities instead of wasting your holidays.

✓ Hands-on training programs. Attending hands-on training programs are the traditional way to learn new technologies and develop skill sets. You can undergo summer training programs or winter training programs that are offered by reputed institutions to spend your vacations in a productive manner. You can choose the training programs as per your requirements and improve your skills accordingly.

✓ Building projects. You can improve your skill sets by building projects on latest technologies to understand them better. Learning while building projects will be an effective approach for you to gain necessary skills. These experiences will also add values to your profile and you can use those to impress your recruiters, as most companies prefer candidates who are more hands-on. You can also try several online project building courses.

- ✓ E-learning. This is also an effective way for you to learn and gain new skills. There are various online platforms available for you, through which you can enroll for some certification courses. Top mentions include NetEase Open Course, icourse163, Tencent Classroom, etc. The advantage of doing these courses is that you can just log in anytime, anywhere and learn without any hassle. You will also be awarded a certificate of completion which you can use to showcase skills to recruiters.

Exercises

1. Work in groups to do a survey of top companies in ECE and then write a survey report, which should include:

(1) Name of the company; Time of foundation; The CEO; Number of employees.

(2) Major businesses; The organization structure.

(3) Recruitment plan; Training and development; Welfare.

2. Work in groups to do a survey of online education resources in ECE and then write a survey report, which should include:

(1) Type of the resource; Name of the resouce; Acess approach.

(2) Content introduction; Duration.

(3) Expected results; Fee.

Unit 18 Communication Skills

There are many qualities and skills an individual needs to become an effective engineer and to have a successful career. Beyond strong technical knowledge, engineering employers place a high value on effective communication and interpersonal skills, such as contributing to discussions, making presentations or public speaking, reading and synthesizing information, writing different types of documents, and so on. Engineers with strong communication skills can succeed in job interviews, and work effectively with colleagues, clients, suppliers or the public.

But many more technically minded people are not naturally adept at speaking in front of people or communication in general. Communication skills need to be improved in practice. This unit gives some practical suggestions for engineering undergraduates.

1. Techniques for Listening

Listening is equal in importance to oral and written communication skills. This skill is essential for engineers to understand problems and issues clearly. Here are some useful tips.

- Show attentiveness. Taking written notes can help you remember everything, and shows others that you value what they are saying. In addition, don't be distracted by mobile phones or tablets.
- Seek understanding. Misunderstanding can be avoided when partners in a conversation ask sincere questions and paraphrase each other's statements.
- Invite dialogue. Engineers should ensure that clients feel that their concerns are truly valued. To engage in a productive conversation, it often takes some work to issue a sincere invitation for everyone to exchange ideas frankly.
- Observe and adapt. While listening, observe the other peoples' body language and adjust your own words. Make sure you pay attention to your audience, watch their faces.

If you are planning a meeting, you may need to brief other attendees about what to discuss in advance, and follow up after the meeting if necessary.

2. Writing a Professional E-mail

Obviously, e-mail has become our most frequent form of correspondence in work. Distributed engineering teams use e-mail to share documentation and information. Work e-mail should include a salutation, a closing, a signature, a specific subject line, and the body, as shown in Figure 18.1.

A well-composed e-mail provides the recipient with a friendly, clear, concise and actionable message. Learning how to write an e-mail that meets all of these criteria takes practice. Consider the following tips to help you write an effective, professional e-mail: identify your goal, consider your audience, keep it concise, proofread your e-mail, use proper etiquette and remember to follow up.

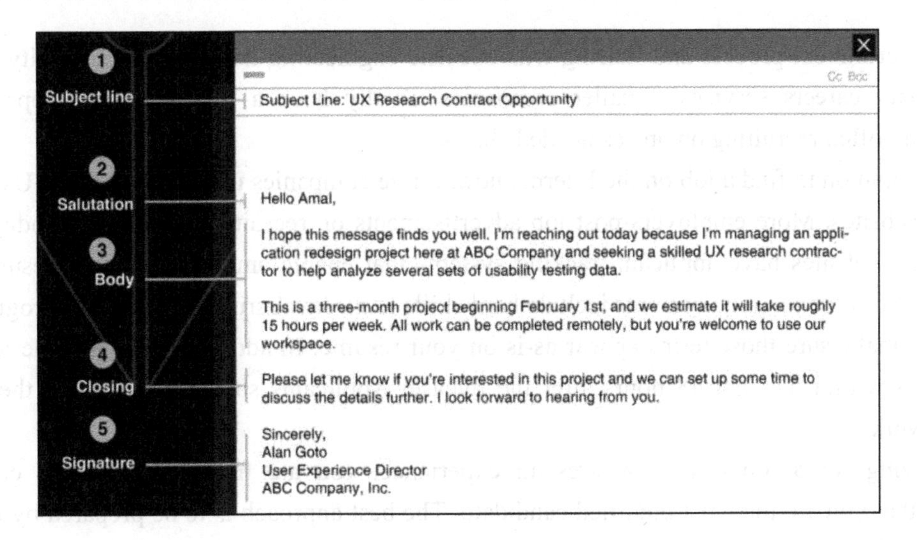

Figure 18.1 Professional e-mail format

Professional e-mail often needs to be forwarded. The engineer should provide enough context and the appropriate level of formality so that a colleague or supervisor can forward the e-mail without rewriting.

3. Presentation Skills

Effective public speaking and presentation skills enable engineers to share their findings, plans, and projects with peers who understand technical language as well as with potential clients in business meetings. Here are some advices.

- Analyzing the audience and setting. The background of the audience will influence both the content and delivery of the presentation. Similarly, the environment setting of the speech is also an influencing factor which need to be considered in advance.
- Overcoming stage fright. Interaction with the audience, by asking questions and watching the reactions, can reduce the anxiety commonly felt in public speaking.
- Designing each slide carefully. Especially for engineers, graphs are a perfect tool to illustrate technical matters. Keep statements, pictures and lettering clear and readable. Avoid long sentences.
- Making and using notes. The notes should contain the basic information you want to present, but should not contain the complete sentences you want to read, because the written language is not suitable for the spoken environment.

If you can get a conversation started off well, then the rest of the conversation is likely to go smoothly. During your presentation, make sure your self introduction is smooth and confident. Put in work to improve your presentation skills, which will become the most powerful skills you have.

4. Application and Interview Skills

There are a variety of methods to conduct your job search. You might focus on networking, meeting with employers on campus, or attending career fairs, among other opportunities.

Large companies usually approach institutions directly, visiting campuses as part of the graduate recruitment process and liaising with specific engineering departments or faculty as well as university careers services. Smaller companies usually have a less structured approach to recruitment, often recruiting on an "as needed" basis.

It is common to find a job on the Internet now. Large companies usually have "Join Us" pages on their websites. More employers post job advertisements on recruitment websites. Today many recruitment websites have applicant tracking systems that enable employers to scan resumes for keywords. So if the job requirements include hard skills (e.g., JavaScript or Python for programmer positions), make sure those terms appear as-is on your resume. In addition, the best time to apply for work is within the first 48 hours after the release. Applicants should often check the newly released work.

Applying for a job can be a stressful experience: you are trying to convince complete strangers that you are the most qualified candidate. The best approach is to be prepared by making sure your written documents (resume and application letter) stand out by making them easy to read and highlighting what makes you different from all the other applicants. There are many good resume and application letter templates on the Internet.

Many students apply for numerous positions with little or no revision to differentiate each resume or application letter. This will cost them most of the opportunities. Effective application materials target not only the desired qualifications listed in a job ad but also the culture of the organization to which they are being submitted. Most organizations state their values on their websites, you need to modify your application materials accordingly. This will greatly increase your chances of getting an interview.

Once you get an interview, you'll need to be prepared. Research the company, research what the job entails, and practice your interview questions and answers. Most large organizations can hold up to three interviews. Following the first interview, successful applicants will then need to attend a technical and an HR interview. Smaller organizations will usually accept resumes and will then hold a first interview and sometimes a second interview. The trend is for employers to have earlier deadlines, partly due to the competition for graduates and partly because the recruitment process is completed in time for students to concentrate on exams. It's a good tip to start thinking about applications in early September, gathering essential information on employers you are interested in and noting their application deadlines.

All engineering interviews have a technical aspect. This may be a separate interview or integrated as part of a first or second interview. Technical questions are not always about having the correct answers and knowledge, but more about having the opportunity to demonstrate your understanding of basic technical concepts and engineering principles and your confidence in applying them. Interviewers will also want to assess your ability to communicate technical ideas and information. If caught out by a difficult question, it is important to be able to admit that you don't know enough to answer, but are interested and willing to learn more.

Today, phone and video conferencing are common for both interviews and meetings. Here are some suggestions for interviews by conference call.

- Before your interview:

- ✓ Find out exactly who you will be talking to.
- ✓ Check whether they will be able to see you or just hear you.
- ✓ Check the date, the time, and the right code to access the online conference.
- ✓ Read your resume and application letter again.
- ✓ Practice answering questions you might be asked.
- ✓ Prepare questions to ask the interviewers.

- During your interview:
  - ✓ Don't be late.
  - ✓ Use your tone of voice to sound confident and enthusiastic.
  - ✓ Do not shuffle papers (this will make a noise).
  - ✓ Sit in a comfortable position, don't move about too much.
  - ✓ Speak very clearly, facing the microphone.
  - ✓ When the interview is over, thank the interviewers and end positivily.

In a world of rapid scientific and technological advancement, new fields are constantly emerging, engineer is a job that requires lifelong learning and skills updating. Whether you are a recent graduate with an engineering degree or a fully licensed and experienced engineer, you still need to find work in similar ways.

You need to stay on top of what's going on in your field. Read trade publications and industry journals, join professional social media groups, and follow social media feeds of companies and industry influencers.

Exercises

1. Read the following e-mail and reply to book a workshop you are interested in.

From: Carla Lin, Training Officer
To: All staff
Subject: Workshops in May

Hi everyone,

Please let me know which workshop you'd like to attend next month and which day you would like to go. Places are limited, so please contact me before 30th April.

Workshops available:
Security procedures: 1 day, 13th or 14th May
Website design: 1 day, 15th or 16th May
Setting up a network: 1 day, 20th or 21st May

Best wishes
Carla

2. Match tough interview questions below with the possible answers.

(1) What kind of personality do you think you have?	A. My managerial experience should be improved.
(2) Are there any weaknesses in your education or experience?	B. I think I am quite extroverted and can work well with others.
(3) What is important to you in a job?	C. As long as my position here allows me to learn and to advance at a pace consistent with my abilities.
(4) If we hire you, how long will you stay with us?	D. Why don't we discuss salary after you decide whether I am right for the job?
(5) What are your salary expectations?	E. Feeling a sense of accomplishment.

3. Understanding job ads.

Telecom Engineer

Responsibilities:

1. Participate in telecom system design for oil, gas, chemical and pharmaceutical projects.
2. Support project estimation, proposal preparation, procurement and construction.
3. Coordinate with other disciplines, i.e., piping, structure/architecture, electric, equipment, etc.
4. Coordinate with vendor/client in engineering design.

Requirements:

1. Major in telecom engineering with bachelor degree or above.
2. 3+ years experience in telecom, communication or network engineering field.
3. Working experience in Class A Chinese design institute is preferred.
4. Good command at English.
5. A team player with strong interpersonal skills.
6. Able to work under pressure to meet challenging objectives on mega-projects.

4. Work in groups to do a survey of online resume templates and then create your own resume.

References

[1] ALEXANDER C K, SADIKU M N O. Fundamentals of electric circuits[M]. 6th ed. New York: McGraw-Hill Education, 2017.

[2] GLENDINNING E H, MCEWAN J. Oxford English for electronics[M]. Oxford: Oxford University Press, 1993.

[3] JIMÉNEZ M, PALOMERA R, COUVERTIER I. Introduction to embedded systems: using microcontrollers and the MSP430[M]. New York: Springer, 2014.

[4] FIORE J M. Semiconductor devices: theory and application[M]. Utica: James M. Fiore, 2018.

[5] OPPENHEIM A V, WILLSKY A S, NAWAB S H. Signals and systems[M]. 2nd ed. Upper Saddle River: Prentice-Hall, 1996.

[6] OPPENHEIM A V, SCHAFER R W. Discrete-time signal processing[M]. 3rd ed. Upper Saddle River: Prentice-Hall, 2009.

[7] GONZALEZ R C, WOODS R E. Digital image processing[M]. 4th ed. New York: Pearson, 2018.

[8] ZIEMER R E, TRANTER W H. Principles of communications: systems, modulation, and noise[M]. 7th ed. Hoboken: John Wiley & Sons, Inc., 2015.

[9] CARLSON A B, CRILLY P B. Communication systems: an introduction to signals and noise in electrical communication[M]. 5th ed. Boston: McGraw-Hill, 2010.

[10] FRENZEL L E. Principles of electronic communication systems[M]. 4th ed. New York: McGraw-Hill Education, 2016.

[11] PROAKIS J G, SALEHI M. Digital communications[M]. 5th ed. Boston: McGraw-Hill, 2008.

[12] JOHNSON D H. Electrical engineering (johnson)[M/OL]. Rice University via Connections. https://eng.libretexts.org/Bookshelves/Electrical_Engineering/Introductory_Electrical_Engineering/Electrical_Engineering_(Johnson).

[13] PETERSON L, SUNAY O, DAVIE B. Private 5G: a systems approach[M]. Tucson: Systems Approach LLC, 2023.

[14] VALDAR A. Understanding telecommunications networks[M]. 2nd ed. London: The Institution of Engineering and Technology, 2017.

[15] INAN U S, INAN A S, SAID R. Electromagnetic engineering and waves[M]. 2nd ed. Harlow: Pearson, 2013.

[16] MAO Z Y. Basic principles of microwave communication[J/OL]. Journal of physics: conference series, 2021, 1885(2021):022062[2021-03-26]. https://doi.org/10.1088/1742-6596/1885/2/022062.

[17] TANENBAUM A S, WETHERALL D J. Computer networks[M]. 5th ed. Boston: Pearson Prentice Hall, 2011.

[18] LITTMAN M L, AJUNWA I, BERGER G, et al. Gathering strength, gathering storms: the one hundred year study on artificial intelligence (AI100) 2021 study panel report[R]. Stanford: Stanford University, 2021.

[19] HURWITZ J, KIRSCH D. Machine learning for dummies[M]. Hoboken: John Wiley & Sons, Inc., 2018.

[20] HURWITZ J, NUGENT A, HALPER F, et al. Big data for dummies[M]. Hoboken: John Wiley & Sons, Inc., 2013.

[21] KUMAR J A, KULKARNI G P, MUNAVALLI J R. Recent trends and developments in computer networks: a literature survey[J]. International journal of advances in electronics and computer science, 2019, 6(9): 2394-2835.

[22] SINGH A, DEV K, SILJAK H, et al. Quantum internet—applications, functionalities, enabling technologies, challenges, and research directions[J]. IEEE communications surveys & tutorials, 2021, 23(4):2218-2247.

[23] PAUK W, OWENS R J Q. How to study in college[M]. 10th ed. Boston: Wadsworth, Cengage Learning, 2011.

[24] HOUSE R, LAYTON R, LIVINGSTON J, et al. The engineering communication manual[M]. Oxford: Oxford University Press, 2017.

[25] THIEL D V. Research methods for engineers[M]. Cambridge: Cambridge University Press, 2014.

[26] OLEJNICZAK M. English for information technology, level 1[M]. Harlow: Pearson Longman, 2011.